Software Innovations in Clinical Drug Development and Safety

Partha Chakraborty
Cognizant Technology Solutions, India

Amit Nagal
GVK Bioscience, India

A volume in the Advances in Medical Technologies and Clinical Practice (AMTCP) Book Series

Medical Information Science
REFERENCE
An Imprint of IGI Global

Managing Director:	Lindsay Johnston
Managing Editor:	Keith Greenberg
Director of Intellectual Property & Contracts:	Jan Travers
Acquisitions Editor:	Kayla Wolfe
Production Editor:	Christina Henning
Development Editor:	Courtney Tychinski
Cover Design:	Jason Mull

Published in the United States of America by
Medical Information Science Reference (an imprint of IGI Global)
701 E. Chocolate Avenue
Hershey PA, USA 17033
Tel: 717-533-8845
Fax: 717-533-8661
E-mail: cust@igi-global.com
Web site: http://www.igi-global.com

Library of Congress Cataloging-in-Publication Data

Software innovations in clinical drug development and safety / Partha Chakraborty and Amit Nagal, editors.
p. ; cm.
Includes bibliographical references and index.
Summary: "This book is a comprehensive resource analyzing the integration of software engineering for the purpose of drug discovery, clinical trials, genomics, and drug safety testing"--Provided by publisher.
ISBN 978-1-4666-8726-4 (hardcover) -- ISBN 978-1-4666-8727-1 (ebook)
I. Chakraborty, Partha, 1972- , editor. II. Nagal, Amit, 1979- , editor
[DNLM: 1. Drug Discovery. 2. Software. 3. Clinical Trials as Topic. 4. Genomics. 5. Pharmacovigilance. QV 26.5]
RS122.2
615.10285--dc23

2015019899

This book is published in the IGI Global book series Advances in Medical Technologies and Clinical Practice (AMTCP) (ISSN: 2327-9354; eISSN: 2327-9370)

British Cataloguing in Publication Data
A Cataloguing in Publication record for this book is available from the British Library.

For electronic access to this publication, please contact: eresources@igi-global.com.

Advances in Medical Technologies and Clinical Practice (AMTCP) Book Series

Srikanta Patnaik
SOA University, India
Priti Das
S.C.B. Medical College, India

ISSN: 2327-9354
EISSN: 2327-9370

Mission

Medical technological innovation continues to provide avenues of research for faster and safer diagnosis and treatments for patients. Practitioners must stay up to date with these latest advancements to provide the best care for nursing and clinical practices.

The **Advances in Medical Technologies and Clinical Practice (AMTCP) Book Series** brings together the most recent research on the latest technology used in areas of nursing informatics, clinical technology, biomedicine, diagnostic technologies, and more. Researchers, students, and practitioners in this field will benefit from this fundamental coverage on the use of technology in clinical practices.

Coverage

- Medical Informatics
- Clinical Data Mining
- Clinical High-Performance Computing
- Diagnostic Technologies
- Clinical Nutrition
- Telemedicine
- Biomechanics
- Biometrics
- Biomedical Applications
- Nutrition

IGI Global is currently accepting manuscripts for publication within this series. To submit a proposal for a volume in this series, please contact our Acquisition Editors at Acquisitions@igi-global.com or visit: http://www.igi-global.com/publish/.

The Advances in Medical Technologies and Clinical Practice (AMTCP) Book Series (ISSN 2327-9354) is published by IGI Global, 701 E. Chocolate Avenue, Hershey, PA 17033-1240, USA, www.igi-global.com. This series is composed of titles available for purchase individually; each title is edited to be contextually exclusive from any other title within the series. For pricing and ordering information please visit http://www.igi-global.com/book-series/advances-medical-technologies-clinical-practice/73682. Postmaster: Send all address changes to above address.

Titles in this Series

For a list of additional titles in this series, please visit: www.igi-global.com

Modern Concepts and Practices in Cardiothoracic Critical Care
Adam S. Evans (Icahn School of Medicine at Mount Sinai, USA) Gregory E. Kerr (Weill Cornell Medical College, USA) Insung Chung (Icahn School of Medicine at Mount Sinai, USA) and Robin Varghese (Icahn School of Medicine at Mount Sinai, USA)
Medical Information Science Reference • copyright 2015 • 1111pp • H/C (ISBN: 9781466686038) • US $415.00 (our price)

Recent Advances in Assistive Technologies to Support Children with Developmental Disorders
Nava R. Silton (Marymount Manhattan College, USA)
Medical Information Science Reference • copyright 2015 • 425pp • H/C (ISBN: 9781466683952) • US $210.00 (our price)

Advanced Technological Solutions for E-Health and Dementia Patient Monitoring
Fatos Xhafa (Universitat Politècnica de Catalunya, Spain) Philip Moore (School of Information Science and Engineering, Lanzhou University, China) and George Tadros (University of Warwick, UK)
Medical Information Science Reference • copyright 2015 • 389pp • H/C (ISBN: 9781466674813) • US $215.00 (our price)

Assistive Technologies for Physical and Cognitive Disabilities
Lau Bee Theng (Swinburne University of Technology, Malaysia)
Medical Information Science Reference • copyright 2015 • 321pp • H/C (ISBN: 9781466673731) • US $205.00 (our price)

Fuzzy Expert Systems for Disease Diagnosis
A.V. Senthil Kumar (Hindusthan College of Arts and Science, India)
Medical Information Science Reference • copyright 2015 • 401pp • H/C (ISBN: 9781466672406) • US $265.00 (our price)

Handbook of Research on Computerized Occlusal Analysis Technology Applications in Dental Medicine
Robert B. Kerstein, DMD (Former clinical professor at Tufts University School of Dental Medicine, USA & Private Dental Practice Limited to Prosthodontics and Computerized Occlusal Analysis, USA)
Medical Information Science Reference • copyright 2015 • 1093pp • H/C (ISBN: 9781466665873) • US $475.00 (our price)

IGI GLOBAL
DISSEMINATOR OF KNOWLEDGE
www.igi-global.com

701 E. Chocolate Ave., Hershey, PA 17033
Order online at www.igi-global.com or call 717-533-8845 x100
To place a standing order for titles released in this series, contact: cust@igi-global.com
Mon-Fri 8:00 am - 5:00 pm (est) or fax 24 hours a day 717-533-8661

Table of Contents

Detailed Table of Contents

The overall process of getting a drug to the market is a long one and takes 10-15 years and costing close to a billion dollar. The success rate as the compound travels from the initial discovery phase to clinical and then through to the market is about 1 in 10,000. The two key phases which together contribute the most to the cost and timeline are clinical development and pharmacovigilance. These two phases together also account for the maximum number of failures. In this chapter, we will look in detail at these two phases with a focus on the business process and process areas which have application of computer systems. The chapter will focus on looking at the various phases of clinical development and their endpoints. Clinical development is the process of testing a drug for safety and efficacy in human subjects. Clinical trial is conducted in 3 phases with the 4th phase which is ongoing post approval which forms an important part of the pharmacovigilance process. These phases will be elaborated in detail.

This chapter contains as well as illustrates different innovations that is changing the way Clinical Drug Development and Safety organizations perceives IT and the ways and means through which these innovations are facilitating the change in business itself. The main content contains illustrations of two structurally different means to create data warehouses, the benefits of the approaches and the difficulties. It also explains the importance of data virtualization technology when implemented in the Clinical and Safety Organizations.

The current digital age is primarily driven by four technology forces namely, Social Media, Mobility, Analytics and Cloud computing. These technologies continue to evolve and shape the digital world, giving people and businesses newer experiences and opportunities that they were not exposed to in the past. Digital technology has the potential to change the world significantly which in turn has a disruptive impact in the world of business. Hence, 'digitizing' its business must be one of top priorities in the medium and long term of every business to ensure a successful future. This chapter begins with by defining each of the four technologies, its benefits and what it means to the key stakeholders in the healthcare business. It also covers many use cases of SMAC with a specific focus on clinical development and pharmacovigilance. The later part of the chapter lays the foundation for setting up a SMAC organization including key strategies, conceptual framework, technology and regulatory compliance considerations.

Collaboration is defined as the actions for individuals and teams to work together for a common goal. There are several bottlenecks to an efficient and effective collaborative model of clinical trial including: the lack of a centralized, consistent, globally accessible platform to manage and store essential study related documentation; inconsistent or incomplete work assignments; inefficient notification of key events requiring follow-on action; and incomplete, missing, expired, or redundant documentation and training activities and need to maintain multiple credential to access various system, Removing these barriers is an important part of establishing an environment that fosters collaboration among all constituencies involved in managing clinical trial keeping them connected, informed, and on task by providing access to everyone at any time, from anywhere.The case study below introduces need of an integrated clinical collaboration platform, addressing key functionality of such an platform and describes the architecture & design consideration to industrialize such a platform. The intended audience of this case study is the architects & designers of similar systems. The clinical trial activity for a drug in research is approximately 70% of the overall drug development cost. It is estimated that 4% of the cost of a trial is in 'rework' involving communication, regulatory issues, patient enrollment, document review and replacement of patients. The integrated clinical collaboration platform has potential to eliminate significant amount of cost of re-work, which is in order of $3.5M per trial.

The pharmaceutical and medical manufacturing sectors have entered a period of disruptive transformation in the way regulatory affairs are conducted globally. The global clinical and regulatory landscape is evolving more quickly in this decade than ever before. The advent of adaptive trial designs, rolling submissions for indications, as well as the impact of regulatory policies in emerging markets, are influencing Pharma's ability to secure approvals efficiently and effectively and with required emphasis on safety and compliance. The impact of these changes on Regulatory Information Management can be significant over the next

5-7 years. Companies are rightfully asking what the transformation in business processes and technology might look like and what types of innovations they can adopt now to prepare them for the future state. The case study below introduces the need for an integrated Regulatory Information Management (RIM) platform, addressing key functionality of such an environment and describes the architecture & design consideration to industrialize such a platform.

DNA sequencing is the process to identification of nucleotides order in genome which developed from very broad history, also it is derived from version of the Sanger biochemistry. SOLiD, 454 and Polonator sequencing based on emulsion PCR to amplify clonal sequencing with in-vitro construction of adaptor-flanked shotgun library, PCR amplified in the context of a water-in-oil emulsion. Solexa technology relies on bridge PCR to amplify clonal sequencing features. At the conclusion of the PCR, each clonal cluster contains ~1,000 copies of a single member of the template library. This chapter focused on next-generation sequencing technologies methods, capabilities and clinical applications of DNA sequencing technologies for researchers in molecular biology and physician scientists. This will also provide the power of these novel genomic tools and methods to use personal diagnostic at molecular level.

Pharmacogenomics deals with drug responses in individual based on genetic variation in genome. Based on genetic variations, drugs may produce more or less therapeutic effect, and same way in side effects also. Physicians can use information about your genetic makeup to choose those drugs and drug doses to get better therapy. Optimizing drug therapy and rational dose adjustment with respect to genetic makeup will maximize drug efficacy and minimal adverse effects. This broken traditional 'trial and error' method of 'one drug fits all', and 'one dose fits all' which contributing to 25–50% of drug toxicity or treatment failures. This will contribute to improve the ways in which existing drugs are used, genomic research will lead to drug development to produce new drugs that are highly effective without serious side effects. This approach to bring personalized medicine more practice and drug combinations are optimized for each individual' genetic makeup.

Epigenetics is the study of changes in organisms caused by modification of gene expression rather than alteration of the genetic code itself. ChIP-seq, is a method used to analyze protein interactions with DNA. It is a type of epigenetic analysis technique. Chromatin immunoprecipitation coupled with massive parallel sequencing (ChIP-seq) is gaining popularity day by day because of its clinical significance. It is a very effective tool in diagnosis of disease such as cancer. ChIP-seq is found to be very effective tool in understanding basic regulatory mechanism, cell differentiation study and studying disease processes with the decreasing cost of sequencing, ChIP-seq has become an indispensable tool for studying gene regulation and epigenetic mechanisms. The Present review explores epigenetic methods, pipeline and its role in cancer.

Recent advances in human exome sequencing and the associated advantages have made it a technology of choice in various domains. The savings in time, cost and data storage compared with whole genome sequencing make this technology a potential game changer in clinical research settings. Recent advances in NGS have made it feasible to use exome sequencing in clinical research for identifying novel and rare variants that can lead to change in protein structure and function which may finally culminate into a totally different phenotype. If whole exome is not desired the same technology can be used for studying target exonic regions to investigate causative genes for a specific phenotype associated with disease. Exome sequencing has emerged as an effective and efficient tool for the translational and clinical research. There is a demand for systematically storing variant information in large databanks. Meaningful information from the exome-seq data can be combined with other data. This can be correlated with clinical findings within a clinical trial setting for a better study outcome.

Given that Agile software development is preferred methodology for products and services in life science industry, in this chapter we will describe how to adopt Agile software development process and still be compliant. We will focus on few Agile methodologies and provide details on what design controls we can adopt in order for the product and process to be compliant. We will also focus on some of the tools that can be used to help put such design and process control in place where we can have complete transparency and traceability.

Foreword

The Pharmaceutical industry is fast evolving with changing guidelines (US FDA, EMEA) aimed at driving higher standards on patient safety. Every year with new drafts/revisions, the Standard Operating Process (SOP) undergo revisions/changes. As a consequence, IT systems, software, hardware, devices need validation as per revised/new norms that drive compliance.

This book describes in easy language the terms, definitions, usage, criteria and the processes from a 'Computer System Validation (CSV)' perspective. This chapter aims to drive better understanding of computer system validation, its deliverables, associated risks, documentation required, including Safe Harbor and Good Practices. It has also dealt with electronic submissions, applications that are hosted on the cloud and their treatment.

Highly recommended for students, professionals who are setting a CSV environment, or to help prepare for a mock audit or submission, quality organizations and IT companies.

Arindam Dey is an exponent on this area and has carefully chosen to simplify and yet highlight the importance of CSV in Pharmaceutical and Med Device organizations.

KV Subrahmanyam
Senior Vice President, Head of Life Sciences & Healthcare, HCL Technologies

Preface

The increasing cost of medicine and healthcare is subject of significant concern throughout the world. Even though significant improvement for medicinal sciences over last one and half centuries, the debate around universal provisioning of healthcare and its quality, spiraling cost of medicines and safety issues associated with them gained momentum in last two decades. Inability to control development of chronic conditions and manage them well, inability to build patient centric model and manage costs though building coalition of the private, public, academic corporate and governmental stakeholders are primary set of challenges for this sector. Existing model of drug discovery and development accentuates these challenges. Life sciences industry, associated with drug discovery and development, is challenged with patent cliff, the significantly smaller pipeline and concerns over compliances & safety. In response to these challenges, healthcare & life sciences industry has started reflecting on the internal inefficiencies of the sector and external change agents, such as change in disease patterns & life style, changing context of urbanization & standard of living, lack of public private partnership, slower adoption of new age technology. Various organizational, institutional and technological initiatives & interventions are seen in different countries. One of these initiatives is considerable adoption of information technology and software application to improve functioning of healthcare & life sciences core process.

Adoption of information technology & software application in healthcare & life sciences is still an emerging phenomenon. There is a need to formalize this to an academic discipline. Few initiatives are taken towards launch discipline like Pharmacoinformatics, Biostatistics etc. However, there is need of significant initiative to consolidate them by theorizing, conceptualizing and defining structures and foundation of a discipline. Emphasize needs to be laid on potential in improving effectiveness & efficiency of the core process along with patient centric services & outcomes as well as improving decision making at the operational, strategic and scientific levels.Focus needs to be given on structured training & learning of this emerging discipline.

This publication aims to create a foundation of a comprehensive guide of software engineering for the purpose of drug discovery, clinical trial, genomics, life science and drug safety. It includes various areas of application of computer; their architecture & design patterns, information models, building search techniques, guidance of implementation of various regulations in software code including US FDA 21 CFR Part 11, security & data privacy, computer system validation.

This publicationaims to address the above considerations in three sections.The first section describes the business processes of clinical trial & drug safety, area of improvement through software innovation. It describes the information framework required to build effective software system to streamline the workflow in the business process and to better analytics for decision making. This section has six chapters Sowmyanarayan describes overview of clinical trial & drug safety process & application of computer system in the first chapter. In the second chapter Kanishka discusses data warehouse & data virtualization techniques, which can be adopted to create information store, required as foundation to form decision insight. In the third chapter, Chandrakant talks about metadata repository and master data management in clinical trial and drug safety.Semantic technologies have gained prominence over the last several years. Semantic technologies & semantic integration are explored in detail in the fourth chapter. Ramin et al describes safety signal detection in drug development process in the fifth chapter. Cloud, analytics, mobile and social media have gained prominence in this decade and show a lot of potential in improving the efficiency of clinical trial & drug safety process area. The section ends with discussion from Manu on application of Social, Media, Analytics& Cloud.

The second section includes topics on software innovation in drug discovery process. This section has four chapters.DNA sequencing is the process to identification of nucleotides order in genome which developed from very broad history. Udaya raja describes personal diagnostics using DNA sequencing in the seventh chapter. In the next chapter, he explains topic of pharmacogenomics; genome wise association with clinical studies. Pharmacogenomics is widely use in drug discovery where drug response is measured with genomic level. Amit explains role of epigenetic in cancer genomics which is very popular modern technique in cancer research. The main aim of this technique is to identify novel drug targets in cancer. Anu et al describes impact of human exome sequencing on clinical research and how it leads to personalize medicine & therapy. Exome sequencing is very popular technique use in drug resistance gene identification which helps in drug discovery process.

The third section reflects computer system validation and agile methodology of software development and how to remain compliant. Subbu, Avik and Arindam describes details of the methodology of computer system validation in the life sciences industry and implication of FDA 21 CFR part 11. This section describes case study on how to remain compliant while adopting agile methodology in software development in the area of clinical research.

Partha Chakraborty
Cognizant Technology Solutions, India

Amit Nagal
GVK Bioscience, India

Chapter 1
Overview of Clinical Trial and Pharmacovigilance Process and Areas of Application of Computer System

Sowmyanarayan Srinivasan
Accenture Services Pvt Ltd, India

ABSTRACT

The overall process of getting a drug to the market is a long one and takes 10-15 years and costing close to a billion dollar. The success rate as the compound travels from the initial discovery phase to clinical and then through to the market is about 1 in 10,000. The two key phases which together contribute the most to the cost and timeline are clinical development and pharmacovigilance. These two phases together also account for the maximum number of failures. In this chapter, we will look in detail at these two phases with a focus on the business process and process areas which have application of computer systems. The chapter will focus on looking at the various phases of clinical development and their endpoints. Clinical development is the process of testing a drug for safety and efficacy in human subjects. Clinical trial is conducted in 3 phases with the 4th phase which is ongoing post approval which forms an important part of the pharmacovigilance process. These phases will be elaborated in detail.

INTRODUCTION AND BACKGROUND

The overall process of getting a drug to the market is a long one and takes 10-15 years and costing close to a billion dollar. The success rate as the compound travels from the initial discovery phase to clinical and then through to the market is about 1 in 10,000. The two key phases which together contribute the most to the cost and timeline are clinical development and pharmacovigilance. These two phases together also

DOI: 10.4018/978-1-4666-8726-4.ch001

Figure 1.

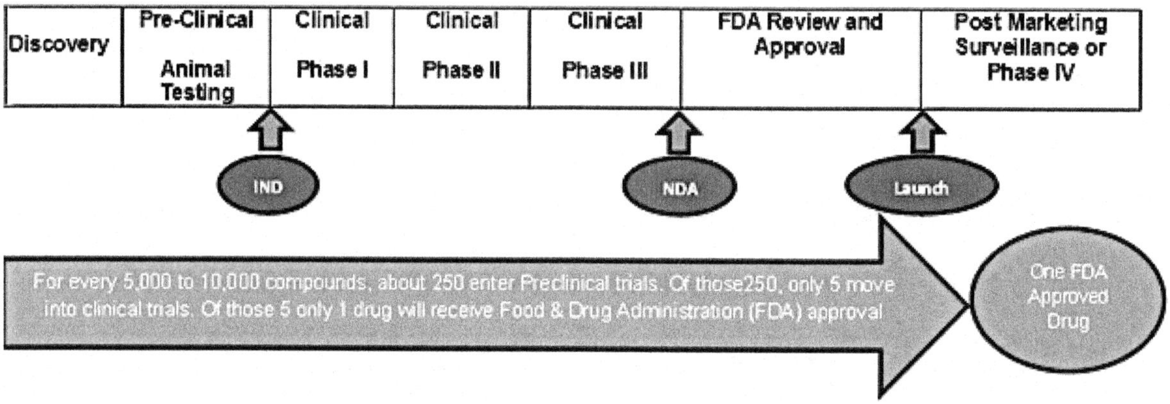

account for the maximum number of failures. In this chapter, we will look in detail at these two phases with a focus on the business process and process areas which have application of computer systems.

Figure 1 illustrates the value chain of bringing a drug to the market starting from understanding of disease to commercialization of the drug for the disease. (R&D Pipeline Management, University of Wisconsin, n.d)

The entire process of drug discovery and development starts *in silico* (in the computer), moves to *in vitro* (in the laboratories) and then finally *in vivo* (inside living beings like animals and humans). This is keeping with the spirit of ensuring safety to all living beings. As the molecule progresses to becoming a drug it graduates from computer to laboratory to living beings.

The initial phase of drug discovery is focused on understanding the disease better and potential drug targets that are relevant for the disease which is largely driven by biology. This part primarily focuses on:

- understanding the mechanism of diseases
- identifying potential targets for therapeutic intervention
- evaluating potential drug candidates

Once the disease is better understood and potential targets identified, the next phase focuses on various aspects of the molecule that can become potential drugs and is largely driven by chemistry. This part addresses the following needs:

- Inventing or identifying safe & effective chemical entities that will become potential drug candidates.
- These drug candidates will act as leads to act on the target disease

These two initial phases of biology and chemistry are largely *in silico* and *in vitro*.

One of the key qualifying parameter for a molecule to become a drug is its medicinal properties. This science is called medicinal chemistry and this characterizes the molecule before it is used in any living being.

Medicinal chemistry is a scientific discipline involved with designing, synthesizing and developing pharmaceutical drugs.

Medicinal chemistry involves the identification, synthesis and development of new chemical entities suitable for therapeutic use. It also includes the study of existing drugs, their biological properties, and their quantitative structure-activity relationships (QSAR). One of the key set of parameters that defines the molecule is referred to as ADMET (Absorption, Distribution, Metabolism, Excretion, and Toxicity).

- **Absorption:** Refers to movement of drug into the body from the site of administration. Before a compound can exert a pharmacological effect in tissues, it has to be taken in to the bloodstream — usually via mucous surfaces like the digestive tract (intestinal absorption). Uptake into the target organs or cells needs to be ensured, too.
- **Distribution:** Refers to movement of drug from intravascular to extravascular compartment. The compound needs to be carried to its effector site, most often via the bloodstream. From there, the compound may distribute into tissues and organs, usually to differing extents.
- **Metabolism:** Refers to chemical transformation of drug. Compounds begin to be broken down as soon as they enter the body. The majority of small-molecule drug metabolism is carried out in the liver by redox enzymes, termed cytochrome P450 enzymes.
- **Excretion:** Refers to removal of drug from the body. Compounds and their metabolites need to be removed from the body via excretion, usually through the kidneys (urine) or in the feces. Unless excretion is complete, accumulation of foreign substances can adversely affect normal metabolism.
- **Toxicity:** Adverse effects occurring as a result of repeated daily dosing of a drug compound, or exposure to the drug compound, for part of an organism's lifespan. With experimental animals, the period of exposure may range from a few days to 6 months. (Clinical Pharmacology I, MCCQE Review Notes, 2002)

ADMET is an extremely important set of parameter since about 50% of the drug candidates fail due to unacceptable side effects and toxicities (ADMET). Some of these failures occur much later in the value chain resulting in additional spend that is avoidable. So it is critical that a compound needs to have a favorable ADMET profile for it to become a drug

Due to this very important nature of ADMET life sciences companies have started monitoring drugs for ADMET across the value chain. ADMET is measured all the way from *in silico* to *in vitro* to *in vivo*. The tests and the end points measured may vary. *in silico* ADMET is also popularly called Predictive ADMET. (Gombar VK[1], Silver IS, Zhao Z., 2003)

Once the medicinal property and safety of the drug is established the molecule becomes a potential drug and is taken up for testing in animals and this phase is called preclinical.

A preclinical study is done to test promising compounds in animals. Companies conduct extensive toxicological tests and animal studies to determine whether the drug is likely to be safe and effective in humans.

If data from the above preclinical study prove the drug's safety and effectiveness, the pharmaceutical research organization submits all the results and files an Investigational New Drug Application (IND) in Food and Drug Administration (FDA) for US and other relevant regulatory authorities for other countries to test drug candidates in humans.

Primary Objectives of Preclinical Studies

- Support for human pharmacology
- Prediction of human pharmacokinetics
- Support for human toxicology
- Support in screening new dosage forms and formulations

Once a drug is determined to have a therapeutic use and safety has been tested in animals the drug becomes a candidate for development. Since drug development is so expensive, compound are carefully considered based on their market potential, availability of other treatments, and difficulty of bringing the drug to market. This phase is called clinical trials and is conducted over four phases as described below.

Clinical Phase 1 Trials

- In phase I, researchers test an experimental drug or treatment in a small group of people (20-80) for the first time to evaluate its safety, determine a safe dosage range, and identify side effects
- Phase 1 studies are designed to determine
  - Metabolic and pharmacologic actions of the drug in humans
  - Side effects associated with increasing doses
  - Early evidence on effectiveness
- Phase 1 studies evaluate
  - Drug metabolism
  - Structure-activity relationships
  - Mechanism of action in humans

Clinical Phase 2 Trials

- In phase II trials, the experimental study drug or treatment is given to a larger group of people (100-300) to see if it is effective and to further evaluate its safety
- Helps to determine the common short-term side effects and risks associated with the drug and to determine optimal dosing (dose and regimen). This is done by subjecting the volunteers/patients through various strengths of dose and arriving at the optimal dose
- The trials done are typically well controlled and closely monitored

Clinical Phase 3 Trials

- In phase III trials, the experimental study drug or treatment is given to large groups of people (1,000-3,000) to confirm its effectiveness, monitor side effects, compare it to commonly used treatments, and collect information that will allow the experimental drug or treatment to be used safely
- Intended to gather the additional information about effectiveness and safety that is needed to evaluate the overall benefit-risk relationship of the drug
- Helps in establishing that the current drug works better than other drugs currently available
- Provide an adequate basis for extrapolating the results to the general population and transmitting that information in the physician labeling

Table 1.

	Number of subjects (approx.)	Duration (approx.)	Purpose	Percentage drugs successfully tested
Phase I	20 to 80	Up to several months	Primarily safety dose (based on tolerance)	70%
Phase II	100 to 300	Up to 2 years	Effective dosage, short term safety/ side effects	33%
Phase III	1000 to 3,000	Up to 4 years	Efficacy and comparison to commonly used treatments, long term safety	25 – 30%

The three phases of clinical trials are followed by submissions and approvals. There are designated organizations in various countries that review and approve drugs in various countries.

An NDA is filed following the successful testing on humans. A New Drug Application is the submission of documentation of safety and efficacy of the new drug to seek approval for marketing. It is also known as a "Marketing Application". The FDA in US reviews the NDA and approves the product for manufacturing and launch.

Table 1 summarizes the typical activities and success ratio across the 3 phases of the clinical trials

In the last several years with drugs showing toxic side effects post approval, a new phase called Phase 4 studies or Post marketing surveillance has been introduced. Post-marketing surveillance ensures that drugs once launched commercially, the safety of a drug is monitored closely. Both regulatory authorities and life sciences companies share responsibility for post-marketing surveillance. Here are some additional details on this phase.

- Post-marketing surveillance (also known as Phase IV trial) is the practice of monitoring the safety of a pharmaceutical drug or device after it has been launched in the market and forms an important part in retrospective or observational studies. Post-marketing surveillance is a part of Pharmacovigilance, an area that we will cover in subsequent sections. All Post-marketing surveillance studies are phase IV trials; however, the converse is not true
- Since clinical trials are conducted in a comparatively controlled environment on fewer patient subjects, there is quite a chance to miss out studying the impact of the drug on medical conditions that might prevail in general populations, other than the participating subjects; hence, post-marketing surveillance can further refine, or confirm or deny, the safety of a drug after it is used in the general population by large numbers of people who have a wide variety of medical conditions in a neutralized environment
- Post-marketing surveillance is a mandate imposed by the FDA for drug safety. Various approaches are used to monitor the safety of licensed drugs such as spontaneous reporting databases, prescription event monitoring, electronic health records, patient registries and record linkage between health databases. These data are reviewed to highlight potential safety concerns in a process known as data mining
- In certain instances, a licensed drug's labeling might need to get updated, its indication may need to be limited to particular patient groups, or in rare cases, the drug might need to be withdrawn from the market completely

Table 2.

Process	Typical Activities
New Entity	1. Molecular modification 2. Organic synthesis 3. Medicinal chemistry
Pre-Clinical Studies	1. Pharmacokinetics 2. Pharmacology/Toxicology 3. ADMET 4. Pre-Formulation/Formulation 5. Manufacturing control
IND	1. Submission 2. FDA review and approval
Clinical Trials	1. Phase 1 2. Phase 2 3. Phase 3 4. Formulation 5. Manufacturing controls
NDA	1. Submission 2. Review, Inspection and Approval
Post Marketing Surveillance	1. Product defect reporting 2. Adverse event reporting

The entire process of drug discovery and development and the associated phases can be summarized in table 2. (Michigan Institute for Clinical and Health Research, n.d; Cern Foundation, n.d; ClinicalTrials.gov, n.d)

Regulations Involved in Drug Development

The process of clinical trials and pharmacovigilance is highly regulated and governed by FDA in the US, EMEA in Europe and MHRA in the UK and other such authorities in different countries.

Here are some reasons why regulations and regulatory authorities are critical for success.

- Ensuring Good practices during the Laboratory research phase, Clinical phase and the Manufacturing phase
- Ensuring that the consumer (patients, clinical research subjects) are informed and their consent is obtained
- Ensuring that the identity of patients / clinical research subjects are protected
- Ensuring Security and Authenticity of e-data submitted to the FDA (21 CFR Part 11)
- Ensuring that all the Life Sciences processes and applications perform their functions "as intended" (Computer Systems Validation)
- Ensuring Safety, Efficacy and Quality of the end-product

Relevance of Computer Systems in Clinical Trials

Clinical trials are structured and conducted in a very scientific manner to ensure hypothesis generation and validation using statistical techniques. The clinical trials generate data over a period of time which

Table 3.

Study design and planning	Site initiation	Patient enrollment and recruitment	Study conduct	Analysis, reporting and study close out
1. Study planning 2. Protocol development 3. Investigator selection	1. Site selection 2.. Site initiation 3. Contract management	1. Patient screening 2. Patient recruitment 3. Informed consent 4. Patient enrollment for specific studies	1. Subject registration and randomization 2. Data entry in case report forms (CRFs) 3. Data validation and cleansing 4. Query generation and resolution	1. Data analysis, reports for submission 2. Database lock 3. Data retention and archiving

needs to be stored, processed and submitted in pre-defined formats and these are the areas where computer systems are playing a big enabling role to make the overall process efficient and reliable.

Most common experimental designs implemented in clinical trials across phases are randomized controlled trails (RCT) but there is a new emerging area called adaptive design that is being used frequently nowadays. Adaptive design is to make the design flexible to account for findings from previous stage of clinical trials. Adaptive design will be explained later in this chapter. The key aspects of RCT are as follows.

- Randomization: Subjects are assigned to intervention vs. Control treatment in random process to overcome confounding factors and any imbalance in the assignment.
- Blinding: None of the participants or researchers knows about the specific treatment details. This is done to reduce bias and unwarranted special care to any patients as that may manipulate the results
- Stratification: Patients are stratified to create balance among the stratifying variables such as gender, age and so on. This is done to achieve balance with respect to stratifying variables and also gain knowledge about sub groups in the patient population.

The various activities within the clinical trials process can be outlined as in Table 3. Each trial has one of more studies as part of the trial process. Each study involves enrolment of sites typically hospitals or clinics and a principal investigator in the site who in turn recruits patients along with his supporting staff within each of these identified sites.

Each of these activities involves set of start and end points that are inter related and come together to ensure a successful trial.

There are several computer systems that support the activities outlined above. Some examples of such systems are as defined below.

Electronic Data Capture Systems (EDC): These systems help capture all the patient information in a structured format so that they can be easily extracted and analyzed at various points during the conduct of the studies. This is set up for every study

Clinical Trial Management Systems (CTMS): These systems help capture trial related information like sites, investigators and other non-patient related information which are important to the conduct of the trials

Clinical Supplies Management Systems (CSMS): These systems capture information related to supply of materials used in the conduct of the trials

Clinical Data Repository (CDR): These systems help bring together information from multiple studies/ trials and make them available for analysis to arrive at insights and conclusions that are then used to file for approval with the regulatory authorities

Usually these systems do not operate in isolation. The systems are integrated and data flows across them thereby enabling easy understanding of the status of trials and insights on the progress of the trial and the ability to meet the expected end point.

Relevance of Computer Systems in Pharmacovigilance

Pharmacovigilance is the science of collecting, monitoring, researching, assessing and evaluating information from life sciences companies, healthcare providers and patients on the adverse effect of medications, biological products, etc. with a view to:

- Identifying new information about hazards associated with medicines
- Preventing harm to patients

Why Do We Need Pharmacovigilance:

- Unexpected adverse reactions
- Drug-Drug interactions
- Long term efficacy
- Quantify and recognise risk factors
- Adverse drug Reactions (ADRs)
- Limitations of pre-marketing data:
  - Pre-Clinical data: insufficient predictive value
  - Clinical Trial data provide provisional evidence of safety but: Too small (patient number limited),limited duration, do not represent the real world

The Pharmacovigilance process can be outlined as in the table 4 below. The process starts from the receipt of an adverse event (AE) and moves from the intermediate processing steps to determine the nature of the event eventually leading to a time bound submission to the health authorities. (Uppsala, n.d)

The above processes are supported by a set of computer systems that enable proactive monitoring and reporting. Some such systems are explained below.

Adverse Events Reporting Systems (AERS): These systems help capture adverse events and have workflows that enable processing and eventual submission to health authorities

Table 4.

Process	Activities
Adverse event case receipt	1. Receive AE case 2. Document receipt 3. Index, file source documents
Adverse event case triage	1. Identify duplicate adverse event cases 2. Assign priority 3. Enter other case data into adverse event reporting system (AERS) 4. Perform preliminary quality assessment of data entered
Event assessment	1. Prepare company narrative for review 2. Assess case from medical perspective 3. Perform final review of case for report ability
Processing follow up information	1. Identify additional information required to analyze and report the case 2. Follow up to obtain additional information 3. Update additional information in AERS
Risk/Benefit Analysis	1. Perform risk/benefit analysis based on AERS data and data available with regulatory agencies 2. Prepare analysis reports
Regulatory submission	1. Prepare safety report 2. Facilitate final review by regulatory affairs 3. Submit report to regulatory agency 4. Track submission date of report and follow up, if any

Safety Datawarehouse: These systems bring adverse events information from various different sources and store it in a single location to enable reporting. Various health authorities have different requirements and frequencies for reporting safety adverse events.

One of the key benefits of a safety warehouse is also the ability to run signal detection. Signal detection is a pro-active approach to identifying potential adverse events and correlations.

Signal Detection: Signal detection aims at finding causality between intervention(s) to adverse event(s) that have not been earlier found or refuted. These findings which can be beneficial or harmful require\deserve immediate attention from the concerned groups. The findings deviate from what is expected from the drug or medical entity. Signal detection is traditionally done by examining the medical evaluation of case reports. But a more comprehensive and practical way of analyzing is through data mining. Data mining allows for a systematic evaluation of the adverse events. Generally a score is assigned based on statistical computation and the higher magnitude of the score reflects stronger association between the adverse event and the drug. Ultimately these scores act as a guideline for providing safety alerts. Based on previous knowledge and understanding of the current data, a threshold is chosen for these scores. Scores exceeding this threshold requires attention. The thresholds for detecting safety signals are a trade-off between sensitivity and specificity. Too stringent threshold results in missing significant signals and thereby effecting sensitivity and too relaxed threshold results in too many false positives impacting the specificity. (Uppsala, n.d)

FUTURE RESEARCH DIRECTIONS

Innovations in Clinical Development

Clinical development is also evolving with technology as new techniques like simulation, predictive modeling, and process automation mature and is available to be used in the design of better trials. The concept of personalized medicine and diagnostics is the new focus as safety events make it evident that one drug does not suit all patients.

Adaptive Trial Design

An adaptive clinical trial is a clinical trial that evaluates patients' reactions to a drug beginning early in a clinical trial and modifies the trial in accordance with those findings. The adaptation process continues throughout the trial. Modifications may include dosage, sample size, drug undergoing trial, patient selection criteria and "cocktail" mix. In some cases, trials have become an ongoing process that regularly adds and drops therapies and patient groups as more information is gained. The aim is to more quickly identify drugs that have a therapeutic effect and to zero in on patient populations for whom the drug is appropriate. A key modification is to adjust dosing levels. Traditionally, non-adverse patient reactions are not considered until a trial is completed.

The FDA issued draft guidance on adaptive trial design in 2010. In 2012, the President's Council of Advisors on Science and Technology (PCAST) recommended that FDA "run pilot projects to explore adaptive approval mechanisms to generate evidence across the lifecycle of a drug from the premarket through the post market phase." While not specifically related to clinical trials, the Council also recommended that FDA "make full use of accelerated approval for all drugs meeting the statutory standard of addressing an unmet need for a serious or life threatening disease, and demonstrating an impact on a clinical endpoint other than survival or irreversible morbidity, or on a surrogate endpoint, likely to predict clinical benefit."

Adaptive trials have resulted in accelerated trials and are supportive of personalized medicine and sub population strategies explained in the next section below. (FDA, n.d)

Translation Medicine and Personalized Medicine

Translational Medicines: Translating Laboratory Discoveries to New Therapies

Translational research is typically a three step process where studies performed in a lab are translated into treatments and/or therapies that are effectively integrated into medical practice.

The first step of translational research is investigating a topic such as the cause of a medical condition or a specific treatment. Researchers then look for new ways to approach some of medical science's most confounding issues.

Discoveries made in the laboratory are used to prompt clinical trials where research findings are given a practical application. During this phase, study volunteers and participants are exposed to the findings from the lab and data about participants' responses, feelings and reactions is recorded. If a positive outcome is achieved, the third step is to put the finding into practice. Doctors implement the researcher's findings, supported by the results from the clinical trials, and patients receive a new form of treatment or therapy.

This three step circular process is often described as taking a "bench to bedside" approach where researchers take what they learned at the research bench and apply it to patients at their bedside.

The process being circular starts at different points and applies the techniques to find effective drugs. (Clinical and Translational Science Institute, University of California Los Angeles n.d)

Personalized Medicine

Personalized medicine refers to tailoring the drugs based on the genetic profile of the patients. The focus is on how patients with different genetic variations respond to the therapeutic treatment and thereby design the drugs accordingly. The advent of genomics technologies like next generation sequencing technologies eases and facilitates personalized medicine. Most of the clinical trials fail at later stages of trials such as Phase III. The main reason is the inappropriate selection of targets. This can remedy pharmacogenomics which is a major tool for personalized medicine. This ensures whether variations in molecular makeup of patients attributes to inefficacy and it can be proved that the intervention can be beneficial to patient populations as determined by the pharmacogenomics.

Centers conducting clinical trials should use real world clinical data obtained from medical institutions, hospitals and academic centers and use them as basis in clinical trial design as most of the clinical trials are controlled designs gathering information from real world data does help. Genomes or exomes of subjects participating in clinical trials needs to be sequenced and subjected to therapeutic intervention to undermine risk benefits of drugs based on genetic profiles. Personalized medicine also helps determine optimal does for the patients. (FDA, n.d)

Simulations and Human Disease Models

Computer simulation of clinical trials has evolved over the past two decades from a simple instructive approach to complete simulation models yielding pharmacologically sound, realistic trial outcomes. The need to make drug development more efficient and informative and the awareness that many industries make extensive use of simulation in product development have advanced considerably the use of simulation of clinical trials in pharmaceutical product development over the past decade.

The availability of data through various open industry initiatives and the reducing research productivity is driving significant adoption of simulation models and making it an integral part of the drug development process.

The other aspect of simulations is complete disease models. These are human models built with the understanding of biology and replicate the behavior of the disease in the computer human model built using experimental and derived data. These models provide researchers with a tool to both understand the mechanisms driving untreated disease progression and to predict physiological responses to therapeutic interventions, including variability across different patient phenotypes. (FDA, n.d)

Innovations in Pharmacovigilance

Pharmacovigilance is moving from a reactive reporting focus to a proactive model where monitoring of safety events start very early and multiple channels including social media are being effectively leveraged to track and report such incidents.

Increasing Use of Social Media to Capture Safety Data

The healthcare industry uses digital media for a number of purposes, including raising disease and treatment awareness, safety issues of drugs, clinical trial enrolment, recruitment and patient support. With people engaging on health issues more and more online, it is essential that companies track this information and derive important insights out of it to have a better vigilance on patient safety

CONCLUSION

Viewpoint: Future of Drug Development

Drug development is a very old process and over the period as the complexity of diseases has increased, most of the known diseases have been addressed, is facing increasing challenges around innovation and more importantly productivity.

There are several late stage failures of drugs and this attrition is causing an increase in the cost of every successful drug that comes to the market. Clearly there is a need to re-look at the approaches being used in drug development and very rapidly bring in changes that allow for more focused way to improve the techniques used and increase innovation.

As explained in the new innovations in this chapter, there are several new techniques and technologies that have emerged over the last decade which raise hopes of a dramatic change in the way drug development is done. Also the move from one drug fits all to a more focused targeted approach also enables bringing new therapies to the market faster.

One other key trend that will continue to evolve is the use of computer systems across the drug discovery and development processes to increase chances of success as the drug moves through the value chain. This is now supported by increasing availability of data which is a key for success of any computer based approach.

In summary computer systems are becoming increasingly important for drug development and their use will increase steadily. We will study more details in the subsequent chapters more details on their usage and applications.

REFERENCES:

Clinical Pharmacology I. (2002). MCCQE Review Notes. Retrieved from http://www.ucl.ac.uk/anaesthesia/education/Pharmacology

Clinical Trial Phases. (n. d.). Cern Foundation. Retrieved from https://cern-foundation.org/?page_id=292

Food & Drug Administration. (n. d.). fda.gov.

Glossary of common terms. (n. d.). ClinicalTrials.gov.

Glossary of terms used in Pharmacovigilance. (n. d.). Uppsala. Retrieved from http://who-umc.org/Graphics/24729.pdf

Gombar, V.K., Silver, I.S., Zhao, Z. (2003). Primary Role of ADME characteristics in drug discovery and their in silico evaluation: in silico screening of chemicals for their metabolic stability. Retrieved from http://www.ncbi.nlm.nih.gov/pubmed/12769701

R&D Pipeline Management. (n. d.). University of Wisconsin. Retrieved from http://maravelias.che.wisc.edu/?page_id=23

University of California Los Angeles. (n.d). Clinical and Translational Science Institute, UCLA.

What is a clinical trial? (n. d.). Michigan Institute for Clinical and Health Research. Retrieved from https://www.michr.umich.edu/about/whatisaclinicaltrial

KEY TERMS AND DEFINITIONS

Clinical Study: A research study using human subjects to evaluate the effect of interventions or exposures on biomedical or health-related outcomes. Two types of clinical studies are interventional studies (or clinical trials) and observational studies.

IND: Investigational New Drug.

NDA: New Drug Application.

Pharmacovigilance (PV): The science and activities relating to the detection, assessment, understanding and prevention of adverse effects or any other drug-related problem.

Chapter 2
Data Warehouse and Data Virtualization

Kanishka Mukherjee
GVK Biosciences Private Limited, India

ABSTRACT

This chapter contains as well as illustrates different innovations that is changing the way Clinical Drug Development and Safety organizations perceives IT and the ways and means through which these innovations are facilitating the change in business itself. The main content contains illustrations of two structurally different means to create data warehouses, the benefits of the approaches and the difficulties. It also explains the importance of data virtualization technology when implemented in the Clinical and Safety Organizations.

INTRODUCTION

Life Sciences Companies are not facing the lack of data as a problem any longer; their problem is absence of '*right*' data at '*right*' time. Business leaders not only require '*right*' data, they require it at '*right*' *(real)* time in order to effectively define the strategic direction of the organization. The problem is, most of the currently available of applications and/or software was designed for functionality, most of the data warehouses and the associated means of information dissemination are based on reports, with a built in time-lag. This is in order to either execute complex algorithms on voluminous data or due to traditional ETL approach for data transportation.

The innovative future generation of applications will require real time access to data and timely analytics as well. While strategically this would require a change of corporate mindset; technically it will require establishment of correct data structures, in order to not only analyze internal data but also to absorb external (structured/unstructured) data, so that algorithms can execute seamlessly to provide uniformly right output to help in deducing right strategy.

In this chapter, we will go through the details of innovative means and mechanisms of utilizing optimal Clinical and Safety aligned data structures; from which and to which, appropriate data can be extracted and retained, for various forward looking as well as past action analysis. We will also look very briefly

DOI: 10.4018/978-1-4666-8726-4.ch002

into importance and integration aspects of external unstructured data (including but not limited to social media) which not only provides voluminous relevant information but also how can they be segregated and analyzed with various parameters to provide extremely worthy and pin-pointed inputs about geographies, markets, competition, products, usage, user satisfaction, adverse reactions etc. We will also discuss to some detail on methods used for data virtualization and the advantages of data virtualization

The chapter is divided into multiple sections and subsections describing various innovations in data warehouses in Clinical and safety areas. First two sections of this chapter deals with Business Case and Resources (human and infrastructure/software etc.) though these two topics are not must for this chapter, I felt, it is important to understand these topics for innovations and successful implementation of a data warehouse project.

1. DEVELOPMENT OF BUSINESS CASE

Multiple studies has shown historically Data Warehouse project don't have high success rate. A study published by Amin & Arefin (2010), list top 20 reasons on why 50% of the Data Warehouse projects fail, 6 out of 20 reasons listed, deals with failure to acquire right Business Case and failure to perform right stakeholder mapping at the start of the project. The main reason to include this section in detail in this chapter is to put further emphasis on the fact when building an innovative application where results could be radically beautiful on one side, on the other side, chances of failure are also high. In such conditions, the business case assumes greater importance.

The most important factor, of the whole sequence, would be for the organization and its leaders to jointly decide, on exactly what they perceive as outcome and what the measurable benefits of the upcoming system are and hence a business case should be developed. It is essential that along with the objectives of the project the Business Case includes as key success criteria; the right accountability structure within the organization as a whole (*business owner*) and for the project organization (*project manager*), right business needs, cost, and estimated business benefit to be derived, on a clear time scale. Who (*business teams*) gets what (*functionality, reports etc.*) when (*timeline*), for how much (*cost, resource/ SME allocation*) should be documented. Finally the risks associated with the achievement of the project objectives is crucial to be identified and documented albeit at a high level at the very beginning, since if kept fluid, this is observed to be the single most important cause of failure of data warehouse projects. A feasibility study could be undertaken during the development of Business case. It allows for a more detailed analysis and assessment of the business problem/s to be dealt by the project team. It also support the team to identify risks and visualize possible pitfalls in advance and finally it present options to mend the solution design envisaged for the project.

It is also vital that one keeps in mind the right template of the Business case which aids to record the essentials, I will not delve into the details and the formats of the business case templates used in the industry however the reader should remember that all the relevant information gets identified, agreed upon, documented and signed off before starting to plan the project execution. One has to remember that this document is used as reference multiple times during the project. At each Quality Gateway the Business Case is used to validate if the benefits, costs and risks currently projected are matching to what was projected in the Business Case. During the Project Closure Meeting the Business Case is discussed with all the stakeholders and the Business Realization Metrics (BRM) is base lined; in this regard, the

Figure 1. Components of Business Case

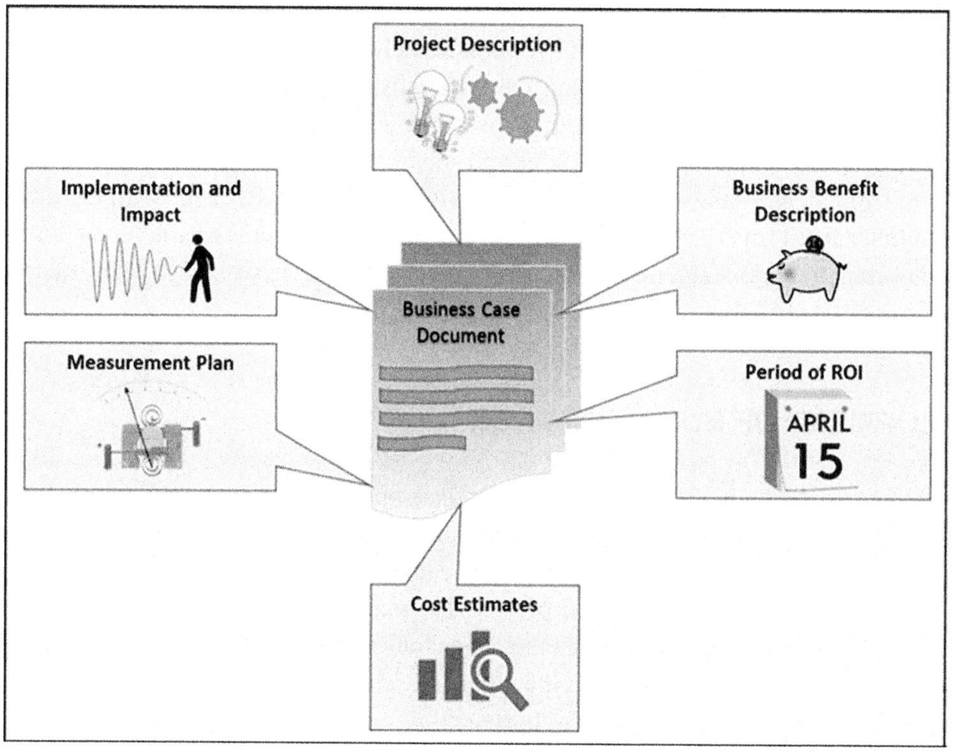

success of the project is measured against the ability of the project to deliver the criteria outlined in the Business Case.

I would like to illustrate the importance of identifying the right business owner and getting their buy-in, using an example case study *(the names of key stake holders and organizations are blinded to protect privacy and maintain confidentiality). In this case, a pharmaceutical organization in China wanted to implement a data warehouse for its clinical data for phase IV and V & BA/BE studies for further analysis. The project was driven by the IT organization of the Pharmaceutical Company using the Marketing Company's own budget, where no key stakeholder from the medical organization was involved with a defined ownership to sponsor the project either during conceptualization or during the development of business case. Business case was developed by the IT organization with their own understanding of the business problem under the guise that "Business didn't understand IT" and business case was approved with neither the key business owner nor the project manager identified. The project was awarded to one of the preferred vendors who was very capable in clinical domain and had prior experience in building clinical data warehouse but had very little presence in China; without any RFP.*

Project went on hold within few weeks since the business refused to provide requirements and kept stonewalling the vendor. After escalations, when the Medical and Clinical organizations ultimately agreed to allocate one SME each to provide requirements, they (SMEs) showed no interest in meeting the project timelines or budget. There were multiple requests for prototypes and samples of static reports, requests for reports generated using mock data for multiple dimensions and also a request for providing training to the larger business team in the concepts of data warehouse in Chinese etc. The IT organiza-

tion and the vendor both didn't anticipate this risk was simply incapable of handling it. It was found that the staffs of the medical and clinical organizations were genuinely not geared up for the project. They really didn't understand the technology, were not articulate enough to provide the requirements to a vendor mostly staffed by non-Chinese personnel and perceived the predicted business benefits too simplistic and not comprehensive.

The project was shelved within 3 months.

Here we see that, the Marketing Company's IT organization made a classical mistake of presuming a business project as an IT project and not involving the key stakeholders at the start. They also didn't perform a feasibility study, which surely would have brought out majority of the issues faced during the execution.

In another case, a European Pharmaceutical company engaged a vendor to build a clinical data warehouse and operationalize the functionality of clinical data harmonization using systematic means, a large, complicated and highly innovative (first of its kind in industry) project. The management of the pharmaceutical company identified a business sponsor and a program manager from the business team, who created a right business case and involved the right stakeholders from various departments including IT to set up the program structure and involved the vendor at an early stage in order to maintain right level of transparency. Even though the program team faced data standards related challenges during the end stage and the program was closed before complete benefit realization was accomplished; the program was concluded to be fairly successful.

Here we see that stakeholder management was excellent right from start. The business case was prepared well and the program governance involving the key stakeholders was set up well in advance. The roles and responsibilities were identified and the accountability structure established. And even though the data issues were identified at late stage, earlier POCs were successful since all involved were aware of the whole problem, acrimonious situation was avoided and the program was closed amicably.

Irrespective of the type of project a business case and the establishment of a business owner are the most important criteria towards the success of the project. Since the success of Data Warehouse projects not only lies in the success of its implementation but primarily in its usage; and involves lot of organizational change management and other political aspects, the presence of an identified business sponsor/owner and well accepted business case plays a larger role.

2. IDENTIFYING RIGHT RESOURCES

In this section, I will start by distinguishing the phrase "Right Resources" as human resources and other resources such as infrastructure – software /hardware, connectivity etc. I will talk about the human resources element first, since human resources are the sources of inspiration and innovation.

Outcome of any project or program depends on multiple factors, some which we will cover during the course of this chapter. However, one of the most critical is, right resource identification for every key roles. Here, I am also including the fact that attributing accountability and authority to each key roles could be equally critical to the success of the program.

Normally, the first resource to be identified for the project is, the project manager. Many organizations including the SI vendors make the mistake of conceptualizing a project and planning its deliverables using highly technical and business resources alone, while it may work in some cases, it certainly isn't a best practice. After all the accountability of the delivery of the entire project lies with the project manager, if that person can be trusted with such a high responsibility, it is only fair that the person is authorized to have a say during the project conceptualization, business case development and planning phases. Another key factor influencing the success of implementation of data warehouse, could be to identify the project manager preferably in-house (though this is not often practiced) as the application owner. This encourages a level of ownership to justify the existence of warehouse not just build it.

In the case of a data warehouse (non-clinical) project involving a European pharmaceutical company in India a PM from a vendor SI organization was chosen purely on the basis of seniority and technical experience and no attention was paid to his ability to lead, communicate, building relationship with the customer. Everything went smooth for the first few months but once the timeline pressure and requirement volatility crept in the PM's lack of project management skills were clearly exposed. Various team members were blamed upon, there were several high level escalations, teams were overworking, change requests were not signed off, and on the whole project status was "red".

In another case a PM of a project from a SI vendor, told a senior customer stakeholder from QA department, during a meeting, that the vendor organization is known for bidding at low cost and the customer stakeholder should be happy with less spend and not bother about the quality processes followed during the project. It caused major embarrassment and all parties had to answer some hard questions.

In another case a transition manager informed his PM about the delay in transition due to consistent non-availability of the customer SMEs. The PM reacted quickly and officially escalated the matter to his counterpart in the customer organization and other senior key stakeholders about the impending delay in transition activity and resulting impact on the timeline and cost. The matter was resolved immediately and vendor team was appreciated for standing out at the right time.

It is not true that every time a good, experienced PM would succeed, since there are various other points of failure, but the probability of success increases with a good PM.

The critical resources such as architects, domain experts, business analysts/SMEs, project leads should also be identified in conjunction with the project manager. The organizations should take care of fact that project complexity, project size, risk assessment, business impact and decide on the resources based on their experience, past performances, interpersonal skill before loading the project team. This might sound idealistic but it is true that a resource with excellent technical skill and very poor interpersonal skills could be as detrimental to the project as a technically poor resource in key role. Another factor which should influence key resource selection is the resource's ability to lead and take accountability of his/her actions and create an environment of trust around them. If the leader cannot take accountability the people will not trust their leader and when that happens the system collapses. The business sponsor, project manager and others of the governance team should pay key attention to this aspect during the course of the execution of the project too and should take immediate action when they see any sign of project organization disruption.

Next step for the PM is to draw out a clear RACI chart. This should include and is to clearly communicate his/her expectations of various stake holders and not only the core project team. This exercise should be accomplished before the start of the project and communicated to all the stakeholders and acquire their agreement. The RACI should include key activities to be accomplished during the course of the project in one axis and key stakeholders in another axis. The metrics should map the RACI for each stakeholder with the activity. Ideally the outcome of the RACI should provide the governance structure too at various levels.

Apart from Human Resources, the PM should also assume the ownership to identify the other key resources for the project. Having the business case handy, the PM should take various 'make/use or buy decisions' in regards to the Infrastructure components such as Servers, Desktops/Laptops; the location of the infrastructure – hosted within the organization domain such as local data centers or hosted outside as a part of a third party cloud – dedicated or multi-tenancy. Each decision needs careful evaluation since they would have impact on the outcome of the project and are strategic in nature.

In one case an APAC market division of a pharmaceutical company took a strategic decision to host its data warehouse (non-clinical) out of its Singapore data center instead of the local market. By using the shared infrastructure, they not only saved on the cost of procuring new infrastructure, the success of the data warehouse project actually encouraged multiple small APAC markets to pay for building GUI interfaces/reports in their local language and capture local data in the same warehouse. It brought down the individual market spend on the warehouse significantly.

While this project was highly successful. But in many cases the PM will have to take these decisions based on non-functional requirements (NFR) such as performance expectations, security, compliance and data privacy etc. There are multiple cases where a Data Warehouse project didn't reach its potential or fail because the performance was bad because of poor infrastructure sizing.

Before I talk in detail about the innovations in Clinical and Safety Data Warehouse Development in recent years, I wanted to cover and emphasize the above two points. As Steve Jobs (Fortune, 1998 and "TIME digital 50" in TIME digital archive (1999)) rightly said, *"Innovation has nothing to do with how many R & D dollars you have. When Apple came up with the Mac, IBM was spending at least 100 times more on R & D. It's not about money. It's about the people you have, how you're led, and how much you get it."* Key resources of a project, the leaders, encourage everyone involved in it including themselves to come up with new, innovative solutions. It really doesn't always require a sophisticated computer lab to build new software. It was also Steve Jobs (The Seed of Apple's Innovation" in BusinessWeek (2004)) who said, *"The system is that there is no system. That doesn't mean we don't have process. Apple is a very disciplined company, and we have great processes. But that's not what it is about. Process makes you more efficient. But innovation comes from people meeting up in the hallways or calling each other at 10:30 at night with a new idea, or because they realized something that shoots holes in how we've been thinking about a problem. It is ad hoc meetings of six people called by someone who thinks he has figured out the coolest new thing ever and who wants to know what other people think of his idea. And it comes from saying no to 1,000 things to make sure we don't get on the wrong track or try to do too much. We're always thinking about new markets we could enter, but it's only by saying no that you can concentrate on the things that are really important."* At the end, it is the people and their passion which matters most, being human is possibly the biggest innovation.

3. CLINICAL AND SAFETY DATA WAREHOUSE DEVELOPMENT AND MAINTENANCE

In this section, I have highlighted multiple recent innovations made in the data warehousing area in the recent times. I am also going to cover using two different case studies to showcase how these innovations are being used in Clinical and Safety areas. In the first case I have described, how the organizations are developing clinical and safety data warehouses, instead of depending on a large enterprise data warehouses. In the other case, I have tried to display, how few organizations are using federation of data warehouses and data bases for to cater to newer areas of discovery as well as clinical development. There can be multiple ways one can design a data warehouse. Choices are Federated, Centralized or De-Centralized. They have their pros and cons. My objective here is not distinguish one over the other but show case the probabilities to the readers.

3.1. Innovations in Data Warehouses applicable to Clinical and Safety Areas

In this section, I have included few of the innovative trends (in brief) in the data warehousing and related technologies which are fast getting adapted in the Clinical and Safety area.

3.1.1. Migration of Clinical and Safety Data Warehouses to Cloud

Many organizations are developing Clinical and Safety Warehouses, which are hosted on Cloud. These applications are available to the pharmaceutical organizations through Private, Multi-tenant and even public cloud options through various commercial models. Obviously, compliance, security, cost of purchase, cost of operation, cost of data storage, flexibility, performance of the applications, time and cost to scale up/down and other parameters are to be kept in mind. However, there certainly is a demand in the market for such applications. Increased dataflow is increasing cost of infrastructure and maintenance, cost of continuous enhancements of warehouses are forcing many pharmaceutical companies to weigh these innovative options for multiple critical warehouses.

3.1.2. Configuration and Enhancement of Data Warehouses to Hold Data from Social Media, Mobile Devices and Other Devices (Internet of things)

The new generation of patients are connected. Social media carries enormous amount of information today about every facet of their life in general, their opinions, and their activities. This sometime includes information regarding medicines they intake, how they feel about it and if there are any discomforts due to the medicinal drug. Pharmaceutical companies are in need of storing and analyzing this information.

One of the European Pharmaceutical companies have built a tool to interface with the patients during the clinical trial. One of the major SI vendors is also working on developing a mobile interface for the patients to communicate with the sponsors.

In another example, the same Pharmaceutical company and SI vendor together built and implemented field monitoring application for the investigators, using a mobile device.

Major mobile application producer companies are also building mobile specific clinical apps to help the cause.

In another case a US based Pharmaceutical Major is partnering with multiple US based pharmacies to integrate the out of medical devices such as blood pressure machines (sphygmomanometers providing digital output) with their in-house data center. This is allowing the pharmaceutical company to monitor the patients more closely.

These mobile technology based applications are proving to be extremely productive tools to not only execute an effective trial but also to bring down the R&D expenses. In the days of spiraling R&D and Healthcare cost, this indeed is a good news.

As mentioned above, while innovations in the examples, are good for clinical development area, the excellent news is that the Data Warehouses are getting better qualified to store and analyze the data.

3.1.3. Integration of newer technologies which can handle large quantity of data for analytics

As I wrote at the start of this chapter, analytics with 'Right Data' at 'Right Time' is playing a major role in today's business. In the above sub-section, there is an awesome torrent of data that is being generated by Social Media, Mobile devices and others. Few years back such volume of data would have been unheard off. To meet the reality though it absolutely necessary for the Data Warehouses to evolve and integrate with software that are adept to handle large volume of data to analyze the same and provide real and correct data to the decision support systems.

The US division of one of the European pharmaceutical majors in alliance with one of the premier educational institute has developed few tools to handle Big Data during discovery and drug development phases.

3.1.4. New Data Management Approaches Such as Data Virtualization

The traditional data warehouses were mostly used as vast data storage or archiving. The usage of data was tactical, limited to certain regular managerial reports. It was in between late 1990s and early 2000, the concept of decision support system using business intelligence tools with data warehouses and data marts being the source became popular. However, not all data warehouses could be used for business intelligence and not all business intelligence tools required data warehouses. While traditional data warehouses were quiet useful for historical analysis of data on a time scale, elusive component was the 'single source of truth'. Since the data definition used in various source or data collection systems were not standardized, there was no one source of truth across discovery, pre-clinical, clinical, safety and other downstream applications. This led to discrepancy in reporting and comparative analysis. And it is in my opinion one of the most important reasons why the usage of the data warehouses were not being maximized.

With the advent and increased popularity of Data Virtualization approach across industry the usage of data warehouses are increasing significantly. The best definition so far is by Rouse (2011) where she defined it as "Data virtualization is any approach to data management that allows an application to retrieve and manipulate data without requiring technical details about the data, such as how it is formatted or where it is physically located." Another way it can be defined as "Data virtualization is process of integrating data from many disparate sources in real-time or near real-time basis." It involves integration (with source systems and support system (*such as MDM etc.*), transformation (*standardization using Enterprise Service Bus Views using Canonical Model*) and data delivery hub (*using web services etc.*) to support applications and processes to support various business requirements. This approach requires creation of standardized data definition across all data elements.

While data virtualization can bring out the 'single version of truth' and makes it possible for multiple applications accessing and modifying data from various locations while keeping the sanctity of the data definition intact there making the comparative analysis of historical data possible in several dimensions; operationally to implement it in an organization is extremely difficult to say the least.

We will discuss this topic in detail again in later sections.

3.1.5. Integration of Various Data Mining Tools

Many of the pharmaceutical companies, I have worked with in the past or helped building their clinical data warehouse either had known data existed but didn't know where or didn't even know the data existed at all. One should not undermine the immense power of good data. The Clinical and Safety organizations are clearly understanding this trend. And the demand is resulting in few excellent data mining tools (which can also perform pattern matching for chemical/organic molecular drawings including DNA and genomic structures). These tools permit quick time fetches of hundreds and thousands of structures and their details from the database or extract similar structures as the input from among millions of records, which otherwise would have been impossible to achieve manually. These tools enables scientists to analyze the scientific data for structural similarity.

3.1.6. Literature Mining or Biomedical Text Mining (also called as BioNLP)

This is another area of software innovation where the tool supports creation of relationship with texts (especially in the area of Drugs, Diseases, Gene name, Protein etc.) to create a better searchable query in databases such as PubMed.

One way of searching PubMed would be find the name of the drugs used in all publications related to Diabetes and get a result.

The other way of searching using a Literature Mining tool would be to arrive at a list of names of every variant of the disease popularly called Diabetes and similarly arrive at a list the names of all the drugs used for all variants of diabetes.

The result for the second search will be broader and possibly provide richer result.

3.2. Designing and Developing a Clinical and Safety Data Warehouse: An Innovative Approach

One of most innovative approaches adopted by many organizations today is to, build Clinical, Transaction (operational) and Safety data warehouses, instead of one enterprise level warehouse containing all the data. Benefits are clear, better flexibility, better standardization and easier access to relevant data.

The figure above describes the reference architectures of a typical Clinical data warehouse. I will not go into the details of various components of the warehouses. However I would like to highlight the level of complexity of each of the data warehouses and let the surmise the complexity it would produce if they were combined.

The figure above shows the innovative approach of building clinical and Safety data warehouses separately. The reference architecture layout show cases a typical data warehouse of a clinical and safety organization.

The user interface and database (storage) layers are coupled, as is usual for any application. However, the distinction here is, data source for applications or portals are the cleaned, managed and standardized data from the virtualized data marts. The integration layer in this architecture, makes it flexible for different databases existing at different location (cloud or otherwise) to seamlessly integrate with the data warehouse. The architecture segregates data warehouses into Clinical, Operational and Safety data for better control and also significantly increase the speed of access through distributed infrastructure and reduces overall cost of ownership.

Figure 2. Reference Architecture of a Clinical Data Repository

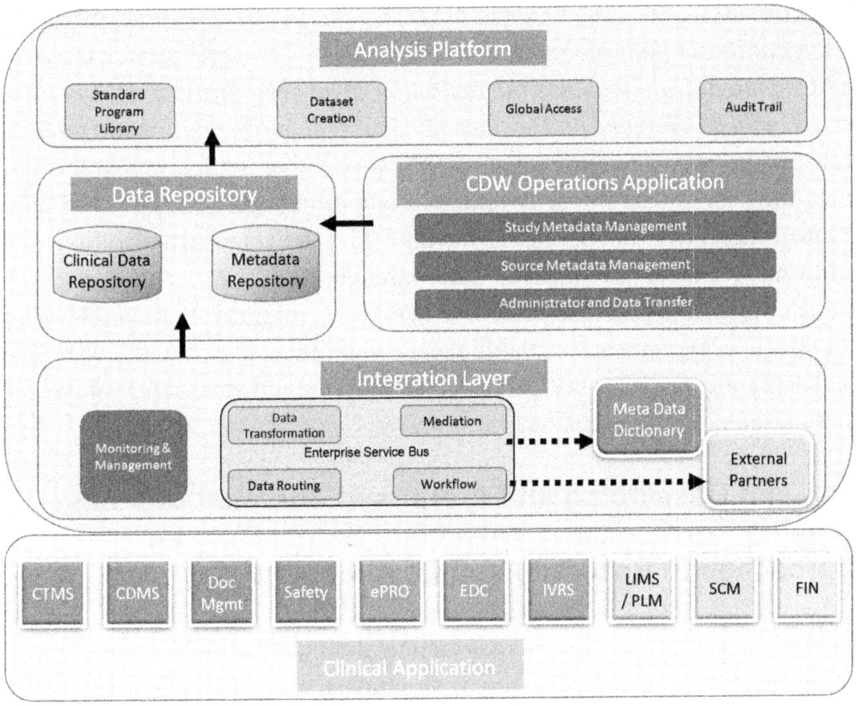

Figure 3. Reference Architecture of an Integrated and Virtualized Clinical, Safety and Operation Data Repository

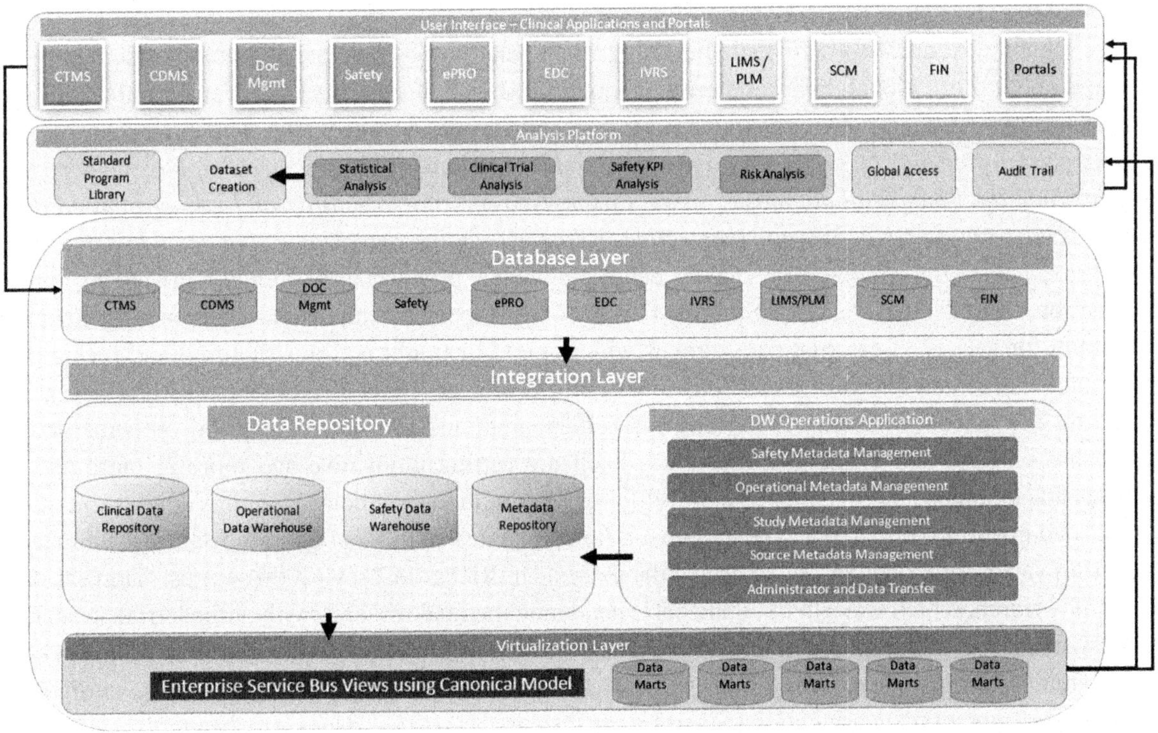

I want to especially highlight the innovative approach of data virtualization at an enterprise level, by the usage of enterprise service bus *(The Enterprise Service Bus (ESB) is basically a software architecture model, used in integration of multiple interacting heterogeneous applications)* views using canonical data model *(The Canonical Data model is the definition of an organization view, of all entities and the relationships between them, designed and implemented for enterprise application integration)*. It in turn enables availability of any data, any time for any application either for analysis or for further processing while maintaining the overall data integrity and not compromising on the speed of data processing or accessing. The Integration layer (not necessarily using ETL) and Data Virtualization are the unique and innovative features not available in traditional data warehouses.

Please note that clinical and operational data can surely be migrated to secure cloud environment. It allows for a cost effective, flexible and scalable means to build and operate the ware houses. However, from security and data confidentiality perspective, and given the sensitive nature of the data, the in-house solution is equally favored by many customers.

3.3. Designing and Developing an Enterprise Federated Database System into using Pre-Clinical, Clinical and Safety Data for alternate approaches to Research and Development

Two of the relatively newer advents in discovery/clinical area are that of Translational Medicine and Drug Repurposing. From a systemic perspective an innovative technological means to achieve better results for both is the introduction of Federated Database Systems. Though Federated Database System or Federation of Databases is not a new concept, usage of it in clinical and safety systems is relatively new. Well, before going any further, the question that can come to one's mind is, what is a federated database system? The best answer I liked so far is, from Sheth and Larson (1990) - 'A federated database system (FDBS) is a collection of cooperating databases that are autonomous and possibly heterogeneous.'

The logical next question here could be, why? What benefit does federated database system provide over and above a traditional data warehouse? I found the answer in the works of Yeung Hall (2007) in Spatial Database Systems: Design, Implementation and Project Management. *The data sources of a database federation may include structured data in relational and object-oriented databases, geometric and attribute data in spatial databases, as well as myriad of semi-structured and non-structured data such as XML programs, flat files and image files. A data warehouse can also be connected to a database federation.* This in my opinion is the biggest benefit and when properly used, highly innovative solution proposition, of Federated Database System. This becomes especially important when one organization is looking for variety of external data sources which might or might not be structured and fitting into the structures created for an organization's own clinical and safety databases or data warehouses. As the entire life sciences industry is looking for better compliance, cost effective and outcome driven healthcare services and medicines; medicines itself are getting innovative and more & more patient centric and personalized. This process of developing 'Personalized Medicine' requires adjustment of the established norms for research and development. It necessitates analysis of genomic data and biomarker data from various publications and external databases. Dr. Richard Casey (2006) writes "Translational medical research, which seeks to integrate basic academic medical research with clinical trials research, will especially benefit from federated systems." As written above another innovative work happening in the area of Clinical Drug Development is 'Drug Repurposing'. Drug Repurposing *is identification of additional potential therapeutics indications of existing marketed drugs or chemical compounds.*

Like Translational medical research, Drug Repurposing is also a data intensive work and what is more, the data is also varied and ranges from data related to drug structures to adverse events to genomic. A federated database system should be apt to build a GRID structure for the purpose of further analytics and outcome for Drug Repurposing.

With the introduction over, I will take you through an in-progress development project, this should help all the readers appreciate the FDBS architecture, the challenges and the innovations that could be brought into the architecture to make it suitable to specific needs.

Business Problem: In this case, the stakeholders is looking to build a platform which is to include 3 different categories of data.

1. Data available from internal as well as sourced from external sources such as public databases, journals etc. e.g. Small Molecules, Activities, Target, Drug, Organization, and Bio-Marker data
2. Data sourced only from external sources e.g. gene expression, pathways, interactome and adverse event data
3. Data received from customers

The expectation from the system is logically and physically store the clean and complete data. Devise a methodology to query multiple data elements for particular set of input parameters and analyze the data further using different algorithms to report possible clues/signs of additional therapeutic indications for the "molecule" supplied by the customer.

Main objective of the system is to bring about a change for Drug Repurposing science by moving it from serendipitous discovery to a data driven approach.

Technical Challenge: Primary technical challenge lied with data, the data sources were quiet varied, ranging from organization's internal data stored in RDBMS, external database, customer provided data to downloaded texts from various PubMed, journals, websites and images. The sheer range of disparate data and the volume made it impossible to convert every data element into a relational database, it was found to be rather easier to store them as is or at the best store as binary image files or as large objects as applicable.

The second challenge was with the completeness of the data fetched from external sources, and the manual cost incurred to cure the data.

The third major challenge was to identify the right data from the internal sources since it was seen to be having duplicate data and also more critically to find and identify right data to use.

Lastly, with extremely high volume of data to be analyzed it was a challenge to develop analytical reports which could be rendered quickly to meet to business requirement

High Level Solution: As a part of the high level solution it was decided to create a Federated database to store and manage all the data elements. Instead of a data cleansing tool it was decided that a set of program, complimenting a GUI would be utilized for the purpose of data cleansing and ensuring data completion. Since, data for same elements are being fetched from multiple sources it was also decided to create clustered autonomous data structures and a multi-layered metadata wrapper based on the ontology with which the data is stored, to not only query data from a single domain but also allows the system to query and compare data from multiple domains. This allowed us to maintain the data integrity of the component schema. The local schema is being utilized to retain virtualized data extracted by the algorithms for further analysis. This 'temporary persistent' layer was to assist quick retrieval of data for various reporting purposes.

Readers please note, I have deliberately abstained from naming the commercially available tools and technologies used for this project. My rational for this are

1. The Case Study is still in progress, and systemic outcome is not available.
2. Essentially, my objective and intention is to showcase the benefits, one could derive, after utilizing federated database in an innovative way in order to obtain desired result, where the business case involves analysis of large volume of disparate data. I leave it to the user's prudence to utilize any technology or tool that could fit into the budget and need to achieve their end goal.

Solution Design and Architecture

The above figure shows the reference architecture of the platform used for the case, I am describing here. For data acquisition and source to sink integration, multiple methodologies are being utilized including manual search and download, depending on the source of the data. All data, including the data fetched from internal data sources, are being stored in tables created in external schema.

The 'filtering process' layer is planned for construction and not developed yet. This layer would have methods and functions to validate the correctness and completeness of the data. The long term plan is to design of the methods and functions are envisaged to be of 'self-learning' in order to improve the programmatic validation over a period of time and solve the second technical challenge completely. For the time being, the data is being curated manually, with a short term plan to provide for a GUI curation tool to the users.

Figure 4. Reference Architecture with Federated Schema

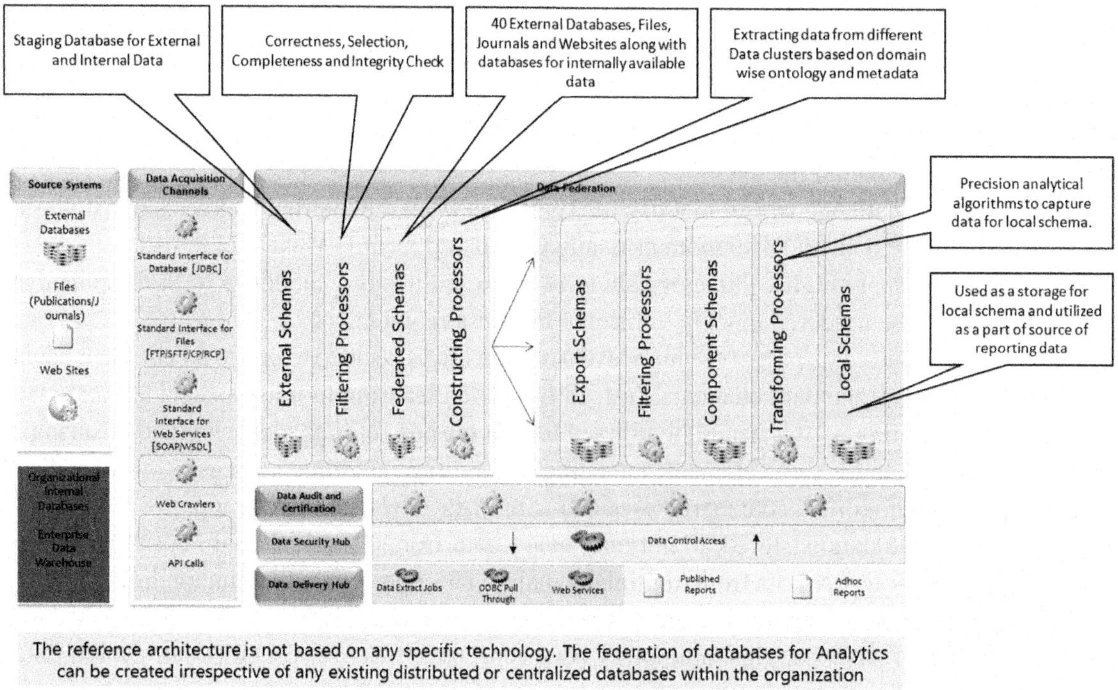

The federated schema consists of multiple autonomous and heterogeneous data bases consisting of various data elements. These databases are logically clustered in order to create a close association with similar domain. While maintaining the autonomy, the data structures are defined for each database following industry standard ontology or standard ontology as defined by the stakeholder following the industry base.

The constructing processor layer is the most critical layer where the metadata definitions are created. The metadata follows the ontological references across the domain there by achieving a cross domain query interface for data delivery. It is critical get/procure data which is not available internally. External data could be unstructured, textual or may not fit the established structures. Development of ontologically aligned metadata based synonym repository or thesaurus significantly reduces the data mapping and gap analysis time, the process could even be automated. This facilitates organizations to align the "external data" with internal data in short time.

The keys to the entire platform, which are the Algorithms (transforming processors), are pointed to the component schema. The result set or the output of the algorithms are stored in the local schema (project based, temporarily persistent database i.e. the database/schema is completely truncated once the project is over). During this process, there is significant usage of a Data Mining tool for structural pattern matching as well as a Literature Mining tool for text augmentation of queries.

At the time of execution of the project, each new run of the algorithm with different parameters would create a fresh table with the output in the local schema. This in turn would be used for further analysis and reporting (batch as well as adhoc). A separate tool is being used for reporting purposes, which points to the local schema for all reports.

Figure 5. Reference Architecture of Federated Database System with Data Delivery Hub

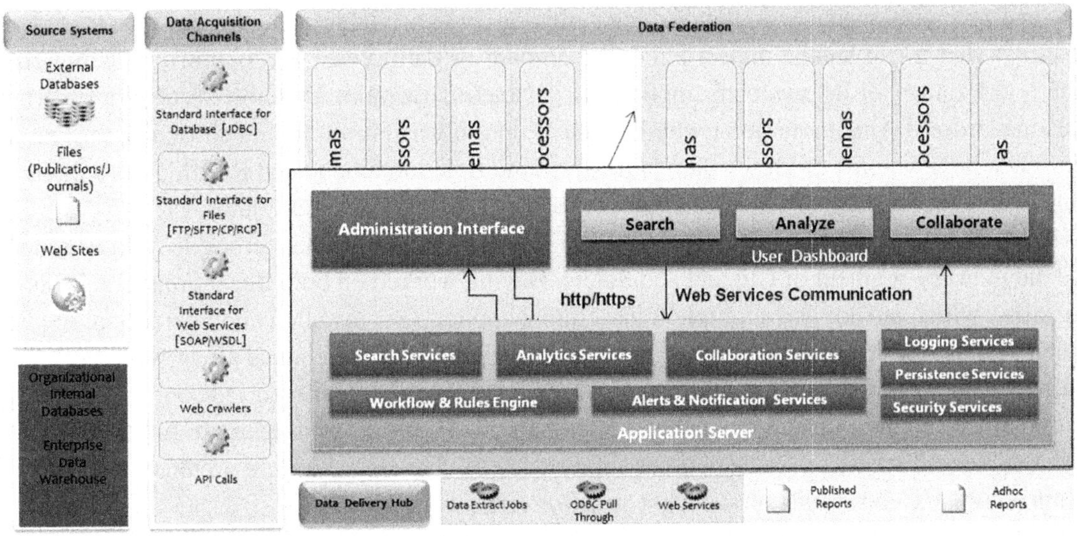

The reference architecture is not based on any specific technology. The federation of databases for Analytics can be created irrespective of any existing distributed or centralized databases within the organization

Note 1: Data Federation has been depicted using a typical FDBS system configuration. Various other models such MRDMS, DDTS, Mermaid etc. can also be used
Note 2: The different schemas and data delivery hub functionalities are described with utilities from Drug Repurposing platform.

Figure 6. Logical View of Reference Architecture of Federated Database System

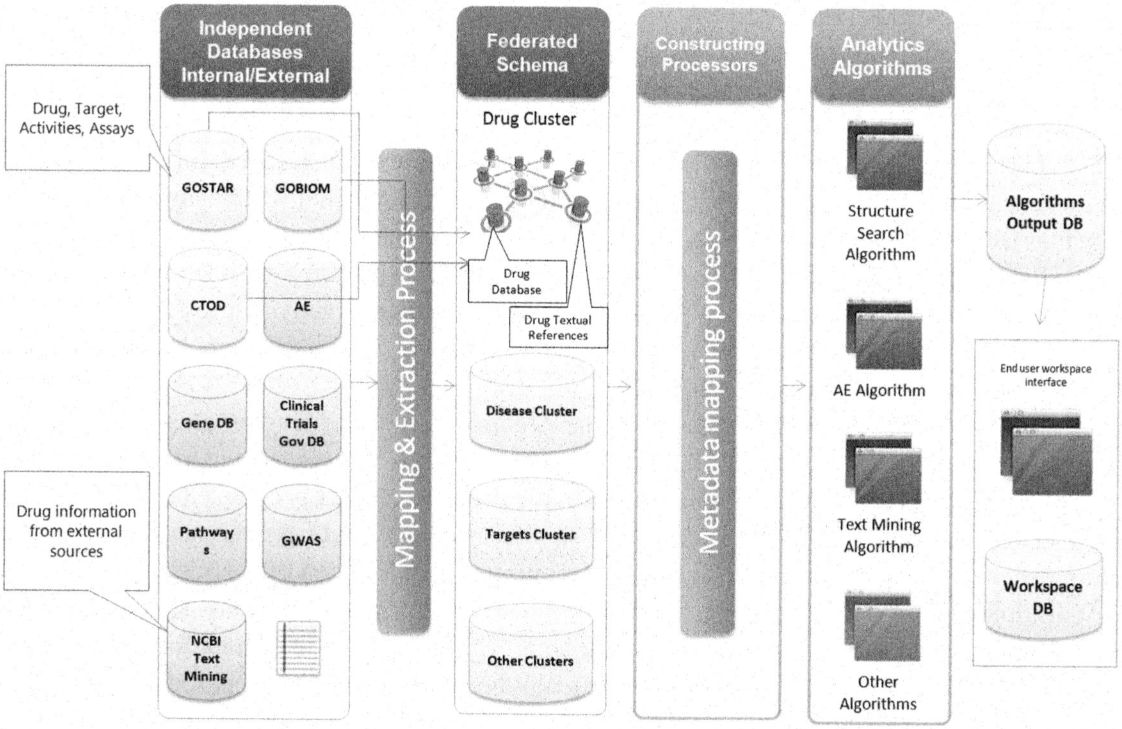

There is separate initiative, beyond this platform, which is being utilized to clinically verify the output of the algorithms.

A Federated Database System or a Federated Data Warehouse as described in this example case study works very well with standardized data (which in this case was available). Another positive for federated databases is that it can handle almost unlimited amount of data. Since this platform is being created for one organization for its own consumption the databases, operating systems, security infrastructure are all standardized. One more point why Federated option was proposed and chosen in this case was the presence large volume of legacy data in the form multiple interconnected RDBMS, as well as large volume of unstructured data (which generally follows the industry standard ontology).

In the end, I would like to conclude, there are several innovations happening in data warehouses, across the industry segment of Clinical and Safety. Having worked on both the approaches, I found, an organization which already has multiple transactional applications installed and operational in clinical and safety landscape it is preferable that they proceed with a data warehouse along with integrated & virtualized data. Whereas if the requirement is to build application/s with varied sources of structured and/or unstructured data a federated database system has been observed to be more suitable. In the two case studies here, I wanted to enunciate the importance of choosing the right architecture (Federated, Centralized or De Centralized) for the right situation and business case.

3.4. Data Virtualization

I have covered Data Virtualization in brief in section 3.1.4. In this section, I would like to elaborate on the topic and describe, how this approach when rightly implemented can effectively make the data available to the end user at 'Right time' and other associated benefits of the approach. As an extension from the previous section, among many reasons behind invention and development of data warehouses; one surely was easy and quick accessibility of required data from large volume of data. Transaction processing and Data entry/modification based applications using DBMS or RDBMS (*Relational or Normalized Data structure*) Databases, has difficulty in churning out data requests, queried by multiple users, quickly, without impacting the performance of the application itself. These applications also has a perennial problem of handling (*updating and querying*) when the data volume grows inordinately large over long period of usage. Introduction and advancement of data warehouse technology (*using 3rd Normal form or more popular De-normalized Data Structure: Kimball Star Schema Dimension Model*) surely solved the problem, to some extent. The reason for mention 'to some extent' is purely due to the fact, while data access became easier and off course the performance impact to the applications (*transaction processing*) was negated; the requirement for comparative and competitive data became more and more complicated. More over as the volume of data grew phenomenally with the growth of business and/or passing years, centralized data warehouse required increased infrastructure to meet the performance need, thereby phenomenally increasing the cost of ownership. Finally, as I mentioned earlier, since the source applications were disparate in nature, from business usage and functionality point of view, definition of data were different, which meant the organization could never get a 'single source of truth'.

Hence, while data warehouses or federated databases could and can retain right data, there surely was a need for innovative means to have quicker access to integrated and standardized data from multiple applications/databases without any hindrances due to physical location. The reason for me to mention federated databases here is to set the context that with the number of semi- and unstructured sources of data growing due to business needs; it has become increasingly difficult to consolidate all of that data into one warehouse and a federation of databases and warehouses are becoming increasingly common.

Data virtualization, as defined earlier, provides an integrated view of completely different data sources irrespective of their physical location or nature of the data stored. The abstracted data created through Enterprise Service Bus Views using Canonical Data Model are stored in the data marts. The data marts can contain structured, un-structured and semi-structured data. Multiple applications can access the data marts using multiple means such as SQL queries, web service calls etc. The abstracted data marts which act as virtualized sources, supplies the data at a real time or near real time.

The figure describes a generic reference architecture of Data Virtualization at an enterprise level. It depicts multiple applications and portals across the organization using data from virtualized data marts and the modified data getting stored in the local databases. Simplistically, the integration layer allows for local databases to get connected to the staging area of the operational data warehouse. In situations, this integration layer allows for seamless communication between source and target systems using various data adapters ensuring smooth flow of disparate data from various locations (*including cloud based applications*) to interact with the staging area of the data warehouse. The data is then validated

Figure 7. Reference Architecture of Data Virtualization

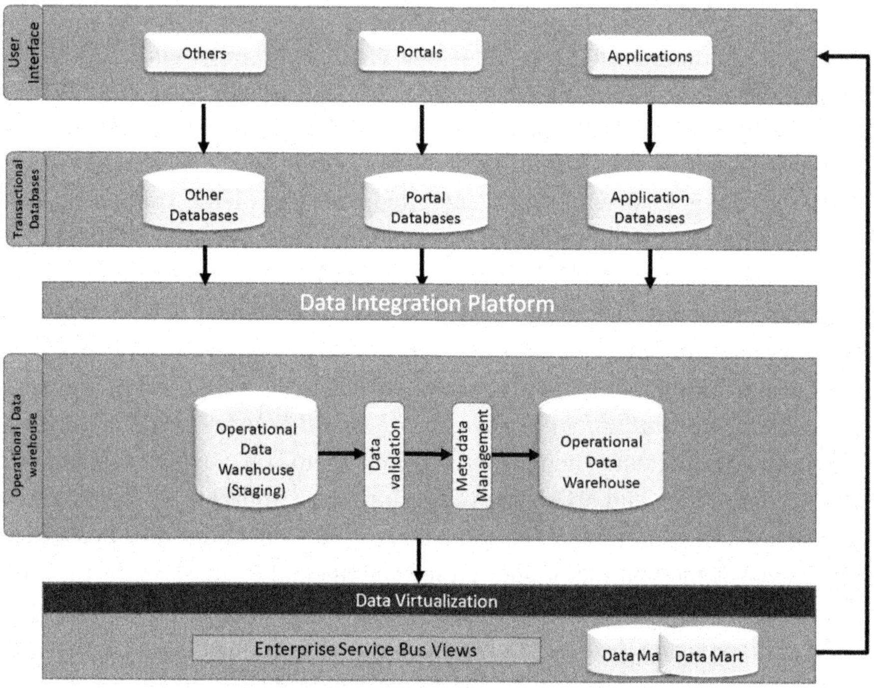

for correctness, so that stored data is secured and complete. The Meta data Management layer is used for the right mapping of the standardized data element across various domains. Finally the Enterprise Service Bus Views using Canonical Data Model allows for the data that can be uniformly accessed by all applications getting stored in the virtualized data marts.

It is quite obvious that data virtualization is a very complimentary methodology adopted by the industry to improve the applicability of data warehouses and federated databases.

Few of the significant benefits of data virtualization that can be derived and measured are as follows

1. **Agility:** Data Virtualization allows for an agility to the enterprises to access cross functional data for real time analysis and planning. This in turn assists in reduced time to market which can be a measureable benefit of great proportion
2. **Secured data access :** The abstracted layer creates a secured gateway through which data is accessed reduces the chances of data corruption
3. **Increased ROI :** Organizations that have invested immense amount of money to build and maintain data warehouses, can invest a fraction of that amount to build the ability of data virtualization. When we allow for the lowered cost to manage cross-functional and non-standard data before the data can accessed and analyzed the ROI can be measurably high. While traditional data warehouses will not lose their ability and importance to retain persistent historical data, the addition of data virtualization ably complements
4. **Usage during Merger and Acquisitions:** Another major utility of data virtualization is during M & A, when the communication and data access between the data stores can be easily made possible by taking the data virtualization approach

5. **Scalability:** Even though many can raise the issue of scalability of the solution and performance concerns, I feel the solution could be complicated to implement due to detailed nature of research required across the enterprise to standardize data definition, the lack of expertise in defining canonical data models for various off the shelf products and the ability to integrate disparate data sources, but scalability should not be a problem

This far we have established that Data Virtualization surely help users to access data at 'Right Time' and also allows the users to compare historical data as well as current data not only from the timeline perspective but allow for a comparative analysis of data from different domains to happen. Lastly, the presence of numerous tools and technologies handling big data along with the ability to compare and analyze different types of data at real time is a real differentiator of this entire approach. It makes the paradigm of 'Right data' at 'Right time' an entirely achievable reality.

I started this chapter talking about the need of the Life Sciences organizations, for the availability of 'Right data' at 'Right time'. During the course of this chapter I have written about the ways and means to anticipate and identify the need of the end users and project stakeholders and how the success of delivery can be measured. I have explained how resources working in the project are the key to success and innovation. In the sections, dealing with design and development of data warehouses, I have showcased, how the software innovations, when integrated with a data warehouse, provide tremendous benefits to clinical and safety organizations. Through multiple case studies, I have exhibited different designs of data warehouses that can be adopted in different situations, in order to store the 'Right data'. And lastly we have seen, how innovative and powerful data virtualization can be, and how this technology can be utilized to obtain 'Right data' at 'Right time' for the business leaders.

ACKNOWLEDGMENT

I would like to acknowledge the help and encouragement of my management and colleagues at GVK Biosciences, without whom many components of this chapter would have been incomplete.

My sincere thanks to Dr. Mahanandeeshwar Gattu (Vice President – GVK Biosciences), who helped me understand various facets of the science of Drug Repurposing and made me see his vision of converting it into a technology enabled data driven process. He also helped me with nuances behind biomedical literature mining.

I would also like to thank GVK Biosciences to allow me use the reference architectures used in the section 3.3 of this chapter to elucidate the Drug Repurposing platform.

My sincere thanks and acknowledgement to Cognizant Technology Solutions for my understanding of Clinical and Safety domain. While I was an employee of Cognizant, I understood the facts of many of the cases, business problems and the reference architectures used in this chapter. The reference architecture used in section 3.2 for Clinical and Safety Data warehouses and the one used along with the case study on Data Virtualization are based on my knowledge gained at Cognizant. I also thank Mr. Partha Chakraborty (Sr. Director – Cognizant Technology Solutions) for his help, backing and persuasion to get this chapter concluded. I have tapped into his knowledge of clinical and safety domains multiple times in past five years enabling me to acquire the courage to write this chapter.

REFERENCES

Amin, R., & Arefin, T. (2010). The Empirical Study on the Factors Affecting Data Warehousing Success.

Casey, D. (2006). Federated Databases in Bioinformatics and Translational Medical Research.

Casey, D. (2006). How Federated Databases Benefit Bioinformatics Research.

Rouse, M. (2011). What is Data Virtualization?

Sheth, A.P. & Larson, J.A. (1990). Federated database systems for managing distributed, heterogeneous, and autonomous databases. *ACM Comput. Surv.* 22(3), 183-236. DOI=10.1145/96602.96604

Yeung, A., & Hall, G. (2007). *Spatial database systems design, implementation and project management* (Vol. 87, p. 566). Dordrecht: Springer.

KEY TERMS AND DEFINITIONS

Data Analytics: It is a process of working with data that results in information which effectively answers a query.

Data Mining: It is a process of extracting information from very large amount of data stored in a database or a collection of databases or a data warehouse.

Data: The basic unit of any information stored in digitalized format.

Innovation: An idea when practically implemented results in increase of effectiveness of the target system or application.

Internet of Things: Network of anything to anything by any means of connectivity to exchange information.

Literature or Text Mining: It is a process of extracting information from very large amount Textual or Literature data stored in file systems or unstructured data repositories.

Chapter 3

Application of Metadata Repository and Master Data Management in Clinical Trial and Drug Safety

Chandrakant Ekkirala
Cognizant Technologies Limited, India

ABSTRACT

This chapter talks about metadata repository, and master data management in clinical trial and drug safety. The chapter begins with the definition of metadata repository and gives an explanation around the same, It talks about a well designed metadata repository and the characteristics associated with the same. A brief around why we need metadata and the reasons for the using the same has also been mentioned. The benefits of a well structured metadata repository was also mentioned in detail. The chapter then gives a detailed explanation on master data management and the usage of MDM in clinical trials. MDM solutions for clinical trials management is also explained in detail.

INTRODUCTION

Metadata Repository

A *Metadata repository* is a database created to store metadata. Metadata itself is information about the structures that contain the actual data. Metadata is often said to be "data about data", but this is misleading. Data profiles are an example of actual "data about data". Metadata is one layer of abstraction removed from this - it is data about the structures that contain data. Metadata may describe the structure of any data, of any subject, stored in any format. Thus Metadata becomes the crux of the data in varying structure, subject or format.

DOI: 10.4018/978-1-4666-8726-4.ch003

A well-designed metadata repository typically contains data far beyond simple definitions of the various data structures. Typical repositories store dozens to hundreds of separate pieces of information about each data structure.

Comparing the metadata of a couple data items - one digital and one physical - will help us understand what metadata really is.

First, digital- for data stored in a database we may have a table called "Patient" with many columns, each containing data which describes a different attribute of each patient. One of these columns may be named "Patient_Last_Name". What is some of the metadata about the column that contains the actual surnames of patients in the database? We have already used two items: the name of the column that contains the data (Patient_Last_Name) and the name of the table that contains the column (Patient). Other metadata might include the maximum length of last name that may be entered, whether or not last name is required (can we have a patient without Patient_Last_Name?), and whether the database converts any surnames entered in lower case to upper case. Metadata of a security nature may show the restrictions which limit who may view these names.

Second, physical- for data stored in a brick and mortar library, we have many volumes and may have various media, including books. Metadata about books would include ISBN, Binding_Type, Page_Count, Author, etc. Within Binding_Type, metadata would include possible bindings, material, etc.

Definition

The metadata repository is responsible for physically storing and cataloging metadata. The physical storage and sequential catalogue is best done in a metadata repository. Data in metadata repository should be *generic, integrated, current, and historical*. Each of the terms, generic, integrated, current and historical is explained with respect to metadata as follows. *Generic*: meta model should store the metadata by generic terms instead of storing it by an applications-specific defined way, so that if your data base standard changes from one product to another the physical meta model of the metadata repository would not need to change. Thus a change in the database standard across various products would not change the metadata model repository. This is one of the important characteristic of the data in a repository. *Integration* of the metadata repository allows all business areas metadata in an integrated fashion covering all domains and subject areas of the organization. The overall coverage of the domains and subject areas function in the integrated approach of the repository. The metadata repository should have accessible *current and historical* metadata. Metadata repositories used to be referred to as a data dictionary.

With the transition of needs for the metadata usage for business intelligence has increased so is the scope of the metadata repository increased. Business intelligence has been an important aspect in the industry. That being requiring the need of metadata, has increased the scope of the repository proportionally. Earlier data dictionaries were the closest place to interact technology with business. Data dictionaries are the universe of metadata repository in the initial stages but as the scope increased business glossary and their tags to variety of status flags emerged in the business side while consumption of the technology metadata, their lineage and linkages made the repository, the source for valuable reports to bring business and technology together and helped data management decisions easier as well as assess the cost of the changes. The metadata thus impacted the data management decisions proportionally and assessed the change in case inversely, making easier decision as well as changes in cost.

Metadata repository explores the enterprise wide data governance, data quality and master data management (includes master data and reference data) and integrates this wealth of information with integrated metadata across the organization to provide decision support system for data structures, even though it only reflects the structures consumed from various systems.

Reasons for Use

Metadata repository enables all the structure of the organizations data containers to one integrated place. Thus all data under one data container. This opens plethora of resourceful information for making calculated business decisions. This tool uses one generic form of data model to integrate all the models thus brings all the applications and programs of the organization into one format. And on top of it applying the business definitions and business processes brings the business and technology closer that will help organizations make reliable roadmaps with definite goals. With one stop information, business will have more control on the changes, and can do impact analysis of the tool. Usually business spends lots of time and money to make decisions based on discovery and research on impacts to make changes or to add new data structures or remove structures in data management of the organization. With a structured and well maintained repository, moving the product from ideation to delivery takes the least amount of time (considering other variables are constant). To sum it up:

1. *Integration* of the metadata across the organization.
2. Build relationship between various *metadata types*
3. Build relationship between various *disparate systems.*
4. *Define business* golden copy of definitions.
5. *Version* control of the changes at structure level.
6. Interaction with *Reference data*
7. Link view to *master data.*
8. Automatic synchronization with various authorized metadata source systems.
9. More *control* to business decisions.
10. *Validate* the structures by overlapping the models
11. Discovering *discrepancies, gaps, lineage, and metrics* at data structure level.

Each database management system (DBMS) and database tools have their own language for the metadata components within. Database applications already have their own repositories or registries that are expected to provide all of the necessary functionality to access the data stored within. Vendors do not want other companies to be capable of easily migrating data away from their products and into competitors' products, so they are proprietary with the way they handle metadata. CASE tools, DBMS dictionaries, ETL tools, data-cleansing tools, OLAP tools, and data mining tools all handle and store metadata differently. Only a metadata repository can be designed to store the metadata components from all of these tools

Design

Metadata repositories should store metadata in four classifications: ownership, descriptive characteristics, rules and policies, and physical characteristics. Ownership, showing the data owner and the application owner. The descriptive characteristics, define the names, types and lengths, and definitions describing business data or business processes. Rules and policies, will define security, data cleanliness, timelines for data, and relationships. Physical characteristics define the origin or source, and physical location. Like building a logical data model for creating a database, a logical meta model can help identify the metadata requirements for business data. The metadata repository will be centralized, decentralized, or distributed. A centralized design means that there is one database for the metadata repository that stores metadata for all applications business wide. A centralized metadata repository has the same advantages and disadvantages of a centralized database. Easier to manage because all the data is in one database, but the disadvantage is that bottlenecks may occur.

A decentralized metadata repository stores metadata in multiple databases, either separated by location and or departments of the business. This makes management of the repository more involved than a centralized metadata repository, but the advantage is that the metadata can be broken down into individual departments.

A distributed metadata repository uses a decentralized method, but unlike a decentralized metadata repository the metadata remains in its original application. An XML gateway is created that acts as a directory for accessing the metadata within each different application. The advantages and disadvantages for a distributed metadata repository mirror that of a distributed database.

Design of information model should include various layers of metadata types to be overlapped to create an integrated view of the data. Various metadata types should be stitched with related metadata elements in a top down model linking to business glossary.

Layers of Metadata

The layers of Metadata are as follows:

1. Business Glossary: contains recursive relationship to Business terms.
2. Business tags: Contains various affiliation to that term or terms.
3. Data Dictionary: contains information from data model tools for the definition of metadata elements and their technical definitions provided by data or enterprise architecture.
4. Conceptual data models
5. Logical data models
6. Physical data models
7. Databases
8. validation rules and data quality rules
9. ETL, business rules and their relationship to attributes and entities
10. Reports
11. Source to target mapping artifacts (relationships)
12. Reporting requirements (relationships)
13. business processes and their relationship to technology
14. people hierarchy and their relationship
15. owner relationship

With the evolution of clinical data standards, metadata management has become all the more important and finding ways to better store, access, and manage metadata is regarded as a priority by most life sciences companies.

The old – and inefficient – way of managing metadata was through spreadsheets, a process still used by many companies. But spreadsheets are not robust enough for this purpose and in general do not provide rigorous controls, which makes it difficult to leverage metadata to support and update standards, map processes and support relationships between data elements.

On the other hand, an efficient, well implemented MDR has the potential to improve the submission process while reducing costs, improving data quality and compliance, improve business processes, and drive automation. By having the ability to populate the MDR with metadata from clinical studies, many of the manual processes of managing data become automated, streamlining processes.

Metadata management is important for the development, implementation, maintenance and administration/governance of standards. Implementation of an MDR is not simply an implementation of the tool itself but also the institution of a governance structure to administrate standards development, the development of processes that formally require and manage usage of those standards, the adaptation of any tools/systems important to the business functions to utilize and propagate those standards, and re-education of both management and operational teams.

Unfortunately products that support the concept of an MDR to manage metadata and support the processes have only recently been released to the market and there is more maturation to be expected from all of the offered solutions.

One of the problems has been that many existing tools on the market have tended to manage metadata as an afterthought, with the majority of tools being data-centric rather than focusing their key capabilities around storing, managing, and consuming the metadata. Moreover, few tools have been robust enough to handle the relationship between various standards, such as ODM, SHARE, and HL7 standards. With this need and growing expectations, some MDR solutions are now emerging that are more truly metadata driven.

As regulation and the number of standards increase, so does the time and effort it takes to develop, manage, and distribute these standards in a way that truly results in the benefits intended. By binding systems to robust common domain terminology produced and disseminated by a standards development organizations (i.e., Semantics Management), industry can more efficiently implement end-to-end data lifecycle processes.

Additionally, regulatory data submission guidelines, reporting mandates, internal business efficiency improvements and merger rationalization efforts are driving life science organizations to replace their existing manual, inefficient management and implementation processes with fully interactive metadata governance platforms.

For most established and emerging pharmaceutical, biotech and medical device companies, there is no mechanism for managing and communicating metadata standards. These organizations are challenged to develop a governance process, or global standard to convey to internal and external data stakeholders.

The absence of standards and the inherent process inefficiencies translate to compromised timelines, data integrity issues and an inability to scale, resulting in potential delays in the delivery of the product to patient population. Is the industry simply looking for a replacement for spreadsheets with better control? Are we considering all the other aspects of metadata independent of the clinical data such as process metadata? Can the industry take full advantage of the power of semantic models for metadata management? These questions have yet to be answered and it will be interesting to see how this area evolves over the coming years.

Master Data Management

Master data management (MDM) is a comprehensive method of enabling an enterprise to link all of its critical data to one file, called a master file, that provides a common point of reference. When properly done, MDM streamlines data sharing among personnel and departments. In addition, MDM can facilitate computing in multiple system architectures, platforms and applications. At a basic level, MDM seeks to ensure that an organization does not use multiple (potentially inconsistent) versions of the same master data in different parts of its operations, which can occur in large organizations. Master data management of disparate data systems requires data transformations as the data extracted from the disparate source data system is transformed and loaded into the master data management hub. To synchronize the disparate source master data, the managed master data extracted from the master data management hub is again transformed and loaded into the disparate source data system as the master data is updated. As with other Extract, Transform, Load-based data movement, these processes are expensive and inefficient to develop and to maintain which greatly reduces the return on investment for the master data management product. Processes commonly seen in MDM include source identification, data collection, data transformation, normalization, rule administration, error detection and correction, data consolidation, data storage, data distribution, data classification, taxonomy services, item master creation, schema mapping, product codification, data enrichment and data governance.

The selection of entities considered for MDM depends somewhat on the nature of an organization. In the common case of commercial enterprises, MDM may apply to such entities as customer (customer data integration), product (product information management), employee, and vendor. MDM processes identify the sources from which to collect descriptions of these entities. In the course of transformation and normalization, administrators adapt descriptions to conform to standard formats and data domains, making it possible to remove duplicate instances of any entity. Such processes generally result in an organizational MDM repository, from which all requests for a certain entity instance produce the same description, irrespective of the originating sources and the requesting destination.

The tools include data networks, file systems, a data warehouse, data marts, an operational data store, data mining, data analysis, data visualization, Data federation and data virtualization. One of the newest tools, virtual master data management utilizes data virtualization and a persistent metadata server to implement a multi-level automated MDM hierarchy.

The benefits of the MDM paradigm increase as the number and diversity of organizational departments, worker roles and computing applications expand. For this reason, MDM is more likely to be of value to large or complex enterprises than to small, medium-sized or simple ones. When companies merge, the implementation of MDM can minimize confusion and optimize the efficiency of the new, larger organization.

For MDM to function at its best, all personnel and departments must be taught how data is to be formatted, stored and accessed. Frequent, coordinated updates to the master data file are also essential.

MDM in Clinical Trials

Leading pharmaceutical companies are reinventing themselves. They're adapting to tremendous pressure to launch blockbuster drugs more quickly, frequently, and cost effectively while minimizing risks of noncompliance. In doing so, they gather more clinical trials data, more often, at every stage of research and development.

But clinical trials data and compound data are often poor in quality. When this data is inaccurate and inconsistent across multiple systems, clinical trials fall behind schedule and go over budget, revenue opportunities are missed, and compliance with regulations becomes difficult without a great deal of manual effort. To address these challenges, pharmaceutical companies require robust technology to manage their clinical trials data and compound data.

MDM Solutions for Clinical Trials Data and Compound Data Management

The MDM solution for clinical trials data and compound data management is required to offer a flexible data model that adapts to evolving research requirements, enables faster implementation, and accelerates time to value.

The MDM solution for clinical trials data and compound data management should have the following capabilities:

- Flexible data model for all clinical trials data
- Extensible data model to adapt to evolving research requirements
- A 360-degree view of clinical trials data including data about drugs and compounds, institutions, and people, via visualization and predefined relationships
- An intuitive dashboard to give business decision makers and data stewards an easy way to maintain clinical trials data and compound data

Persistent data quality to ensure the ongoing trustworthiness of clinical trials data and compound data, correct errors, and eliminate duplicates

Market-leading enterprise data integration to effectively integrate clinical trials data and compound data from operational and analytical applications, whether on premise or in the cloud

With drug pipelines looking less and less promising, and a challenging political climate, pharmaceutical companies are scrambling for new ideas to stay profitable during these troubling economic times. Many have turned to cost cutting and productivity improvements in order to stay competitive. While these approaches are suitable short-term responses, they do not address the root cause of the problem, namely the challenge of accurately identifying promising therapies and accelerating the research and clinical trials process to more rapidly bring these therapies to market.

It is common knowledge that drug introduction is costly and time-consuming. The average drug is the result of 12 to 14 years of effort and millions, if not billions, of dollars spent. Yet despite these steep investments, thousands of therapies are discarded for every one successful drug market entry largely because of the numerous and multi-step processes that are required.

In response to this reality, researchers and those responsible for clinical trials management are searching for innovative ways to accelerate these processes and reduce the size of the slope. One new approach involves improving the capture, management, and sharing of data related to compounds, research, and trials. In order to achieve this goal, companies are turning to the emerging software category called master data management (MDM).

A typical drug pipeline involves numerous stages and processes. By capturing, managing, and sharing data related to compounds, research, and trials, organizations can speed time to market. (Source: Siperian Inc.)

Figure 1.

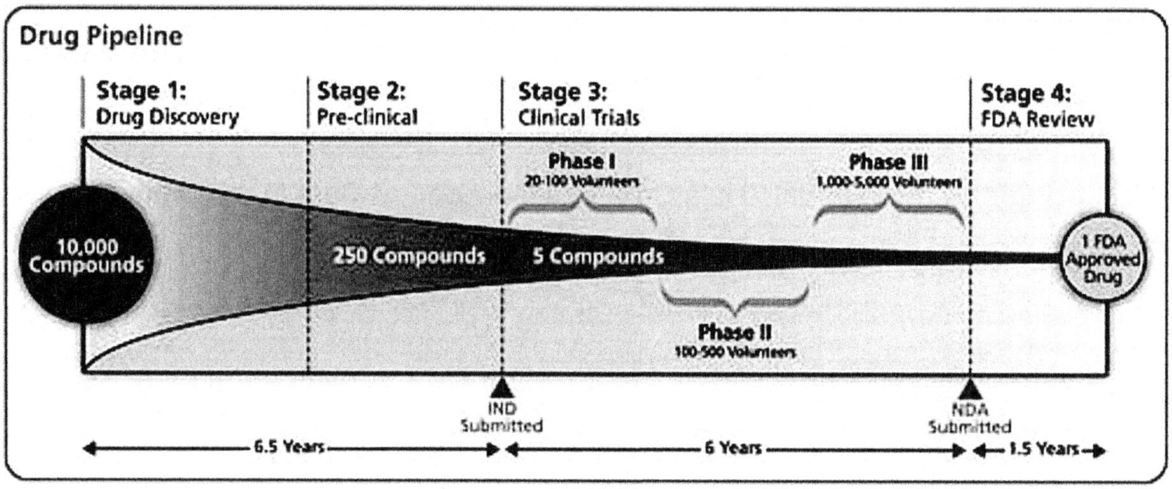

The Data Challenge

A deeper look at each step of the drug development process reveals a recurring theme of missing, incomplete, or erroneous data that wastes time and resources and adds months or even years to the drug introduction process.

For example, one drug manufacturer once lamented, "We're pretty sure that we've found the cure for cancer. We just lost it." This statement, underscores the difficulty researchers have in managing the immense number of compounds that are tested across the enterprise. One cause of this complexity is that compounds are often housed across multiple business units in separate systems, where they are given differing names and identifiers. As a result, successes achieved in one therapy are often invisible to other parts of the business. Worse, side effects and setbacks are also not always shared, meaning the same failed compound can re-enter the drug pipeline multiple times across different therapies, thereby wasting precious time and resources on research that would not have been conducted had full institutional knowledge of this compound been properly shared and managed across the enterprise.

It is well known that not all sites perform in a similar fashion and that some variance can exist for specific sites based on the particular demographics that are being recruited or the investigators involved in the study. Yet most institutions do not have access to the data that would enable them to clearly identify and recommend those sites that would be best suited to conduct the trial based on the compound that is being tested. Better insight into sites and their performance across a number of important metrics could dramatically reduce the time and cost of clinical trials, while maximizing the reliability of the trial itself.

In the same way that all sites do not perform equally, not all investigators can be relied upon. The ability to properly manage the list of investigators and describe their performance against key metrics could also allow clinicians to drive down the time and cost typically associated with clinical trials.

When an organization identifies a lack of promising new drugs for a therapeutic area, many routinely turn to academia for new promising compounds. Scouts seek out these key researchers and establish relationships in an attempt to leverage whatever successes might result from their early-stage research.

Figure 2.

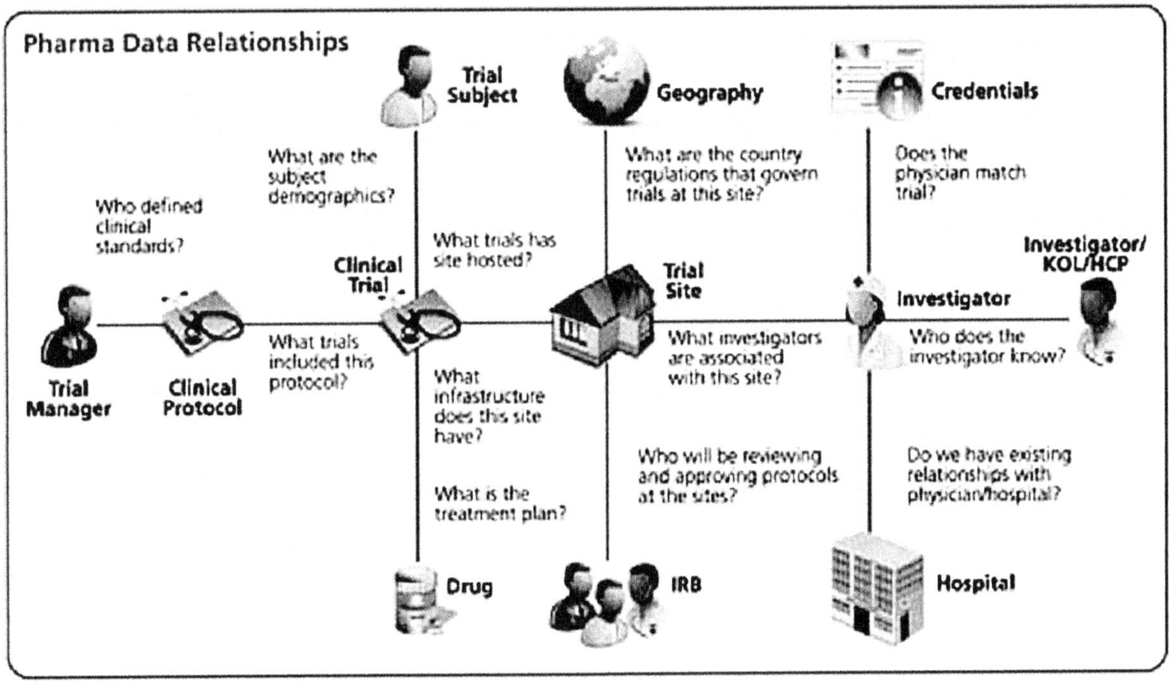

However, most companies lack clean, reliable, centralized data that can identify these key resources, discoveries and compound inventories or they lack the ability to access it.

The reality is that, in addition to understanding each of these data elements independently, understanding the relationships among them is just as critical. By institutionalizing processes that depict these relationships and by enabling easy access to the data itself, organizations can radically reduce the time of each phase of drug development. Organizations can also benefit from internally generated data such as employees, clinical protocols, and clinical trials in addition to the third-party data that is mentioned above.

Understanding the relationships among data types and entities can greatly accelerate drug discovery time and development and help to reduce waste. (Source: Siperian Inc.)

Discovering Master Data Management

MDM is a growing trend among pharmaceutical institutions looking to accelerate drug discovery. In short, master data management aims to manage, integrate, and reconcile disparate data across the various information technology landscapes. MDM requires a combination of process and organization to ensure success, but pivots on important technology enablers. In evaluating a solution for the types of challenges outlined above, several key components of an MDM technology are critical and should be included:

A flexible data model. The solution must accommodate the different types and forms of master data such as people, compounds, clinical trials, and locations, in order to fully address the different data needs across the drug discovery process.

Data cleansing, standardization, and enrichment. Given the state of data that is typically found across an enterprise, it is critical that an organization leverage a solution that can continually integrate and cleanse data as opposed to relying on a one-time only data clean-up effort.

Robust matching and survivorship rules. Establishes a "best version of truth" or "golden record" of data and resolves inevitable data discrepancies in a rules-based manner, which in turn dramatically reduces the need for manual effort.

Flexible relationship management. Relationships among clinical trials, clinical protocols, compounds, sites and investigators are just as important as the individual items themselves. It is quite possible, for instance, that an investigator performs quite well in one site, but poorly in another. It is also important to understand the relationships between compounds and trials for compliance and covigilence. To be effective, the proper solution must allow for these complex relationships to be maintained.

Robust data stewardship capabilities. MDM by its very nature will dramatically reduce the work effort associated with data management. However, an ideal solution also allows humans to interact with the data as needed and perform manual data management and exception handling to resolve conflicts and ensure maximum reliability and performance. This is something MDM experts and influencers refer to as "data stewardship." It is also related to data governance, which is a process that applies a definition and the enforcement of rules and procedures as part of data maintenance.

Flexible integration framework. While the ability to create reliable data is important, the ability to access and retrieve these data, both through a Business Intelligence solution or through operational systems is just as critical. The ideal solution must allow for flexible integration in batch and real time with the business applications, where research and research-related management decisions are made.

Ensuring a Successful MDM Outcome

Given the challenges of a weakening pipeline and tough economic conditions, innovative companies are looking for new ways to accelerate the drug development process and improve the bottom line. Early adopters are already starting to see improved performance in R&D and clinical trials management, shaving weeks and even months off of the difficult drug introductory process. The best way to ensure a strong outcome is to create and maintain reliable data that is capable of supporting the company's current and future business requirements. Achieving a healthy MDM journey will reap many rewards provided that considerations are identified and research is performed right from the start.

Clinical Data Management

CDM is defined as "the development, execution and supervision of plans, policies, programs and practices that control, protect, deliver and enhance the value of data and information assets" in the clinical trial arena. With its diverse connectivity, cross-functional features, and a wide range of responsibilities, CDM has come a long way in the past two decades, and is a recognized profession with increasingly realized importance within and outside biopharmaceutical research and development. As complex and dynamic as the profession is, CDM globally continues to grow into a firmly established discipline in its own right, focuses on managing clinical trial-related data as a valuable resource, and is becoming a career that requires multiple skill sets, such as a background of sound clinical skills, scientific rigor, information technology, systems engineering, and strong communications ability. With the continued global harmonization of clinical research and introduction of regulatory-mandated electronic submission in the industry, it is critical to understand, appreciate, work within the framework of global clinical development, and apply standards in the development and execution of architectures, policies, practices, guidelines, and procedures that properly manage the full clinical data lifecycle needs of an enterprise. This definition is fairly broad and encompasses a number of professions which may not have direct technical contact with lower-level aspects of data management, such as relational database management. Many other topics, processes, and procedures are also relevant, including:

- Data governance, such as standards management, SOPs, and guidelines
- Data architecture, analysis, and design including data modeling for potential clinical data repository or warehouse
- Database management including data maintenance, administration, and data mapping across related clinical or external systems
- Data security management including data access, archiving, privacy, and security
- Data quality management including query management, data integrity, data quality, and quality assurance
- Reference and master data management including data integration, external data transfer, master data management,

Clinical Data Management Perspectives

CDM has evolved and will continue to develop in response to the special cross-functional needs and according to the particular strengths of e-clinical research advances due to much enhanced clinical harmonization, global standardization, and expected clinical systems interoperability initiatives.

The future is not what it used to be, and will undergo many anticipated reality checks. CDM professionals once optimistically predicted that EDC technology would radically increase efficiency by reducing the amount of paper documentation associated with clinical trials, and streamline the CDM process considerably. Indeed, some sponsor companies have realized some claimed clinical efficiencies with planned long-term cost savings, but not all of them do so well. It is not uncommon to see sponsor companies spending a large resource and investment to establish an electronic documentation system, such as Electronic Documentum, to store study-related documents while still maintaining a concurrent manual paper filing system. It seems a reasonable reality that the current clinical studies are operated in both traditional PDC-based and EDC-supported environments by sponsors and/or CROs with differential levels of automation. The speed at which paper mountains accumulate may have been reduced by some sponsor companies; however, adoption of an electronic document management or clinical trial management system seems unable to eliminate the document piling. Therefore, successful implementation and integration of EDC technology with other key clinical systems depends as much on managing change as it does on clinical science and technology itself, and changes, especially organizational ones, have never been easy for sponsor e-clinical solutions implementation.2 To realize the full potential of EDC technology in e-clinical research, both sponsor and site personnel need to make logistic reorganizational changes in their offices and surroundings, in entering and retrieving clinical information, in managing the issuance or closure of queries, in interacting and dealing with others stakeholders such as colleagues, CROs, and study subjects, and, most importantly, in gaining an understanding of the technology advantages and limits to achievement of business objectives.

Electronic Solutions in Clinical Data Management

Technology-driven strategies and initiatives have the potential to alleviate the significant pressure to market a medicine as early in the patent life as possible to maximize the period without competition, both to increase total revenue and to shorten the time to market sales. The increase in regulatory requirements and competition seen in the recent years, coupled with reforms in health care services, has presented extreme challenges for the biopharmaceutical industry, suggesting the need for sponsor companies to invest significantly in technological solutions and add an additional emphasis on business process re-engineering and improvement to engender long-term clinical efficiencies and cost benefits. In this environment, the effectiveness of the clinical data management function is crucial to substantiate early approval for a new product launch and subsequent successful marketing. Delay, deficiency, or quality issues in the CDM process can be costly. Further, speed is not enough by itself and success needs to be achieved with other quality attributes. There is an ever-increasing demand for sponsors, including CROs, to strike the right balance between time, cost, process, and quality in conducting all clinical studies.

Applying e-clinical technology, including EDC, in such a context is the anticipated industry trend and will continue to offer superior benefits to sponsors as collaboration, standardization initiatives, and technology innovation are constantly geared towards more and wider technology adoption.

Status of Data Management in Clinical Studies

Slow yet increasing EDC adoption combined with EDC technology improvement has demonstrated the reality and complexity of implementing re-engineered e-clinical processes along with new technology introduction. There is still the presence of PDC in a large number of sponsor firms, especially in Phase I clinical studies or studies sponsored by small-sized or start-up firms. Medium or large biopharmaceutical firms are tending to move into EDC, or have accumulated implementation expertise with the technology and associated e-clinical systems. It is not surprising that the traditional PDC and evolving EDC may coexist for a sponsor or CRO. To address the clinical operational needs, a sponsor firm or CRO may have a different set of procedures, standard work practices, guidelines, or business documents for PDC and EDC. Some sponsors may outsource the PDC data management functions to CROs in a complete fashion. Other sponsors may take a combinational approach whereby they would have an internal core team design the CRFs and come up with varied edit check specifications, but seek CROs to build the database and program those checks. To ensure that a standardized set of forms and edit checks are applied for cross-therapeutic clinical studies, sponsor firms must have the proper oversight and expertise to drive CRO data management or database design deliverables. There also seems to be an evolving trend whereby sponsor firms separate clinical database design (CRF or eCRF) and deployment functions into a specific unit from the CDM group due to the increasing sophistication of technology improvement, innovation, or clinical systems integration. It is also common for a different clinical programming unit to be set up for programming edit checks, listings, or reports for different functional groups. Increasing EDC computerization has enabled a paperless environment where key study variables based on protocols and electronic querying need to be transmitted between a clinic and a sponsor via a web browser entry. An independent CDM organizational unit with data managers designated to various therapeutic areas seems to be more beneficial to sponsors in terms of standardization, systems integration, and process consolidation than multiple CDM units affiliated with different therapeutic functions.

Scope of Clinical Data Management

It is now a known fact that the scope of data capture, CRF design, and CDM activity vary widely between different companies engaging in clinical studies. For small-size entities, traditional data entry from paper CRF at a central location or outsourced CRO may still be the most effective strategy when all factors are taken into consideration. Larger companies have turned to EDC technology to deal with ongoing clinical study challenges, and long-term benefits of pursuing EDC-enabled global strategies are being realized gradually. The associated changes in the CDM process and ensuing reorganizational structuring indicate that the roles of those employed in CDM become increasingly blurred with those of their colleagues in clinical monitoring, quality assurance, and application development.8 Moreover, the pace of technology development or optimization may be so rapid that additional consideration is required for any company planning to invest in new hardware and software for EDC technology in a changing operational environment.

Challenges in Clinical Data Management

Although EDC technology and e-clinical systems have been implemented to enhance various aspects of the data management process, implementation has not been without difficulty nor has it been improved as rapidly as many had anticipated. The pharmaceutical, biotechnology, and medical device industry, as well as academia and the government, have all started to learn about the technology advantages; some have gained implementation expertise in adopting or configuring it as a new data management tool. EDC acceptance seems strong, and there are few instances where sponsors have gone back to PDC studies when they have had the experience of EDC. Although the goal of data management will not change, i.e., assurance of clean data at the end of the study, there is no doubt that data management processes will evolve with the use of EDC and e-clinical systems.

Critical Clinical Form Design with Balancing Needs

There are interdisciplinary eCRF design challenges involving technology, protocol-driven science, standardization, validation, and work-flow usability for both PDC and EDC studies. Ultimately, the final study report, which is the product of sophisticated computer programs and a statistical analysis, is only as good as the data collected in the CRF or eCRF. The whole process from defining the data to be collected, the collecting, checking, analyzing and presenting it, is resource-intensive, utilizing sophisticated technology and employing highly skilled professionals. The competing/ complementary demands made on the CRF or eCRF by site users, sponsors, and/or CROs must be acknowledged and addressed through balancing standards with the individual protocol requirements, considering the preference of the team members and site users, and engaging in collaboration and negotiation of the human issues involved in the process of cross-functional team review. The growing importance of post marketing data collection in large-population safety studies, the economics of drug therapy, and proteomics/ genomics/ pharmacogenomics presents multiple challenges including collecting, storing, integrating, querying, and analyzing growing lists of data sources, such as insurance claims, cost, large size of laboratory datasets, and patient-reported outcome data. It should be emphasized that study designers need to play a key role in driving and achieving core clinical database building. It is mission-critical for a sponsor to recruit a talented pool of professionals who excel in a fluid environment, pay great attention to protocol details, have developed expertise in therapeutic areas and technologies, and are capable of communicating and leveraging their working knowledge of clinical and systems engineering.

REFERENCES

Lu, Z., & Su, J. (n. d.). *Clinical data management: Current status, challenges, and future directions from industry perspectives Metadata Management in Clinical Research*. Retrieved From http://www.octagonresearch.com/assets/files/MDR_whitepaper_2_2012.pdf

KEY TERMS AND DEFINITIONS

Business Intelligence: It is a set of tools and techniques for transformation of raw data into meaningful and useful information for business analysis purpose.

Data Dictionary: Contains information from data model tools for the definition of metadata elements and their technical definitions provided by data or enterprise architecture.

Master Data Management: A comprehensive method of enabling an enterprise to link all of its critical data to one file called a master file.

Metadata Repository: Metadata repository is a database created to store metadata.

Metadata: Data about the structures that contain the data.

Chapter 4
Semantic Interation, Text Mining, Tools and Technologies

Chandrakant Ekkirala
Cognizant Technologies Limited, India

ABSTRACT

Semantic technologies have gained prominence over the last several years. Semantic technologies are explored in detail and semantic integration of data will be outlined. The various data integration techniques and approaches will also be touched upon. Text Mining, different associated algorithms and the various tools and technologies used in text mining will be enumerated in detail. The chapter will have the following sections – 1. Data Integration Techniques • Data Integration Technique – Extraction, Transformation and Loading (ETL) • Data Integration Technique – Data Federation 2. Data Integration Approaches • Need Based Data Integration • Periodic Data Integration • Continuous Data Integration 3. Semantic Integration 4. Semantic Technologies 5. Semantic Web Technologies 6. Text Mining 7. Text Mining Algorithms 8. Tools and Technologies for Text Mining

INTRODUCTION

Data Integration Techniques

Data integration is a fundamental, yet deceptively challenging, component of any organization's business intelligence and data warehousing strategy. Data integration involves combining data residing in different data repositories and providing business users with a unified view of this data. In addition, companies face a challenge of ensuring that data being reported is current and up-to-date. Companies are now increasingly incorporating both traditional batch-oriented techniques for query performance and real-time data integration to eliminate the annoyance of out-of-date data. The top batch-oriented technique that companies utilize is known as ETL while one of the popular real-time techniques is known as Data Federation.

DOI: 10.4018/978-1-4666-8726-4.ch004

Figure 1. Data Integration Techniques – ETL and Data Federation

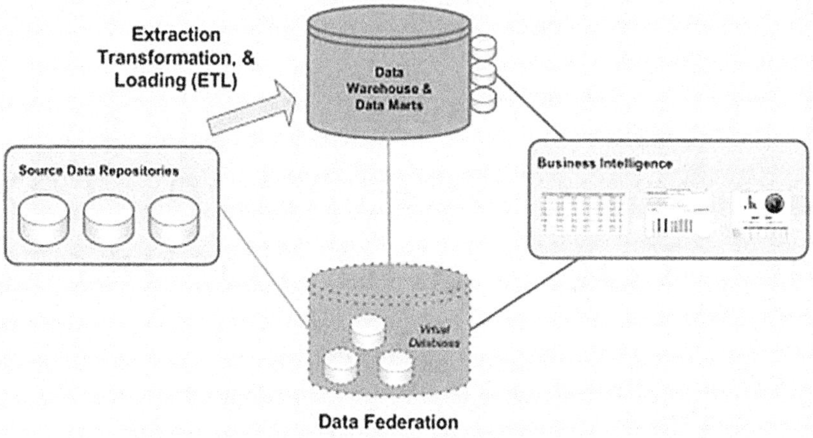

Data Integration Technique: Extraction, Transformation and Loading (ETL)

The term ETL which stands for extraction, transformation, & loading is a batch or scheduled data integration processes that includes extracting data from their operational or external data sources, transforming the data into an appropriate format, and loading the data into a data warehouse repository. ETL enables physical movement of data from source to target data repository. The first step, extraction, is to collect or grab data from its source(s). The second step, transformation, is to convert, reformat, cleanse data into format that can be used be the target database. Finally the last step, loading, is import the transformed data into a target database, data warehouse, or a data mart. A data warehouse holds very detailed information with multiple subject areas and works towards integrating all the date sources. A data mart usually holds more summarized data and often holds only one subject area.

ETL Step 1: Extraction

The extraction step of an ETL process involves connecting to the source systems, and both selecting and collecting the necessary data needed for analytical processing within the data warehouse or data mart. Usually data is consolidated from numerous, disparate source systems that may store the date in a different format. Thus the extraction process must convert the data into a format suitable for transformation processing. The complexity of the extraction process may vary and it depends on the type and amount of source data.

ETL Step 2: Transformation

The transformation step of an ETL process involves execution of a series of rules or functions to the extracted data to convert it to standard format. It includes validation of records and their rejection if they are not acceptable. The amount of manipulation needed for transformation process depends on the data. Good data sources will require little transformation, whereas others may require one or more transformation techniques to to meet the business and technical requirements of the target database or the data warehouse. The most common processes used for transformation are conversion, clearing the duplicates, standardizing, filtering, sorting, translating and looking up or verifying if the data sources are inconsistent.

ETL Step 3: Loading

The load is the last step of ETL process involves importing extracted and transformed data into a target database or data warehouse. Some load processes physically insert each record as a new row into the table of the target warehouse utilizing a SQL insert statement. Whereas other load processes include a massive bulk insert of data utilizing a bulk load routine. The SQL insert is a slower routine for imports of data, but does allow for integrity checking with every record. The bulk load routine may be faster for loads of large amounts of data, but does not allow for integrity check upon load of each individual record.

ETL Tool Providers

Here is a list of the most popular commercial and freeware (open-source) ETL Tools.

Commercial ETL Tools:

- IBM Infosphere DataStage
- Informatica PowerCenter
- SAP Business Objects Data Integrator (BODI)
- SAP Business Objects Data Services
- Oracle Warehouse Builder (OWB)
- Oracle Data Integrator (ODI)
- SAS Data Integration Studio
- Microsoft SQL Server Integration Services (SSIS)
- Ab Initio
- SyncSort DMExpress
- iWay DataMigrator
- Pervasive Data Integrator

Freeware, Open Source ETL tools:

- Pentaho Data Integration (Kettle)
- Talend Integrator Suite
- CloverETL
- Jasper ETL

Data Integration Technique: Data Federation

Data federation is a category of data integration technology that provides the ability to query and aggregate data from disparate sources in a virtual database so it can be used by business intelligence, reporting, or analysis applications in real-time. The virtual database created by data federation technology doesn't contain the data itself. Instead, it contains information or metadata about the actual data and its location. The actual data is physically left in place within its source data repository.

Data federation is used to create virtualized and integrated views of data and allows for execution of distributed queries against multiple data sources (relational databases, enterprise applications, data warehouses, documents, XML) at the same time. Data federation allows for accesses to data without physical movement of data and provides a layer of abstraction above the physical implementation of data.

Data federation is synonymous with other technologies and commonly referred to as …

- Data Virtualization
- Enterprise Information Integration (EII)

Pros of Data Federation

- Access current and transactional data stored in multiple sources
- Does not require movement of data (No ETL)
- Only requested data returned
- Real-time data access
- Quicker development time – supports incremental development
- Reduces data storage and data transfer

Cons of Data Federation

- Still queries against original data sources
- Only contains as much data as source system contains
- If data is archived off source, data is no longer available in federation tool
- Query performance is not as good as a data warehouse
- High system performance transferred to an application server from a database server

Data Integration Approaches

Need Based Data Integration

Analysts begin by identifying the administrative datasets that can be utilized to address a particular business or analytic need at hand. Once the required datasets have been identified, and the necessary programmatic and legal approvals obtained, data is collected, organized, and prepared for analysis.

The key element of this approach is that the data matching effort is undertaken to address a specific business or analytic need, and the data preparation tasks such as data profiling, cleansing, transformation, matching, and linking are performed in the context of the particular business problem or analytic need. For example, a child welfare agency wanting to better assess the needs of its aging out population may be interested in information about living arrangements of its past clients. For this particular analytic need, the agency may want to match its client datasets with those of the homeless system.

Periodic Data Integration

With the Periodic Data Integration approach, enterprises implement a process of collecting, cleansing, organizing, and storing data at pre-determined frequencies such as monthly, quarterly, or annually. Unlike the Need Based Data Integration approach in which cross-agency data matching is carried out after a particular analytic need has been identified, Periodic Data Integration seeks to pre-integrate data for a class of business problems or analytic needs.

With this approach, participating organizations develop data sharing agreements to formalize their data sharing relationships. A data sharing agreement is a formal contract that identifies specific data categories to be shared, the obligations of the participating organizations regarding the purpose for which data can be used, the frequency with which data will be made available, client confidentiality requirements, data security requirements, etc.

Continuous Data Integration

With the Continuous Data Integration approach, participating organizations develop a shared information source that is continuously kept current with the administrative systems. Unlike Periodic Data Integration, in which large datasets are brought together at regular intervals, Continuous Data Integration involves immediate posting and constant mirroring of administrative data in the shared data repositories. The key advantage of the Continuous Integration approach is that it makes integrated data available not only for research and policy purposes but also for casework and operational planning purposes.

Continuous Data Integration must also address certain complexities that result from the real-time processes. In particular, unlike Periodic Integration, there is no batch window available to complete data quality checks, source to target data reconciliations, and other quality control functions. Thus, the data quality checks and measurements must be implemented as pre-defined business rules.

Further, because data is brought into the shared information repositories while it may still be volatile, the IDS must recognize and process any changes when they occur. For example, if an incorrect social security number is entered for a client at a source agency and is then corrected a few hours later, the associated impact on data matching or linking must be determined and corrected through an automated process. Such complexities do not exist in Periodic Integration that primarily deals with non-volatile data.

Semantic Integration Explained

Semantic integration is the process of interrelating information from diverse sources, for example calendars and to do lists, email archives, presence information (physical, psychological, and social), documents of all sorts, contacts (including social graphs), search results, and advertising and marketing relevance derived from them. In this regard, semantics focuses on the organization of and action upon information by acting as an intermediary between heterogeneous data sources, which may conflict not only by structure but also context or value.

In enterprise application integration (EAI), semantic integration can facilitate or even automate the communication between computer systems using metadata publishing. Metadata publishing potentially offers the ability to automatically link ontologies. One approach to (semi-)automated ontology (structural frameworks for organizing information) mapping requires the definition of a semantic distance or its inverse, semantic similarity and appropriate rules. Other approaches include so-called lexical methods, as well as methodologies that rely on exploiting the structures of the ontologies. For explicitly stating similarity/equality, there exist special properties or relationships in most ontology languages. OWL, for example has “sameIndividualAs” or “same-ClassAs”.

Eventually system designs may see the advent of composable architectures where published semantic-based interfaces are joined together to enable new and meaningful capabilities. These could predominately be described by means of design-time declarative specifications that could ultimately be rendered and executed at run-time.

Semantic integration can also be used to facilitate design-time activities of interface design and mapping. In this model, semantics are only explicitly applied to design and the run-time systems work at the syntax level. This “early semantic binding” approach can improve overall system performance while retaining the benefits of semantic driven design.

What makes semantic integration a challenge is two-fold: first, the representation of information and the information itself are often bound tightly together; and second, that information frequently lacks context. Developers often think not of the data itself but rather the structure of those data: schemas, data types, relational database constructs, file formats, and so forth – structures that don’t pertain directly to the information at hand, but rather our assumption of what the data should look like. In tightly-coupled architectures, data structures are absolutely necessary, since they provide systems a way of coping with the information they are being fed.

However, in a standards-based, loosely coupled architecture, when the barriers to application integration are removed, instead of being helpful constructs, these various data structure representations actually get in the way. How information is stored and represented interferes with the meaning of that

Figure 2. Semantic Integration

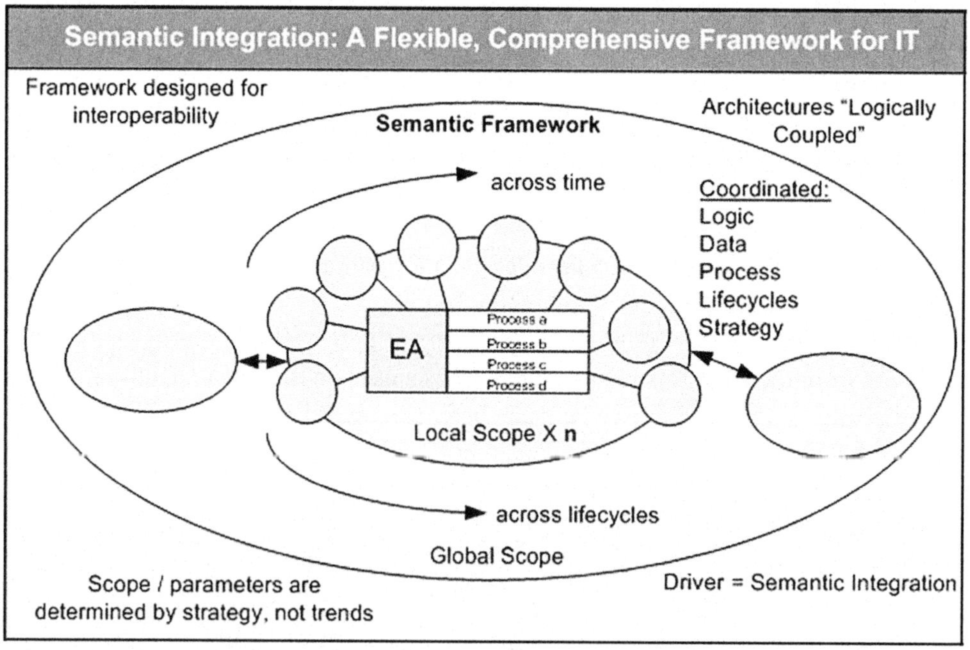

information. To be more precise, the meaning of information and the structure of that information aren't one and the same. For example, "August 7, 2003″ is a date for sure, but whether or not it is stored as a string, date type, or integer shouldn't matter. Yet, developers often needlessly combine the structure and meaning together inextricably. Furthermore, there aren't enough contexts in the structure to understand if the date is a birth date, the date this Zap Flash was written, or any other date.

Thus, when one developer's assumption of a particular structure for some datum conflicts with another's representation, you get an impedance mismatch – in other words, a data integration problem. In order for data to flow unimpeded in a Service-Oriented Architecture, Service providers must isolate requesters from the underlying data structure assumptions. The issue here is therefore one of loose coupling. While we might loosely couple application interfaces through the use of SOAs, if we deal with data the same way we've always done – by imposing the data structures of Service providers on Service requesters, the result is every bit as tightly coupled as previous architectural approaches. In order to provide the promise of seamless data integration, we must transcend simply loosely coupling the application interface and in addition provide loose coupling at the semantic level.

The practice of SOA includes different ways of looking at the problems that architecture solves, known as "views." In the information view, an information architect focuses on the meaning of the information that moves through the company, who is responsible for it, and what people do with it. The work in this view includes identifying how information is created, transported, secured, stored, and destroyed. The data architect takes the data view, in which he or she focuses on the taxonomies (way to group things together) that the company will use. The bulk of the work in this view consists of normalizing the various vocabularies across the enterprise, and understanding and delineating just what data the company wants to use. The end-product of the activities in this view is the schemas and namespaces that the business processes and the business services they contain will reference.

So, here's where data and Services fit together. In an SOA, the Service Model serves as the referee between the business requirements on the one hand, and the technical implementations on the other. And, therefore, the Services represented in the Service Model must provide the layer of abstraction between the data representations on the one hand and how they are consumed by business-level Services on the other. However, in today's early SOA implementations, companies often implement static service definitions, which mean that the Web Services' interface contracts are set at design time. While UDDI and Service-Oriented Management provide the means for dynamic discovery of such Services, those Services are still essentially static.

In order to achieve the sort of semantic data integration we are seeking, we must implement dynamic service definitions. In essence, the definition of the Service interface must change based on the context of the Service requester. As a result, a Service can change its contract between invocations. For example, the fact that a Service provider requires first names to be no longer than 40 characters should not require the requester to know that fact. The contracted Service interface is supposed to provide that isolation. Service interfaces must therefore become much smarter. Instead of having to know ahead of time what specific data requirements are needed by a Service, the Service requester should be able to dynamically discover a Service interface that can not only provide the needed functionality, but also understand the information payload.

In order to follow this "Just-in-time" integration style, for Service requesters to be able to consume data in an SOA, the data must be decoupled from any specific technical assumption (such as a specific data schema or format) so that they can be accessed via discoverable, loosely coupled, dynamically bindable Services. Now this requirement doesn't mean that the data shouldn't have any structure at all, it just means that the Service interface hides the details of that structure from the user, and the Service interface itself is dynamically created based on the context of the Service requester. Sound complicated? Well, it is. Fortunately, we have a bit of a head start: XML provides the technical means to isolate the specifics of a data format from the consumer of the data. Web Services in turn provide the means to discover and understand how to consume the data.

Semantic Technologies

The term semantic technologies represent a fairly diverse family of technologies that have been in existence for a long time and seek to help derive meaning from information. Some examples of semantic technologies include natural language processing (NLP), data mining, text mining, artificial intelligence (AI), category tagging, and semantic search.

Natural-language processing (NLP). NLP technologies attempt to process unstructured text content and extract the names, dates, organizations, events, etc. that are talked about within the text.

Data mining. Data mining technologies employ pattern-matching algorithms to tease out trends and correlations within large sets of data. Data mining can be used, for example, to identify suspicious and potentially fraudulent trading behaviour in large databases of financial transactions.

Artificial intelligence or expert systems. AI or expert systems technologies use elaborate reasoning models to answer complex questions automatically. These systems often include machine-learning algorithms that can improve the system's decision-making capabilities over time.

Classification. Classification technologies use heuristics and rules to tag data with categories to help with searching and with analyzing information.

Semantic search. Semantic search technologies allow people to locate information by concept instead of by keyword or key phrase. With semantic search, people can easily distinguish between searching for John F. Kennedy, the airport, and John F. Kennedy, the president.

Semantic Web Technologies

Semantic Web technologies are a family of very specific technology standards from the World Wide Web Consortium (W3C) that are designed to describe and relate data on the Web and inside enterprises. The standard promotes common data formats on the World Wide Web. By encouraging the inclusion of semantic content in web pages, the Semantic Web aims at converting the current web, dominated by unstructured and semi-structured documents into a "web of data".

These standards include:

- A flexible data model (RDF),
- Schema and ontology languages for describing concepts and relationships (RDFS and OWL),
- A query language (SPARQL),
- A rules language (RIF),
- A language for marking up data inside Web pages (RDFa),

The main purpose of the Semantic Web is driving the evolution of the current Web by enabling users to find, share, and combine information more easily. Humans are capable of using the Web to carry out

Figure 3. Layered Model for Data Integration and Interoperation on Semantic Web

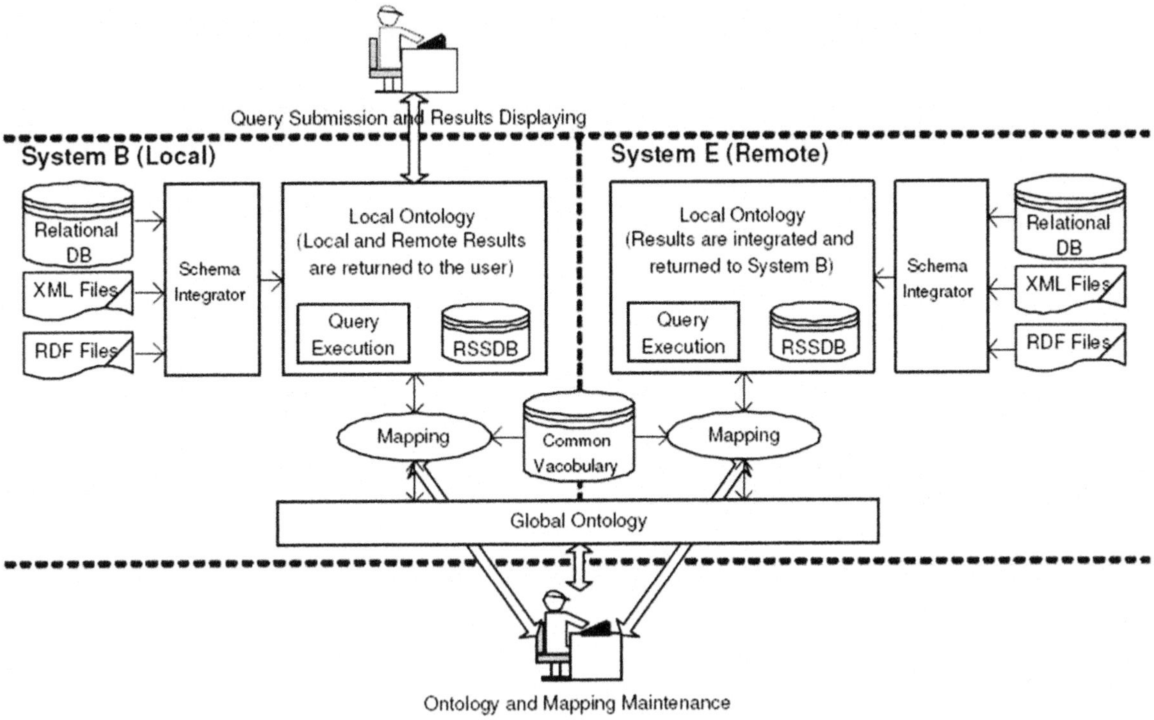

tasks such as finding the German translation for "eight days", reserving a library book, and searching for the lowest price for a DVD. However, machines cannot accomplish all of these tasks without human direction, because web pages are designed to be read by people, not machines. The semantic web is a vision of information that can be readily interpreted by machines, so machines can perform more of the tedious work involved in finding, combining, and acting upon information on the web. It uses metadata as well.

So what precisely is the relationship between semantic technologies and Semantic Web technologies?

In short Semantic Web technologies are a set of technologies that happen to be especially well-suited for implementing semantic technology algorithms and solutions.

Collectively, Semantic Web technologies are a toolbox; as such, they can be used to implement a wide variety of algorithms, solutions, and applications. However, they are particularly appropriate for implementing semantic technologies. Consider the following examples:

- Classifying data can be accomplished very effectively by describing information using the schema and ontology languages that are part of the Semantic Web technology set.
- Semantic search requires a way to describe data conceptually and a way to search via these concepts. The Semantic Web technology stack satisfies both these conditions.
- NLP tools can identify unanticipated relationships between entities in source documents. The flexible graph-based data model that is one of the core Semantic Web standards is an ideal way of capturing all information obtained by NLP technology without the need to discard any data.

Figure 4. High Level Text Mining Functional Architecture

Text Mining

Text mining, also referred to as text data mining, roughly equivalent to text analytics, refers to the process of deriving high-quality information from text. High-quality information is typically derived through the devising of patterns and trends through means such as statistical pattern learning. Text mining usually involves the process of structuring the input text (usually parsing, along with the addition of some derived linguistic features and the removal of others, and subsequent insertion into a database), deriving patterns within the structured data, and finally evaluation and interpretation of the output. 'High quality' in text mining usually refers to some combination of relevance, novelty, and interestingness. Typical text mining tasks include text categorization, text clustering, concept/entity extraction, production of granular taxonomies, sentiment analysis, document summarization, and entity relation modelling (i.e., learning relations between named entities).

Text mining is the automatic and semi-automatic extraction of implicit, previously unknown, and potentially useful information and patterns, from a large amount of unstructured textual data, such as natural-language texts . In text mining, each document is represented as a vector, whose dimension is approximately the number of distinct keywords in it, which can be very large. One of the main challenges in text mining is to classify textual data with such high dimensionality. In addition to high dimensionality, text-mining algorithms should also deal with word ambiguities such as pronouns, synonyms, noisy data, spelling mistakes, abbreviations, acronyms and improperly structured text.

Eighty percent of information in the world is currently stored in unstructured textual format. Although techniques such as Natural Language Processing (NLP) can accomplish limited text analysis, there are currently no computer programs available to analyse and interpret text for diverse information extraction needs. Therefore text mining is a dynamic and emerging area. The world is fast becoming information intensive, in which specialized information is being collected into very large data sets. For example, Internet contains a vast amount of online text documents, which rapidly change and grow. It is nearly impossible to manually organize such vast and rapidly evolving data. The necessity to extract useful and relevant information from such large data sets has led to an important need to develop computationally efficient text mining algorithms [5]. An example problem is to automatically assign natural language text documents to predefined sets of categories based on their content. Other examples of problems involving large data sets include searching for targeted information from scientific citation databases (e.g. MEDLINE); search, filter and categorize web pages by topic and routing relevant email to the appropriate addresses.

Text analysis involves information retrieval, lexical analysis to study word frequency distributions, pattern recognition, tagging/annotation, information extraction, data mining techniques including link and

Figure 5. A text mining approach for extracting sequence variants in biomedical literature

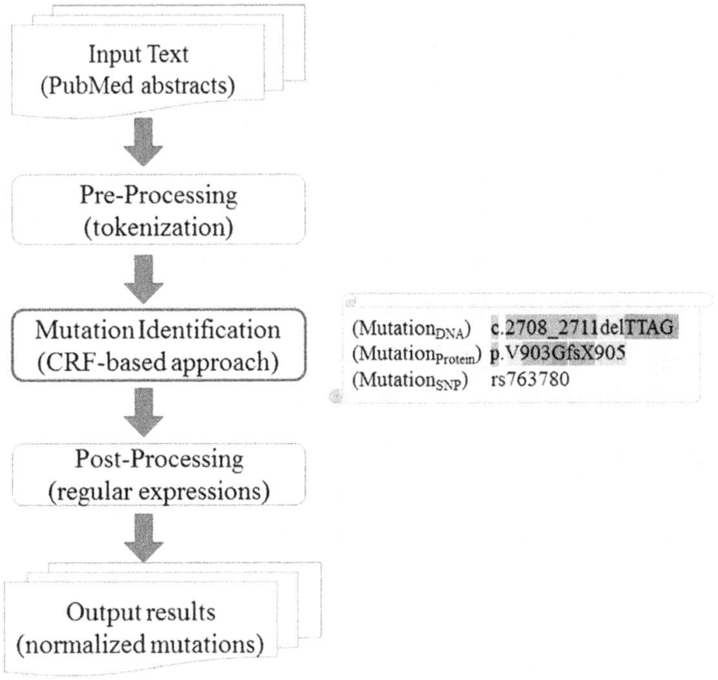

association analysis, visualization, and predictive analytics. The overarching goal is, essentially, to turn text into data for analysis, via application of natural language processing (NLP) and analytical methods.

Text Analysis Processes

Subtasks — components of a larger text-analytics effort — typically include:

- Information retrieval or identification of a corpus is a preparatory step: collecting or identifying a set of textual materials, on the Web or held in a file system, database, or content management system, for analysis.
- Although some text analytics systems apply exclusively advanced statistical methods, many others apply more extensive natural language processing, such as part of speech tagging, syntactic parsing, and other types of linguistic analysis.
- Named entity recognition is the use of gazetteers or statistical techniques to identify named text features: people, organizations, place names, stock ticker symbols, certain abbreviations, and so on. Disambiguation — the use of contextual clues — may be required to decide where, for instance, "Ford" can refer to a former U.S. president, a vehicle manufacturer, a movie star, a river crossing, or some other entity.
- Recognition of Pattern Identified Entities: Features such as telephone numbers, e-mail addresses, quantities (with units) can be discerned via regular expression or other pattern matches.
- Coreference: identification of noun phrases and other terms that refer to the same object.

- Relationship, fact, and event Extraction: identification of associations among entities and other information in text
- Sentiment analysis involves discerning subjective (as opposed to factual) material and extracting various forms of attitudinal information: sentiment, opinion, mood, and emotion. Text analytics techniques are helpful in analyzing sentiment at the entity, concept, or topic level and in distinguishing opinion holder and opinion object
- Quantitative text analysis is a set of techniques stemming from the social sciences where either a human judge or a computer extracts semantic or grammatical relationships between words in order to find out the meaning or stylistic patterns of, usually, a casual personal text for the purpose of psychological profiling etc

Applications

The technology is now broadly applied for a wide variety of government, research, and business needs. Applications can be sorted into a number of categories by analysis type or by business function. Using this approach to classifying solutions, application categories include:

- Enterprise Business Intelligence/Data Mining, *Competitive Intelligence*
- *E-Discovery*, Records Management
- *National Security*/Intelligence
- *Scientific discovery*, especially Life Sciences
- *Sentiment Analysis* Tools, Listening Platforms
- Natural Language/Semantic Toolkit or Service
- Publishing
- Automated *ad placement*
- Search/Information Access
- *Social media* monitoring

Text Mining Algorithms

Text mining algorithms are two types: Supervised learning and unsupervised learning. Support vector machines (SVMs) are a set of supervised learning methods used for classification and regression. Non-negative matrix factorization is an unsupervised learning method.

Supervised Learning

Supervised learning is a technique in which the algorithm uses predictor and target attribute value pairs to learn the predictor and target value relation. Support vector machine is a supervised learning technique for creating a decision function with a training dataset. The training data consist of pairs of predictor and target values. Each predictor value is tagged with a target value. If the algorithm can predict a categorical value for a target attribute, it is called a classification function. Class is an example of a categorical variable. Positive and negative can be two values of the categorical variable class. Categorical values do not have partial ordering. If the algorithm can predict a numerical value then it is called regression. Numerical values have partial ordering.

Unsupervised Learning

Unsupervised learning is a technique in which the algorithm uses only the predictor attribute values. There are no target attribute values and the learning task is to gain some understanding of relevant structure patterns in the data. Each row in a data set represents a point in n-dimensional space and unsupervised learning algorithms investigate the relationship between these various points in n-dimensional space. Examples of unsupervised learning are clustering, density estimation and feature extraction.

Different Text Mining Algorithms

There are a number of Text Mining Algorithms available like Naïve Bayes, Generalized Linear Models, Support Vector Machine, k-Means, Non-negative Matrix Factorization, Apriori, Minimum Descriptor Length and so on.

Details of some of the popular algorithms are given below -

Naive Bayes

The Naive Bayes algorithm is based on conditional probabilities. It uses Bayes' Theorem, a formula that calculates a probability by counting the frequency of values and combinations of values in the historical data.

Bayes' Theorem finds the probability of an event occurring given the probability of another event that has already occurred. If B represents the dependent event and A represents the prior event, Bayes' theorem can be stated as follows.

Prob(B given A) = Prob(A and B)/Prob(A)

To calculate the probability of B given A, the algorithm counts the number of cases where A and B occur together and divides it by the number of cases where A occurs alone.

Support Vector Machines

Support Vector Machines (SVM) is a powerful, state-of-the-art algorithm with strong theoretical foundations based on the Vapnik-Chervonenkis theory. SVM has strong regularization properties. Regularization refers to the generalization of the model to new data.

In the SVM algorithm, a text document is represented as a vector whose dimension is the approximately the number of distinct keywords in it. Thus, as the document size increases, the dimension of the hyperspace in which text classification is done becomes enormous, resulting in high computational cost. However, the dimensionality can be reduced through feature extraction algorithms. An SVM model can then be built based on the extracted features from the training data set, resulting in a substantial decrease in computational complexity.

SVM models have similar functional form to neural networks and radial basis functions, both popular data mining techniques. However, neither of these algorithms has the well-founded theoretical approach to regularization that forms the basis of SVM. The quality of generalization and ease of training of SVM is far beyond the capacities of these more traditional methods.

SVM can model complex, real-world problems such as text and image classification, hand-writing recognition, and bioinformatics and biosequence analysis.

SVM performs well on data sets that have many attributes, even if there are very few cases on which to train the model. There is no upper limit on the number of attributes; the only constraints are those imposed by hardware. Traditional neural nets do not perform well under these circumstances.

Non-Negative Matrix Factorization

Non-negative matrix factorization (NMF), also non-negative matrix approximation is a group of algorithms in multivariate analysis and linear algebra where a matrix V is factorized into (usually) two matrices W and H, with the property that all three matrices have no negative elements. This non-negativity makes the resulting matrices easier to inspect. Since the problem is not exactly solvable in general, it is commonly approximated numerically.

Non-negative matrix factorization is a feature extraction algorithm that decomposes text data by creating a user-defined number of features. NMF gives a reduced representation of the original text data. It decomposes a text data matrix Amn where columns are text documents and rows are attributes or keywords, into the product of two lower rank matrices Wmk and Hkn, such that Amn is approximately equal to Wmk times Hkn. In NMF, in order to avoid cancellation effects, the factors Amn and Hkn should have non-negative entries. NMF uses an iterative procedure to modify the initial values of Wmk and Hkn so that the product approaches Amn. The procedure terminates when the approximation error converges or the specified number of iterations is reached. The matrix decomposition can be represented as:

Amn = Wmk × Hkn,

where,

Amn: (m×n) matrix: Each column of which contains m nonnegative values (word counts) of one of the n text documents.

Figure 6. The Support Vector Machine Algorithm

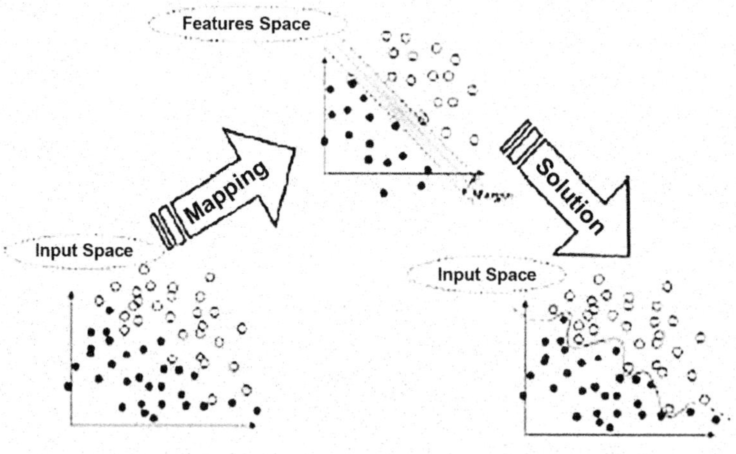

Wmk: (m×k) matrix: k columns of W are called basis document vectors or feature vectors.

Hkn: (k ×n) matrix: each column of H is called encoding or weight column.

Matrix A represents a collection of text documents, where Aij is the number of times the ith word in the vocabulary appears in the jth document. The above equation illustrates the decomposition of the matrix Amn into two lower rank matrices Wmk and Hkn. The columns of the matrix Wmk can be viewed as the underlying basis document vectors. In other words, each of the n columns of the matrix Amn can be built from k columns of Wmk. Columns of the matrix Hkn give the weights associated with each basis document vector. Basis vectors of Wmk are not necessarily orthogonal and can have overlap of topics. Each document of a text collection can be represented as a linear combination of basis text document vectors or "feature" vectors.

Tools and Technologies for Text Mining

Some of the more popular software for Text Analysis, Text Mining and Text Analytics are given below:

- **SAS Text Analytics:** SAS provides a comprehensive Text Analytics software suite to discover and extract information from text content. SAS software uses advanced statistical modeling, natural language processing and advanced linguistic technologies to discover patterns and trends from any text in any format.
- **IBM Text Analytics:** IBM provides natural language processing solutions LanguageWare and IBM Content Analytics. These solutions provide an easy to use environment for capturing the knowledge into dictionaries and semantic rules for re-use, allows customizable Information Extraction and offers entity and relationship recognition.

Figure 7. Non Negative Matrix Factorization

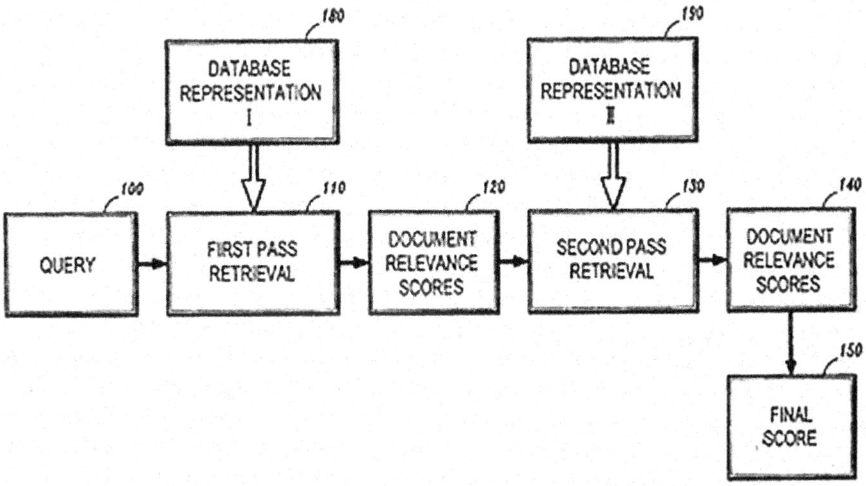

- **SAP Text Analytics:** SAP Data Services is an enterprise solution for data integration, quality, profiling, and text analysis which delivers trusted data to support your critical business functions and improve decision making.
- **Lexalytics Text Analytics:** Lexalytics is a provider of text analytics engine. Lexalytics builds a multi lingual text analytics engine, Salience. Salience supports a number of text processing, natural language processing and text analytics technologies such as Sentiment Analysis, Named Entry Extraction, Context Extraction, Entity Level Sentiment Analysis, Summarization and Facet and Attribute Extraction. Supports multi languages.
- **Smartlogic:** Smartlogic provides content intelligence platform containing commercial text analytics, NLP, rule-based classification, ontology/taxonomy modeling and information visualization.This solution automatically applies the metadata and classification to deliver better search and navigation experience. This solution improves findability of information from text.

A few more tools are ai-one, Provalis research, OpenText, Pingar, AlchemyAPI, Attensity, Clarabridge, Content Analyst etc.

Some of the free software available for Text Analysis, Text Mining and Text Analytics are given below –

- **GATE:** This is an open source toolbox for natural language processing, and language engineering. It is used for all sorts of language processing tasks and applications including voice of customer, cancer research, drug research, information extraction, semantic annotation to name a few.
- **Carrot2:** This does text and search results clustering framework. It is an open source search clustering engine. Apart from two specialized search results clustering algorithms, it also offers ready to use components for fetching search results from various sources like GoogleAPI, BingAPI, eToolsMetaSearch and more.
- **KH Coder:** An application for quantitative content analysis, text mining or corpus linguistics. By inputting the raw texts the searching and analysis functionalities like KWC, collocation statistics, co-occurrence networks self-organizing map, multi-dimensional scaling, cluster analysis and corresponding analysis can be utilized.

A few more similar tools are tm, Gensim, Natural Language Toolkit, RapidMiner, Unstructured Information Architecture, OpenNLP, KNIME Text Processing etc.

REFERENCES

Evans, B.L. (2003). Non Negative Matrix Factorization, Multidimensional Digital Signal Processing. Retrieved from http://www.ece.utexas.edu/~bevans/courses/ee381k/projects/spring03/

Gunn, S. (1998, May 14). Support Vector Machines for Classification and Regression. http://homepages.cae.wisc.edu/~ece539/software/svmtoolbox/svm.pdf

Hearst, M. (1999). Untangling text data mining. http://www.sims.berkeley.edu/~hearst/papers/acl99/acl99-tdm.html

Hearst, M. (2003, October 17). What is text mining? Retrieved from http://www.sims.berkeley.edu/~hearst/text-mining.html

Langville, A., & Meyer, C. (2005, June 9). ALS Algorithms Nonnegative Matrix Factorization Text Mining. Retrieved from http://meyer.math.ncsu.edu/Meyer/Talks/SAS_6_9_05_NmfWorkshop.pdf

Ma, L., Mei, J., Pan, Y. (2007). Semantic Web Technologies and Data Management. Retrieved from http://www.w3.org/2007/03/RdfRDB/papers/ma.pdf

Noy, N.F. (2004, June 25). Semantic Integration – A Survey of Ontology based approaches. Retrieved from http://web.stanford.edu/~natalya/papers/SigmodRecordReview.pdf

KEY TERMS AND DEFINITIONS

Analytics: It is the process of identifying and communicating meaningful patterns in data.The purpose of the same is to predict and improve business performance.

Data Mining: The process of discovering and identifying patterns in large sets of data using computational methodologies.

Ontology: It is the study of entities and their relationships. In other words it is a formal naming and definition of types, properties and interrelations of the entities that exist.

SOA: Service Oriented Architecture, which essentially is a collection of services wherein the services interact with each other. The service here is a well-defined function and also self-contained and does not depend on the context or state of other services.

Taxonomy: It is the science of defining groups of biological organisms on the basis of shared characteristics and identifying the different groups with names.

UDDI: Universal Description, Discovery and Integration is a platform independent extensible markup language based registry by which businesses worldwide can enlist themselves on Internet, and a mechanism to register and locate Webservice applications.

Chapter 5
Safety Signal Detection in the Drug Development Process

Ramin B. Arani
Advanced Analytic Center, USA

Antoni F.Z. Wisniewski
AstraZeneca Pharmaceuticals, UK

ABSTRACT

Drug development is a complex set of inter-linked processes in which the cumulative understanding of a drug's safety and efficacy profile is shaped during different learning phases. Often, drugs are approved based on limited safety information, for example in highly at risk or rare disease populations. Therefore, post approval, regulatory organizations have mandated proactive surveillance strategies that include the collection of reported adverse events experienced by exposed populations, some of whom may have been on treatment for extended periods of time. Analyzing these accumulating adverse event reports to understand their clinical significance, given the limitations imposed by the methods of data collection, is a complicated task. The aim of this chapter is to provide the readers with a general understanding of safety signal detection and assessment, followed by a description of statistical methods (both classical and Bayesian) typically utilized for quantifying the strength of association between a drug and an adverse event.

INTRODUCTION

The successful development, registration and marketed use of a new pharmaceutical entity require regular reappraisal of its risk-benefit profile, throughout its life-cycle. An intimate understanding of the real or potential risks to patients of taking a pharmaceutical intervention will inform how it is formulated, packaged, labeled, prescribed and taken by patients, including how known risks are managed or avoided. A key component of a marketing authorization is the agreed strategy for identifying new risks and how emergent risks will be characterized and managed once the pharmaceutical product is launched; in the context of risk management, what remains *unknown* about the product becomes as important as that

DOI: 10.4018/978-1-4666-8726-4.ch005

which is known. Indeed, a failure to present an adequate strategy to deal with uncertainty in relation to risk is likely to result in the rejection of an application to market a product.

The safety profile of a new pharmaceutical entity evolves over the drug development process and new risks continue to be identified once it is launched onto the market. The information upon which decisions related to safety are made may vary considerably, depending on the life-cycle stage of the product and the methods of detecting and evaluating new risks will differ based on the type, quantity and quality of the available data. In the early stages of drug development, the number of exposed subjects will be small, although the quantity of information on each may be plentiful and the quality high in terms of how the data are collected and recorded. In the latter stages of drug development, the number of exposed patients will generally be much higher although the amount of information collected on each patient may be less. Even in the largest phase 3 trials, the quality of the data is still subject to appraisal and control and it is usually possible to obtain further information to help characterise any emerging safety concerns. Once on the market, patient exposure to a product is likely to grow massively in comparison with that seen in development; even if the quantity of data from these patients is growing exponentially, there will only limited ability to directly influence the quality of the information coming from marketed use and in most cases, the information content itself may be sparse. In this chapter we will explore how a range of more or less innovative and established statistical methods can be employed to enhance risk identification over the life-cycle of product with a particular focus on those that are amenable to automation and computer-aided screening of large datasets as encountered in the post-marketing setting.

BACKGROUND

For the sake of clarity the following definitions are assumed throughout this chapter:

An *adverse drug reaction* (ADR) is `a noxious and unintended response to a medicinal product for which there is a reasonable possibility that the product caused the response.' (International Conference on Harmonisation, 1994)

An *adverse event* (AE) is defined as `any untoward medical occurrence in a patient or clinical investigation subject administered a pharmaceutical product which does not necessarily have a causative relationship with this treatment.'(International Conference on Harmonisation, 1996)

Disproportionality analysis (DPA) is `the application of computer assisted computational and statistical methods to large safety databases for the purpose of systematically identifying drug-event pairs reported at disproportionately higher frequencies relative to what a statistical independence model would predict.'(J. Almenoff et al., 2005)

A *drug-event pair* is the combination of a medicinal product and an adverse event which has appeared in at least one case report entered into a spontaneous report [safety] database.(CIOMS Working Group VIII, 2010)

The *safety profile* of a drug or other therapeutic intervention can defined as the aggregate knowledge of the severity and frequency of adverse drug reactions and other risks related to the use of the intervention.

A *safety signal* is defined as `information that arises from one or multiple sources (including observations and experiments), which suggest a new potentially causal association, or a new aspect of a known association, between an intervention and an event or set of related events, either adverse or beneficial, that is judged to be of sufficient likelihood to justify verificatory action.'(CIOMS Working Group VIII, 2010)

Table 1. Drug-Event scenarios

	Drug D	Other	Total
Adverse Event E	n_{11}	n_{12}	n_{1+}
Other	n_{21}	n_{22}	n_{2+}
Total	n_{+1}	n_{+2}	n_{++}

A *spontaneous report* is an unsolicited communication by health care professionals or consumers to a company, regulatory authority or other organisation that describes one or more suspected adverse drug reactions in a patient who was given one or more medicinal products.(International Conference on Harmonisation, 18 November 2004)

A *statistic* (sometimes a *signal) of disproportionate reporting* (SDR) is a numerical result above a preset threshold generated from any data-mining algorithm using disproportionality analysis applied to a spontaneous report database. An SDR alerts medical assessors to a specific adverse event reported for a particular medicinal product (drug-event pair) that should be explored further. Note: SDRs that originate from spontaneous report databases cannot be interpreted as scientific evidence for establishing causality between medicinal products and adverse events, and thus they are distinct from statistical associations from formal epidemiological studies.(CIOMS Working Group VIII, 2010)

The accepted gold standard for clinical evidence supporting claims related to therapeutic efficacy is the randomised double-blind, placebo-controlled trial. In contrast, with the exception of a few, pre-specified safety endpoints, it is generally not possible to conduct trials of sufficient magnitude to comprehensively characterise the safety profile of an intervention since issues related to `the power to detect' and of multiplicity will become serious limitations. A great deal can be known about the safety profile of an intervention from the data gathered during a clinical development programme. However, given the practical limitations of the randomised double-blind, placebo-controlled trial and the observation that `absence of evidence is not evidence of absence' [...of a safety risk], different approaches to safety signal identification and characterisation are needed in the various stages of a product's life-cycle, not least during the marketed phase.

As well as the textual definitions provided above, the following statistical concepts and definitions are fundamental to the understanding of the quantitative aspects of safety signal detection. Table 1 introduces the concept of the two-by-two contingency table from which much of what follows will flow.

Assuming the total number subjects exposed to drug D is known, as in the clinical trial or epidemiological study setting (ie, n_{+1} and n_{+1} denote total exposed), then

Relative Risk or Risk Ratio (RR): is the probability of event E in subjects exposed to drug D relative to the overall probability of event E, that is

$$RR = \frac{P\left(E|D\right)}{P\left(E\right)} = \frac{n_{11} \div n_{+1}}{n_{1+} \div n_{++}}$$

Odds Ratio (OR): is the ratio of the odd of event E in subjects exposed to drug D to the odd of event E in subjects not exposed to drug D, that is

$$OR = \frac{P\left(E|D\right) \div P\left(\bar{E}|D\right)}{P\left(E|\bar{D}\right) \div P\left(\bar{E}|\bar{D}\right)} = \frac{n_{11} \div n_{21}}{n_{12} \div n_{22}}$$

Where $\bar{D}$ means "not D".

Now assuming the total number of subjects exposed to drug D and experienced event E are unknown, this case refers to post approval of drug where one of the principal sources of new data are from spontaneous reporting of adverse events. Hence, the entries in Table 1 are based on number of reports that mention drug D and event E, as a proxy for actual exposure. In later sections, the equivalent formulae for use in the post-marketing setting are described, given that spontaneous adverse event reports are the main analysis substrate.

QUANTITATIVE SAFETY SIGNAL DETECTION

Over the past several decades there has been a move towards employing more statistically driven methods of signal detection in spontaneous report datasets, in both regulatory and pharmaceutical company setting, as a compliment to more traditional methods such as manual review of case series (J. Almenoff et al., 2005; Lindquist et al., 1999; Van Puijenbroek et al., 2002). The main drivers for this have been:

- Large increases in reports of adverse events due to progressive changes in reporting practices and regulation as well as increasing engagement of the public in AE reporting making comprehensive individual case report assessment unfeasible
- The need for pharmacovigilance departments to appropriately prioritise safety concerns according to potential public health importance
- The aspiration to identify new safety concerns more quickly than traditional pharmacovigilance approaches
- The aspiration to enable objective identification of adverse drug reactions caused by off-target effects not predictable from the known pharmacology of the therapeutic agent (so called idiosyncratic or Type B ADRs)
- Advances in electronic data storage and processing enabling the exploitation of complex computational screening methodology not feasible previously.

To further these ends, a multitude of methods based on the statistical principles described above have been developed (J. S. Almenoff, DuMouchel, Kindman, Yang, & Fram, 2003; Bate et al., 1998; Brown et al., 2011; Cornelius & Evans, 2009; Evans, Waller, & Davis, 2001) in order to identify *statistical* associations between a medicinal intervention and adverse events (i.e, drug-event pairs) commonly referred to as statistics of disproportionate reporting (SDR). These are quite distinct from safety signals that originate from other sources such as assessment of individual case reports and formal epidemiological

Figure 1. Representation of statistical signal, with parameter estimate and corresponding lower and upper confidence bound.

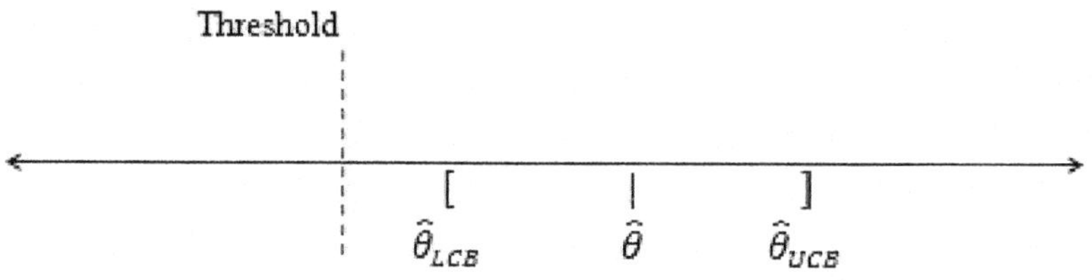

studies. It is important to establish from the outset that SDRs are measures of *statistical* association and should not been used as a measure of the degree of *causal* association between an intervention and the observed effect. It is also noteworthy that different statistical methods will tend to generate different SDRs although, as has been shown recently by Candore et al.(Candore et al., 2015), the relative performance of the various methods is predictable given the same analysis dataset. The common Bayesian and non-Bayesian methods used to generate SDRs are reviewed in latter sections.

To some extent, the specific method of generating SDRs is less important than the threshold or signal level above which further action is triggered. In this context further action will generally mean a review of the underlying case report information within the database from which the SDR is signaling. This in turn may result in a further evaluation involving more advanced analytic techniques and/or of wider scope involving the assessment of relevant information from other sources. Establishing the most appropriate signal threshold is important since there appears to be a direct trade off between the sensitivity and specificity which will result in more or fewer signals requiring evaluation with the resultant impact on the work load of expert resources. In practice, given that this tension between sensitivity and specificity exists, pharmacovigilance organizations do not rely on quantitative methods alone but will have integrated signal detection systems employing both quantitative and qualitative approaches.

Having provided the overall context and the caveats surrounding the use of quantitative methods of signal detection, the technical basis for these methods are now explored.

It is common for a confidence interval approach to be applied to the identification of SDRs, that is disproportionality scores (e.g., RR and OR) are estimated and the corresponding confidence interval is constructed. If lower confidence bound (LCB) is higher than a pre-specified threshold, then strength drug-event association indicates a statistical signal. This can be illustrated as follows:

To validate or confirm a safety signal, following questions need to be addressed:

1. Is the statistical signal (i.e., drug-event association) maintained over time?
2. Can one observe a similar pattern of drug-event association in different databases?
3. Can medical experts explain or justify the drug-event associate?
4. Can causality of AE be established beyond reasonable doubt?

The remainder of this chapter will present both established and emerging statistical methods for safety signal detection along the drug development value chain, starting with clinical development and progressing into the post marketed setting.

QUANTITATIVE SAFETY SIGNAL DETECTION DURING DRUG DEVELOPMENT

Even though randomized controlled clinical trials data provide a gold standard for the assessment of safety, due to the limited number of subjects (i.e, size) and relatively limited follow-up time only commonly occurring adverse events will be adequately characterised and inadequate power to detect rarer ADRs or adverse events with longer onset date.

In a clinical trial setting safety signals are identified in laboratory or adverse event data. In early phase (Phase I and II) studies, AE data are often not mature in terms frequency and follow-up time, therefore there is a heavy reliance on laboratory data for safety analysis. In this setting extreme value modelling can be useful to evaluate the predictive probability of observing selected laboratory parameters outside of the normal reference range in the subsequent stages of a drugs development and may be considered for future investment decisions such as go/no-go decision for Phase III. This approach has been studied extensively by Harry Southworth (Southworth & Heffernan, 2012a; Southworth & Heffernan, 2012b; Southworth, 2014) and an overview of this approach is provided in next sub-section.

In larger Phase II and III studies and where safety data are amenable to aggregation, adverse event data has its great potential to identify ADRs. It should be noted, in order to facilitate summarisation and analysis, adverse events are coded to a standardised medical ontology typically employing a hierarchical framework of similar medical concepts and/or within related organ systems. The universal adverse event coding dictionary used by the pharmaceutical industry and regulatory authorities world-wide is the Medical Dictionary for Regulatory Activities (MedDRA) (ICH MedDRA Maintenance and Support Services Organization (MSSO),). Other ontologies include the World Health Organisation Adverse Reaction Terminology (WHO-ART) (Uppsala Monitoring Centre (UMC),) and SNOMED (International Health Terminology Standards Development Organisation (IHTSDO),). A paper by Berry and Berry (Berry & Berry, 2004) proposed a three-level hierarchical mixed model for the statistical identification of potential adverse drug reactions in clinical trials data that exploits the hierarchical component of an adverse event coding dictionary. Briefly, the first (i.e. at the level of the unique medical concept) level is Preferred Terms (PT) of AEs, the second level refers to the respective body system of PT-AEs in level 1, and the third level is the collection of all body systems. Their analysis allows for borrowing `strength' [of a statistical association] across body systems. The probability that a drug has caused a type of AE is greater if its rate is elevated for several types of AEs within the same body system than if the AEs with elevated rates were in different body systems. This method is explained in more detail in the next sub-section.

Early Phase Development: Extreme Value Analyses Application

Considering continuous endpoints such as blood pressure or laboratory biochemistry parameters such as the liver function test alanine aminotransferase (ALT), most statistical methods focus on modeling the distribution central tendency, such as mean, however extreme value modeling seeks to characterize the upper or lower tail of a distribution of interest.

Generally there are two ways to capture extremes in real data. The first approach considers the maximum the variable takes in successive periods or block of observations, for example weekly or monthly maximum of laboratory measurement per subject. This is referred to as the block maxima method. The second approach focuses on the realizations (i.e., values) exceeding a given (high) threshold $u(>0)$. All realizations exceeding the threshold u constitute extreme events. This method is referred to as the peak over threshold (POT) or threshold exceedance method. The block maxima method is the traditional method, however, the threshold method uses data more efficiently and, for that reason, seems to become the method of choice in recent applications.

Block Maxima Method:

It can be shown that block maxima $M_n = \max\{X_1, \cdots, X_n\}$ as block size $n \to \infty$, then M_n follows a generalize extreme value (GEV) distribution defined by:

$$P(X \le x) = G(x) = \begin{cases} \exp\left(-\left[1+\xi\left(\frac{x-\mu}{\sigma}\right)\right]^{-1/\xi}\right) & \text{if } \xi \ne 0 \\ \exp\left(-\exp\left(-\frac{x-\mu}{\sigma}\right)\right) & \text{if } \xi = 0 \end{cases}$$

Where μ, σ, and ξ are location, scale and shape parameters, respectively.

Note that if $\xi = 0$ then $G(x)$ known as a Gumbel distribution. If $\xi = 1/\alpha > 0$ then the distribution $G(x)$ is referred to as a Frechet distribution, and if $\xi = -1/\alpha < 0$ then the distribution $G(x)$ is referred to as a Weibull distribution.

Of interest often is the value expected every *M* subjects (i.e, block maxima), z_M referred to as the *return level*,

$$P(X > z_M) = 1 - G(z_M) = \frac{1}{M}.$$

For example a 100-subject return level corresponds to the 95% percentile, in which it takes 100 subject to see a value as high as the 95 percentile.

The return value is calculated using an inverse distribution function (i.e., quantile function)

$$z_M = G^{-1}\left(1 - \frac{1}{M}\right) = \mu - \frac{\sigma}{\xi}\left(1 - \left(-\log\left(1 - \frac{1}{M}\right)\right)^{-\xi}\right)$$

Hence $z_M = \mu - \sigma \log\left(-\log\left(1 - \frac{1}{M}\right)\right)$

Note that z_M can be estimated empirically as:

$$\hat{z}_M = \hat{G}^{-1}\left(1 - \frac{1}{M}\right) = \inf\left\{x | G(x) \geq \frac{1}{M}\right\}.$$

Peak Over Threshold (POT) Method

Let us consider the exceedance over the threshold defined as $Y = (X - u | X > u)$. It can be shown that over large values of $u(> 0)$, Y asymptotically follows a Generalized Pareto distribution (GPD)

$$P\left(Y \leq y \middle| X \rangle u\right) = \begin{cases} 1 - \left(1 + \frac{\xi y}{\sigma}\right)^{\frac{-1}{\xi}} & \text{if } \xi \neq 0 \\ 1 - \exp\left(-\frac{y}{\sigma}\right) & \text{if } \xi = 0 \end{cases}$$

where σ, and ξ are scale and shape parameters, respectively. Similar to GEV, there are different types of distribution corresponding to value of ξ. Note that if $\xi > 0$ then the distribution $G(x)$ is referred to as a Frechet distribution with heavy tail behavior, and if $\xi < 0$ then the distribution $G(x)$ is referred to as a Weibull distribution, with upper finite endpoint.

Even though, there is a rich history of research and publication on extreme value modeling, only recently have they been applied to safety issues in clinical trials (Southworth, 2014). The steps in this modeling process are outlined as follows:

Typically, laboratory data are captured longitudinally per subject. Often the baseline value is highly correlated with the post-baseline measurement. Therefore, the first step is to remove the dependence on the baseline value, which is done through a robust regression approach,Where Y_{ij} is the j-th measurement (often only the maximum value Y_{iM} is considered) of i-th subject, $\hat{\theta}$ refers to the estimate of regression parameters, and e_{ij} denotes residuals.

1. For each laboratory parameter, the residuals are calculated and fitted to either a Generalized Pareto Distribution (GPD) or a Generalized Extreme Value Distribution (GEV) as defined earlier.
2. Estimates for the parameters are obtained through the maximum likelihood, or in the case of small sample sizes, the penalized maximum likelihood method. A bootstrap approach is utilized to assess the distributional properties of parameters. Next, appropriate return values are calculated.
3. Note that often laboratory parameters are correlated therefore a multivariate framework seems to be a more reasonable framework. Therefore multivariate extreme value models need to be utilized. However this topic is not in scope of this discussion: but see Heffernan and Tawn for additional details (Heffernan & Tawn, 2004).

The **texmex** R-package (The Comprehensive R Archive Network,) is capable of performing both univariate and multivariate analysis. This package was specially developed to address safety related issues in clinical trials. The example codes and analysis strategy within **texmex** guide may be consulted.

There are other R-packages such as **extremes** (The Comprehensive R Archive Network,) which may be utilized.

Late Phase Development: Hierarchical Method Application

The main purpose of clinical trials is to establish efficacy with acceptable tolerance for safety endpoints. Therefore, in the clinical trial setting, the scope of analysis for adverse events has been limited to reporting their frequency. Typically, no formal hypothesis is tested with respect to any adverse event endpoints. However, the number of unique adverse event preferred terms are relatively large which increases the false positive rate. Berry & Berry (Berry & Berry, 2004) provide an alternative to control the false positive rate through a three-level hierarchical model utilizing the natural hierarchy of MedDRA.

Berry and Berry Hierarchical Method

Suppose there are B SOC levels, and within each body system $i = 1, \cdots, B$, there $j = 1, \cdots, K_i$ adverse events. Out of N_C subjects in control arm X_{ij} experienced adverse event A_{ij} and similarly of N_T subjects in control arm Y_{ij} experienced adverse event A_{ij}. The probabilities of experiencing the adverse event A_{ij} in control and treated arms are denoted by P_{ij}^C and P_{ij}^T, respectively. Let's consider the following logistic transformation.

$\phi_{ij} = \log\left(\dfrac{\boldsymbol{P}_{ij}^C}{1 - \boldsymbol{P}_{ij}^C}\right)$ and $\theta_{ij} = \log\left(\dfrac{\boldsymbol{P}_{ij}^T}{1 - \boldsymbol{P}_{ij}^T}\right) - \phi_{ij}$ The hierarchical prior distributions are,

Stage 1: $\phi_{ij} \sim N\left(\mu_{\phi_i}, \sigma_\phi^2\right)$ for $i = 1, \cdots, B$ and $j = 1, \cdots, K_i$.

Note that the log-ratio of $\theta_{ij} = 0$ means that the probability of experiencing adverse event A_{ij} is same in control and treated arms. Assume θ_{ij} follows a mixture distribution,

$$\theta_{ij} \sim \pi_i I\left[0\right] + \left(1 - \pi_i\right) N\left(\mu_{\theta_i}, \sigma_{\theta_i}^2\right) \text{ for } i = 1, \cdots, B \text{ and } j = 1, \cdots, K_i.$$

The standard approach is to assume a prior distribution to the hyperparameters, which yields the second stage.

Stage 2: $\mu_{\phi_i} \sim N\left(\mu_{\phi_0}, \tau_{\phi_0}^2\right)$ for $i = 1, \cdots, B$ and $\sigma_\phi^2 \sim IG\left(\alpha_{\sigma_\phi}, \beta_{\sigma_\phi}\right)$ and $\pi_i \sim \text{Beta}\left(\alpha_\pi, \beta_\pi\right)$

$\mu_{\theta_i} \sim N\left(\mu_{\theta_0}, \tau_{\theta_0}^2\right)$ for $i = 1, \cdots, B$ and $\sigma_\theta^2 \sim IG\left(\alpha_{\sigma_\theta}, \beta_{\sigma_\theta}\right)$

Similarly, assuming a prior distribution to the hyperparameters yields the third stage.

Stage 3: $\mu_{\phi_0} \sim N\left(\mu_{\phi_{00}}, \tau^2_{\phi_{00}}\right)$ for $i = 1, \cdots, B$ and $\sigma^2_{\phi 0} \sim IG\left(\alpha_{\tau_\phi}, \beta_{\tau_\phi}\right)$

$\mu_{\theta_0} \sim N\left(\mu_{\theta_{00}}, \tau^2_{\theta_{00}}\right)$ for $i = 1, \cdots, B$ and $\sigma^2_{\theta 0} \sim IG\left(, \alpha_{\tau_\theta}, \beta_{\tau_\theta}\right)$

The calculations for this model are carried out using x Markov chain Monte Carlo (MCMC) methods. The complete conditionals and the details of the MCMC methods are presented in the Berry paper (Berry & Berry, 2004).

POST-MARKETING

Once a drug is approved and is launched onto the market, the requirements to monitor safety and continuously communicate changes in benefit-risk become paramount (International Conference on Harmonisation, 18 November 2004). These requirements can be divided into four distinct sets of activities:

- Collecting individual adverse event reports (often called individual case safety reports) from marketed use of a product and reporting a subset of these cases meeting given criteria to the regulatory agencies responsible for licencing the product in that country. Individual case safety reports are received from various sources including: patients or the relatives of patients taking a pharmaceutical product; health care professionals, typically doctors, nurses, dentists and pharmacists; case reports written up in the biomedical literature; pharmaceutical sales representatives; other pharmaceutical companies and regulatory agencies who are recipients of first-hand reports that include mention of a third parties' product. From an analysis standpoint, a fundamental assumption is that the reporting of each case is considered to be a spontaneous occurrence independent of any other. It is well documented that adverse events are massively under-reported in the post marketing setting.
- Carrying out post-marketing safety studies to address specific safety concerns as a condition of regulatory approval or ad hoc behest of a regulatory agency.
- Providing periodic reports at regular intervals containing a cumulative analysis of all new safety data relevant to the product including those from Phase VI clinical or observational post-marketing studies, clinical development programmes in new indications or line extension, reviews of the biomedical literature and aggregate analysis of spontaneous data.
- Semi-continuous analysis of different kinds of safety data for the identification of new safety signals some of which will lead to further evaluation and ultimately specific actions to communicate and, where feasible, mitigate risk. Typically this will be the aggregate analysis of spontaneous reports that are recorded in a dedicated company or regulatory agency database. More recently, the widespread growth of powerful computing environments and sophisticated data mining and screening techniques has prompted a significant interest in using health insurance data and electronic medical records for safety signal detection (in this context, mining these databases for

safety signals in the anticipation of identifying previously unrecognised adverse drug reactions constitutes secondary use of the data which are ostensibly collected for a different purpose). These spontaneous and observational data domains have proven most amenable to novel and advanced statistical treatment in the quest to identify safety signals. Even though the information content of each case report or patient record may be poor (e.g. in the case of adverse event reports from consumers) or badly structured for observational research purposes (e.g. in the case of data from health insurance claims databases), the sheer volume of cases or records in these databases gives power to the analyses that no clinical trial could feasibly achieve.

Spontaneous Reporting Databases

The post-approval spontaneous safety reporting systems maintained by pharmaceutical companies and regulatory agencies contain substantially larger volumes of data than pre-approval databases, which are mainly based on clinical study data. As an example, the US Food and Drug Administration (FDA) Adverse Event Reporting System (FAERS, formerly AERS) is a database that contains information on adverse event and medication error reports submitted to the FDA (U.S. Food and Drug Administration,).

Besides those from manufacturers, reports can be submitted from health care professionals and the public. The original system was started in 1969, but since the last major revision in 1997, reporting has markedly increased as it has for other spontaneous report databases. At the end of 2013Q2, FAERS contained over 5,700,000 individual cases safety reports not originating from clinical studies and after de-duplication. In the European Union, Eudravigilance(European Medicines Agency (EMA), 2014) was established in 2002 and is maintained by the European Medicines Agency (EMA) to support pharmacovigilance of products marketed in the EU. Most national agencies maintain a database of adverse events or adverse drug reaction reports originating from organised data collection schemes in their country. Similarly, pharmaceutical companies are obliged to record reports arising from the use of the products that they market under licence and these databases are more or less global in scope depending on the size, reach and diversity of products they sell.

There are large differences in the size of these adverse event report databases, ranging from a few tens of thousands to the millions in number and thus their potential utility for statistical signal detection screening methods vary accordingly. Also, pharmaceutical company databases will inevitably contain a proliferation of reports from use of products targeted to a more or less limited number of therapeutic indications resulting in potentially skewed distributions of background events and a resultant risk of confounding statistical signal detection techniques.

The largest international database of adverse events is Vigibase™, maintained on behalf of the World Health Organisation by the Uppsala Monitoring Centre (Uppsala Monitoring Centre (UMC),). In a database survey conducted as part of a European consortium project (Wisniewski et al., 2012), by the end of June 2010, it was shown that Vigibase contained almost five and half million spontaneous reports. The sizes (in terms of the number of spontaneous reports they contain) of the other databases that participated in the same survey are shown in Figure 2.

Key to figure: UMC – Uppsala Monitoring Centre; EMA – European Medicines Agency; GSK – GlaxoSmithKline (Pharma); BSP - Bayer Pharma; MHRA - Medicines and Healthcare products Regu-

Figure 2. Comparison of sizes of a number of spontaneous databases based in Europe

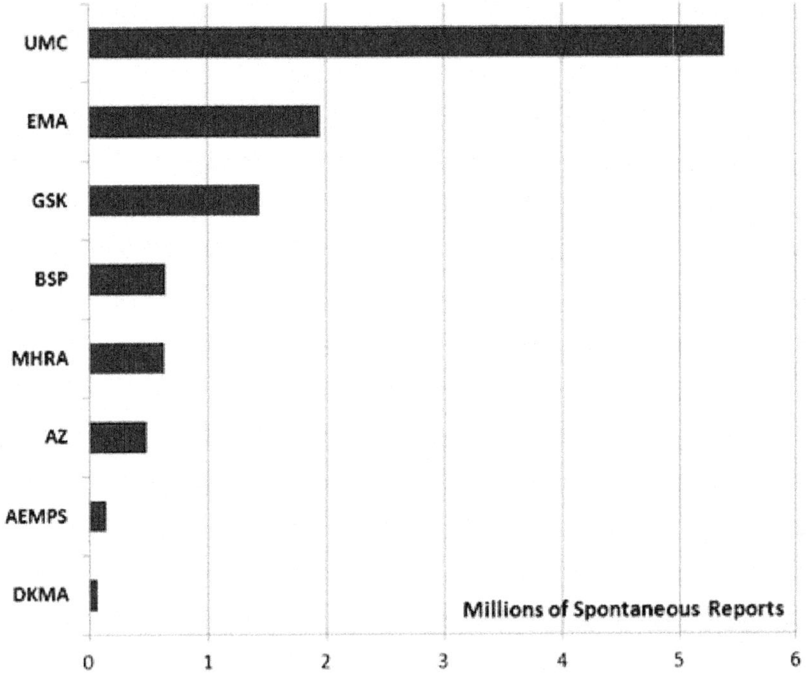

latory Agency (UK); AZ – AstraZenenca (Pharma); AEMPS - Agencia Espaniola de Medicamentos y Productos Sanitarios (Spain); DKMA - Danish Medicines Agency.

Spontaneous report databases are not homogenous entities. The implementation of any statistical signal detection approach requires a good knowledge of the underlying database and its inherent biases so that the chosen signal detection algorithms and the subsequent interpretation and evaluation of the resultant signals can take these biases into account. Recent evidence from the EU sponsored Innovative Medicines Initiative PROTECT project (Innovative Medicines Initiative PROTECT,) indicates that the choice of disproportionality statistic is less important than the policies or thresholds used to identify what is considered to be a signal – in this case, a specified level of (or change in) disproportionality (Candore et al., 2015; Slattery et al., 2013). The study assessed the performance of different signal detection algorithms (i.e. statistical method + signal detection policies/thresholds) in identifying adverse drug reactions labelled in the reference safety information of 220 marketed products and within 2 international, 1 national regulatory and 4 pharmaceutical company databases. This work suggests that there is an inevitable and predictable trade-off between the sensitivity and positive predictive value of a given approach. Figure 3 from the Candore study illustrates this trade-off; it also shows how the choice of signal identification policy/threshold can have a significant impact on performance even for the same method generating the disproportionality statistic.

Key to figure: EB05 - lower bound of the 90% confidence interval for the Empirical Bayesian Geometric Mean; PRR025 – lower bound of the 95% confidence interval for the Proportional Reporting Ratio; ROR025 – lower bound of the 95% confidence interval for the Reporting Odds Ratio; URN1/500 - Based

Figure 3. Performance characteristics of several implementations of a number of disproportionality algorithms

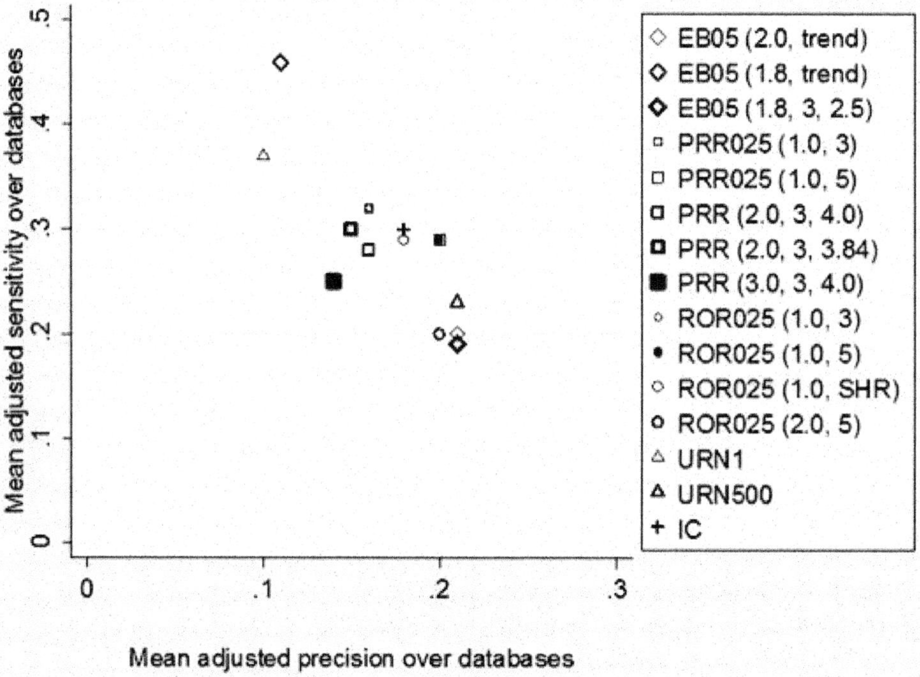

on Fisher's Exact Test the methods require at least 1 or 500 reports to be present for a drug-event pair to be included in the calculation. Number in brackets indicate, in order: the minimum threshold level of the lower bound, minimum report number and minimum mean value if applied; trend indicates that signals are also identified based on a change in disproportionality score over time; SHR – shrinkage.

These findings point to the need to test the performance of an implemented signal detection algorithm and to adjust the signal identification policies/thresholds in order to meet the desired performance characteristics of the system overall. It is worth stressing again at this point that safety signal detection systems based solely on statistical screening methods such as those described here are not recommended and they should be complemented with non-statistical methods since the latter tend to be more sensitive to very rare but clinically important adverse drug reactions and they can have a beneficial effect on precision by e.g. avoiding drug-event pairs that have been reviewed in the past or those that are already established as part of the known safety profile of the drug.

The following are commonly employed statistical methods for the generation of disproportionality scores in spontaneous datasets.

Statistical Methods

The statistical means of generating disproportionality scores can be divided into frequentist and Bayesian. The essential principle common to all methods is the calculation of an observed count over expected count

O/E. The principal difference between the methods lies in the way that the expected value is calculated. See a paper by Gipson (Gipson, 2012) for more details. Table 1 should be referred to in relation to the methods described here.

Frequentist:

Relative Reporting Ratio (RRR): Ratio of – number of reports that mention drug-event pair (D,E) to number of reports that mention drug D – to – number of reports that mention event (E) to total number of reports. The RRR can be expressed as

$$RRR = \frac{P\left(E|D\right)}{P\left(E\right)} = \frac{n_{11} \div n_{+1}}{n_{1+} \div n_{++}}$$

The variance of $\ln\left(RRR\right)$ and 95% confidence interval can be calculated by

$$Var\left(RRR\right) = \frac{1}{n_{11}} - \frac{1}{n_{+1}} + \frac{1}{n_{1+}} - \frac{1}{n_{++}}$$

$$95\%\ CI\ for\ RRR:\ \exp\left[\ln\left(RRR\right) \pm 1.96\sqrt{\frac{1}{n_{11}} - \frac{1}{n_{+1}} + \frac{1}{n_{1+}} - \frac{1}{n_{++}}}\right]$$

Proportional Reporting Ratio (PRR): Ratio of – number of reports that mention drug-event pair (D,E) to number of reports that mention drug D – to – number of reports that mention drug-event pair (not D,E) to number of reports that mention drugs other than D (Evans et al., 2001). The PRR can be expressed as

$$PRR = \frac{P\left(E|D\right)}{P\left(E|\bar{D}\right)} = \frac{n_{11} \div n_{+1}}{n_{12} \div n_{+2}}$$

The variance of $\ln\left(PRR\right)$ and 95% confidence interval can be calculated by

$$Var\left(PRR\right) = \frac{1}{n_{11}} - \frac{1}{n_{+1}} + \frac{1}{n_{12}} - \frac{1}{n_{+2}}$$

$$95\%\ CI\ for\ PRR:\ \exp\left[\ln\left(PRR\right) \pm 1.96\sqrt{\frac{1}{n_{11}} - \frac{1}{n_{+1}} + \frac{1}{n_{12}} - \frac{1}{n_{+2}}}\right]$$

Reporting Odds Ratio (ROR): Ratio of – odd of reports mentioning adverse event E in those report that mention the drug D – to – odd of reports mentioning adverse event E in those report that do not mention the drug D. The ROR can be expressed as

$$ROR = \frac{P\left(E|D\right) \div P\left(\bar{E}|D\right)}{P\left(E|\bar{D}\right) \div P\left(\bar{E}|\bar{D}\right)} = \frac{n_{11} \div n_{21}}{n_{12} \div n_{22}}$$

The variance of $\ln\left(ROR\right)$ and 95% confidence interval can be calculated by

$$Var\left(ROR\right) = \frac{1}{n_{11}} + \frac{1}{n_{21}} + \frac{1}{n_{12}} + \frac{1}{n_{22}}$$

$$95\%\ CI\ for\ ROR:\ \exp\left[\ln\left(ROR\right) \pm 1.96\sqrt{\frac{1}{n_{11}} + \frac{1}{n_{21}} + \frac{1}{n_{12}} + \frac{1}{n_{22}}}\right]$$

Bayesian:

Bayesian Confidence Propagation Neural Network (BCPNN)

Vigibase is the database of adverse drug reactions (ADRs) held by the Uppsala Monitoring Centre on behalf of the 47 countries of the World Health Organization (WHO) Collaborating Programme for International Drug Monitoring. A flexible Bayesian method, Bayesian Confidence Propagation Neural Network (BCPNN) was developed (Bate et al., 1998) to automate the signal detection procedure. The strength of dependency between a drug and adverse event is quantified by the Mutual Information or Information Component (IC),

$$IC = \log_2 \frac{P\left(E,D\right)}{P\left(D\right)P\left(E\right)} = \log_2 \frac{P\left(E|D\right)}{P\left(E\right)} \qquad (1)$$

where $\left(E,D\right) = \text{Joint occurance of event } E \text{ and Drug } D$,
$P\left(E|D\right) = \text{Probabilty of Event given Drug, and } P\left(E\right) = \text{Probability of Adverse event.}$

It follows that

$$IC = \log_2 \frac{P\left(E|D\right)}{P\left(E\right)} = \log_2 RRR$$

Considering Table 1, Equation 1 can be represented in terms of the observed to expected drug-event rates,

$$IC = \log_2 \frac{Observed\ No.of\left(E,D\right)reports}{Expected\ No.of\left(E,D\right)reports} = \log_2 RRR.$$

Note that a positive *IC* value indicates that $\left(E, D\right)$ reported more frequently than expected under the assumption of independence, a negative *IC* value indicates that they co-occur more rarely, and zero value for *IC* indicates there is not sufficient evidence to support a statistically significant dug-event association between the number of reports mentioning the drug and number of reports mentioning the adverse event.

The data consist of n_{++} total number of reports; n_{i+} number of reports mentioning drug *i*; n_{+j}, number of reports mentioning event *j*; and n_{ij}, number of reports mentioning drug *i* and event *j*; n_{i+}., n_{+j} and n_{ij} are binomially distributed random variables with respective parameters p_i, q_j, and r_{ij} .

$$n_{i+} \sim Binomial\left(n_{++}, p_i\right) \tag{2}$$

$$n_{+j} \sim Binomial\left(n_{++}, q_j\right) \tag{3}$$

$$n_{ij} \sim Binomial\left(n_{++}, r_{ij}\right) \tag{4}$$

Assuming a Beta prior distribution for the probabilities in equations (2-4), the posterior distributions also follow a Beta distribution (Casella & Berger, 2001).

$$p_i \sim Beta\left(a_i, b_i\right),\ p_i \mid data \sim Beta\left(a_i + n_{i+}, b_i + n_{++} - n_{i+}\right) \tag{5}$$

$$q_j \sim Beta\left(c_i, d_i\right),\ q_j \mid data \sim Beta\left(c_i + n_{+j}, d_i + n_{++} - n_{+j}\right) \tag{6}$$

$$r_{ij} \sim Beta\left(e_i, f_i\right),\ r_{ij} \mid data \sim Beta\left(e_i + n_{ij}, f_i + n_{++} - n_{ij}\right) \tag{7}$$

In practice, often the parameters are fixed $a_i = b_i = c_i = d_i = e_i = 1 \text{ and } \mathrm{f} = 3$, (Gould, 2003; Gould, 2008).Note that Equation (1), can be written as:

$$IC_{ij} = \log_2\left(\frac{r_{ij}}{p_i q_j}\right) = \left[\log\left(2\right)\right]^{-1}\left\{\log\left(r_{ij}\right) - \log\left(p_i\right) - \log\left(q_j\right)\right\}$$

Bate et al (Bate et al., 1998), using the delta method (i.e., Taylor expansion of IC), and assuming independence between numbers of adverse events and drugs reports, derived approximate posterior mean and variance of IC_{ij}. Therefore, if the lower bound of an approximate 95% credible interval, i.e, $E\left(IC_{ij}\right) - 2\sqrt{Var\left(IC_{ij}\right)}$, for IC_{ij} is greater than zero then ij-th drug-event is flagged as a potential signal.

Box 1. Example of SAS codes for BCPNN method

```
%MACRO BCPNN(NDE,ND,NE,NT,title);
PROC iml;
a1=1; b1=1 ; *Beta prior parameter for drug;
c1=1; d1=1 ; *Beta prior parameter for Event;
e1=1; f1=3 ; *Beta prior parameter for drug-event;

a=a1+&ND; b=b1+(&NT-&ND); *POSTERIOR Beta parameter for drug;
c=c1+&NE; d=d1+(&NT-&NE); *POSTERIOR Beta parameter for Event;
e=e1+&NDE; f=f1+(&NT-&NDE); *POSTERIOR Beta parameter for drug-event;

/** Exact mean IC **/
EIC_exact=(1/log(2))*(digamma(e)-digamma(e+f)-
(digamma(a)-digamma(a+b)+digamma(c)-digamma(c+d)));
/** Exact Variance IC **/
VIC_exact=(1/log(2)##2)*(trigamma(e)-trigamma(e+f)+
(trigamma(a)-trigamma(a+b)+trigamma(c)-trigamma(c+d)));

DRG_AE=&NDE; DRUG=&ND; AE=&NE; Total=&NT; Title=&title;
PRINT DRG_AE DRUG AE Total EIC_exact VIC_exact Title;
RUN;
%MEND;
%BCPNN(900,1300,1100,2000,"BCPNN Analysis");
```

The above calculation can be made more precise by using the exact mean and variance of IC_{ij} as suggested by Gould (Gould, 2003; Gould, 2008). The exact expression for the mean and variance of IC_{ij} are:

$$E\left(IC_{ij}\right) = \left(\log 2\right)^{-1}\left\{\Psi\left(\mathrm{e}_{ij}\right) - \Psi\left(\mathrm{e}_{ij} + \mathrm{f}_{ij}\right) - \left[\Psi\left(\mathrm{a}_{ij}\right) - \Psi\left(\mathrm{a}_{ij} + \mathrm{b}_{ij}\right) + \Psi\left(\mathrm{c}_{ij}\right) - \Psi\left(\mathrm{c}_{ij} + \mathrm{d}_{ij}\right)\right]\right\}$$

and

$$Var\left(IC_{ij}\right) = \left(\log 2\right)^{-1}\left\{\Psi'\left(\mathrm{e}_{ij}\right) - \Psi'\left(\mathrm{e}_{ij} + \mathrm{f}_{ij}\right) - \left[\Psi'\left(\mathrm{a}_{ij}\right) - \Psi'\left(\mathrm{a}_{ij} + \mathrm{b}_{ij}\right) + \Psi'\left(\mathrm{c}_{ij}\right) - \Psi'\left(\mathrm{c}_{ij} + \mathrm{d}_{ij}\right)\right]\right\}$$

where Ψ and Ψ' are digamma and trigamma functions respectively. Both of these functions are built-in functions in many scientific software programs, such as SAS®. The following is an example of SAS code for the BCPNN method.

Multi-Items Gamma Poisson Shrinker (MGPS):

In a seminal paper, DuMouchel (DuMouchel, 1999) in collaboration with FDA, introduced a robust Bayesian method coined as Multi-items Gamma Poisson Shrinker (MGPS). The BCPNN, MGPS and variation of such methods have become the backbone of modern disproportionality analysis.

Let N be the number of report for a drug-event follows a Poisson distribution,

$$f\left(n;\mu\right) = \frac{\mu^{n}\exp\left(-\mu\right)}{n!} \text{ where } \mu = \lambda E .$$

Assume prior distribution of λ follows a Gamma-mixture:The marginal likelihood for N can be used to estimate the parameters

$$f\left(n;\lambda,\alpha_0,{}^{2}{}_{0},\alpha_1,{}^{2}{}_{1},\acute{E}\right)=\prod_{i=1}^{N}\acute{E}f_{NB}\left(n_i;\alpha_0,\frac{{}^{2}{}_{0}}{{}^{2}{}_{0}+E_i}\right)+\left(1-\acute{E}\right)f_{NB}\left(n_i;\alpha_1,\frac{{}^{2}{}_{1}}{{}^{2}{}_{1}+E_i}\right)$$

where

$$f_{NB}\left(n_i;\alpha_k,\frac{{}^{2}{}_{k}}{{}^{2}{}_{k}+E_i}\right)=\left(n_iB\left(\alpha,n_i\right)\right)^{-1}\left(\frac{{}^{2}{}_{k}}{{}^{2}{}_{k}+E_i}\right)^{\alpha}\left(1-\frac{{}^{2}{}_{k}}{{}^{2}{}_{k}+E_i}\right)^{n_i}$$

The posterior distribution of λ follows a Gamma-mixture:where

Note that the posterior distribution is defined explicitly therefore the 5% quantile, denoted by EB05, may be evaluated. Commonly, it is the different threshold of EB05 values that defines different strategies for flagging a pair of drug-event association as a statistical signal.

In the aforementioned formulation of MGPS the unit of analysis is the unique drug-event report which is referred to as 2-dimentional disproportionality analysis. This approach can be easily extended to a 3-dimentional analysis by considering the unit of analysis to be the unique reports of either (drug1, drug2, event) or (drug, event1, event2). Consequently, reported adverse events due to the drug-drug effect can be assess by defining the Interaction Signal Score (INTSS), which is a relative measure of how much excess disproportionality is present in the three-dimensional combination of two drugs and one adverse event term. An INTSS >1 indicates that the CI for the three-dimensional analysis is larger than and does not overlap with the CI from the highest two-dimensional analysis, i.e.

$$INTSS=\frac{EB05\left(D_1,D_2,E\right)}{\min\left\{EB05\left(D_1,E\right),EB05\left(D_2,E\right)\right\}}\;(>1?)$$

There is no gold standard for the choice of specific threshold, however Szarfman (Szarfman, Machado, & O'Neill, 2002) suggested a threshold value of EB05 of 2, meaning that the number of drug-event pairs of interest was present in the reporting database are at least twice as much as the average for all the other drugs and events in the database (i.e. observed/expected >2) with 95% probability that the statistical association is not a false positive one. However the IMI-PROTECT Signal Detection performance work referred to earlier puts this one-fits-all strategy in doubt(Candore et al., 2015; Slattery et al., 2013). For an example of the application of disproportionality analysis in pharmacovigilance, refer to the paper by Baker et al (Baker et al., 2009).

Interpretation of Statistical Signals of Disproportionate Reporting

The importance of having knowledge of the content of the database where automated safety signal screening methods are used has already been stressed – the impacts on signalling are described here. Databases may be subject to a number of sources of potential bias that may result in high signal scores:

- Associations due to underlying disease or indication. High signal scores may be generated for events that are actually related to the underlying disease rather than due to the product itself e.g. reports of breast cancer progression presented as potential signals for aromatase inhibitors used to prevent recurrence.
- Associations due to medications used concomitantly with the drug of interest. High signal scores may be generated for events that are actually causally related to a commonly co-prescribed drug and not the drug of interest. For example high signal scores on 'respiratory depression' may occur for local anaesthetics, an 'innocent bystander', as this event is known to be related with the commonly co-prescribed opioids.
- Database quality bias. High signal scores may be generated for events that have been reported in duplicate from several different sources. This can partly be corrected for by using duplicate detection and removal algorithms. (Tregunno, Fink, Fernandez, & Noren, 2013; Tregunno, Fink, Fernandez-Fernandez, Lazaro-Bengoa, & Noren, 2014).
- Notoriety bias. High signal scores may be generated for events due to over-reporting as a result of media exposure, 'dear doctor' letters advising health care professionals of new safety information, marketing campaigns, mass litigation campaigns by lawyers, sales forces providing awareness of specific safety concerns, and awareness of specific unwanted effects in drugs of the same class or hitting the same target.
- Administrative bias. High signal scores may be generated for events due to regional regulations on the nature of reportable adverse events: e.g. The Japanese Early Post-marketing Phase Vigilance (EPPV) program obtains solicited reports which skews adverse event reporting towards products newly launched in Japan.
- Reporting bias. High signal scores may be generated for events during the initial launch period of a new drug. The product life cycle should be considered for time-dependent reporting patterns. Reporting bias also involves any skewed geographical distribution that may indicate cultural differences in drug usage and/or differences in the availability of competitor products. Low signal score may be generated for events due to underreporting, therefore, certain high profile events may warrant further investigation regardless of signal score (e.g. agranulocytosis, aplastic anaemia, Stevens-Johnson syndrome).
- Cloaking or masking. Low signal scores may be obtained due to cloaking. This may occur by drug (i.e. very high levels of reports of event X for drug A masks a high signal score for drug B with the same event) or by disease/demographics (i.e. high reporting of hip fracture in elderly patients masks a signal in younger patients, etc). This can occur in relation to notoriety bias described above. (Gipson, Schaaf, DuMouchel, Valentino, & Wisniewski, 2010)
- Database differences. Signal scores from external and internal databases may be inconsistent due to differences in the total number of reports and different levels of representation of particular drugs in databases.
- Dictionary structure. High or low signal scores may be generated for events due to the MedDRA structure. In MedDRA, closely related medical terms may be coded to different system organ classes. For example the preferred term 'Hyperkalaemia' occurs in the system organ class `Metabolism and nutrition disorders' where as an almost identical medical concept (from a signal detection perspective) termed 'Blood potassium increased' occurs in the system organ class 'Investigations'. Also, contradictory terms on a lower level may be merged under a common MedDRA term on

high level, for example the preferred terms 'blood pressure increased' and 'blood pressure decreased' both occur under the same High Level Term within the MedDRA hierarchy.

- Misclassification Bias. High or low signal scores may be generated for events due to coding errors or variations in coding conventions (i.e. lumping/splitting of terms) at time of data entry.

The specific implementation of statistical screening algorithms in a database can, to a certain degree, take account of these sources of error where they are recognized. For example, the calculation of signal scores might be adjusted through stratification or subgrouping of by key variables such as country of report source or gender although this is only really feasible in databases of sufficient size to provide adequately sized strata or sub-groups. In the case of masking, strategies have been described for automated assessment of the size of the masking effect inherent in a database (Juhlin, Ye, Star, & Noren, 2013; Maignen, Hauben, Hung, Holle, & Dogne, 2013; Maignen, Hauben, Hung, Van Holle, & Dogne, 2014), which in turn may indicate the need to censor the drug-event pairs causing the masking from the calculations of the disproportionality statistics.

Other Statistical Methods for Safety Signal Detection in Spontaneous Datasets

Over more than a decade, the use of statistical signal detection algorithms based on the PRR, IC and EB05 have predominated in pharmacovigilance organisations. However, new approaches based on logistic regression have been published and interest is growing concerning their potential role in signal detection (Caster, Noren, Madigan, & Bate, 2013). Logistic regression has also been promoted as a means of helping in the assessment of safety signal evaluation. The value of regression techniques in both of these roles is yet to be established.

Statistical signal detection in vaccines has also evolved different approaches in recent years since the exposures are typically short term (often only dose of a vaccine is given), the exposed population is typically healthy, and the administration of vaccines can occur over short but intense periods of time (e.g. winter influenza campaigns) or to a restricted population (e.g. infant vaccines) (Kurz et al., 2010; Van Holle & Bauchau, 2014). Most pharmacovigilance organisations will run signal detection screening methods for vaccines separately from that for small molecules or therapeutic biologics even if the data are held in the same database. The FDA has a separate database for vaccines known as VAERS, distinct from the FAERS database described above (Banks et al., 2005).

QUANTITATIVE SIGNAL DETECTION IN OTHER DATA SOURCES

Observational Data

Even though the randomized controlled trial is the gold standard, in some situations randomization to a control arm may violate ethical principles, or may be impractical in the case of rare events or diseases. In such situations an alternative is the use of observational data such that it approximates to a randomized controlled experiment. Typical examples observation al studies are:

- Case Controlled Study: which groups differing outcomes are identified and the respective supposed causal attribute compared.

- Cross Sectional Study: involves comparing attributes of a population at one specific point in time.
- Longitudinal Study: Examine the observed attributes of a population over long periods of time.
- Cohort or Panel Study: a particular form of longitudinal study where a group of patients is closely monitored over a span of time.

The observational studies are used to:

- assess real-world use of medicines and practice
- identify and characterize safety signals and subsequently update about the benefits and risks profile in the general population
- develop hypotheses to be tested in subsequent clinical experiments.

Sentinel and OMOP

In 2007, recognizing that the increased use of electronic health records (EHR) and availability of other large sets of marketplace health data provided new learning opportunities, Congress directed the FDA to create a new drug surveillance program to more aggressively identify potential safety issues. The FDA launched several initiatives to achieve that goal, including the well-known Sentinel program to create a nationwide data network(U.S. Food and Drug Administration (FDA), 2014). A similar approach has been developed in Europe under the auspices of the EU-ADR network (Eu-Adr Group,).

In partnership with PhRMA and the FDA, the Foundation for the National Institutes of Health launched the Observational Medical Outcomes Partnership (OMOP), a public-private partnership. This interdisciplinary research group tackled a surprisingly difficult task that is critical to the research community's broader aims: identifying the most reliable methods for analyzing huge volumes of data drawn from heterogeneous sources.

Employing a variety of approaches from the fields of epidemiology, statistics, computer science and elsewhere, OMOP took on a critical challenge: what can medical researchers learn from assessing these new health databases, could a single approach be applied to multiple diseases and could their findings be proven? Success would mean the opportunity for the medical research community to do more studies in less time, using fewer resources and achieving more consistent results. In the end, it would mean a better system for monitoring drugs, devices and procedures so that the healthcare community can reliably identify risks and opportunities to improve patient care.

There has been increased interest in using multiple observational databases to understand the safety profile of medical products during the post-marketing period. However, it is challenging to perform analyses across these heterogeneous data sources. The Observational Medical Outcome Partnership (OMOP) provides a Common Data Model (CDM) for organizing and standardizing databases. OMOP's work with the CDM has primarily focused on US databases. As a participant in the OMOP Extended Consortium, we implemented the OMOP CDM on the UK Electronic Healthcare Record database-The Health Improvement Network (THIN).

Additional information regarding current research and available SAS macros and R codes can be obtained at the OMOP web portal (The Observational Medical Outcomes Partnership (OMOP),).

Social Media

A number of initiatives related to the mining of social media for the detection of adverse events are underway or planned, including those under evaluation by the US FDA and the Innovative Medicines Initiative in the EU. Karlin (Karlin, 2014) summarises the current state of the art that focuses in the main social media platforms. The specific challenges to exploiting social media for safety signal detection are:

- The identification of putative adverse events from social internet site postings that include non-medical, common-use language or even slang. The term `proto-adverse event' has been coined to describe raw, un-curated text that appears to describe a medical event. Proto-adverse events will need to be mapped to an established medical ontology such as MedDRA. Similarly, the accurate identification of the pharmaceutical product or device associated with the adverse event is needed.
- Avoiding duplicate reports from re-postings of the same events. The role of automated de-duplication algorithms similar to those in use in some spontaneous adverse event datasets is of interest in this context. A potentially greater challenge exists in identifying posts that describe the same event also reported in other datasets, for example as a spontaneous report from the consumer themselves or their doctor. The issue at stake is to avoid the double counting of events.
- The credibility of the reporting source may be impossible to determine. There is little or no barrier to posting an observation in relation to a drug exposure that may be based on hearsay or describe an experience second-hand. The possibility of postings that falsely describe an adverse drug reaction in order to provoke adverse publicity for a product or pharmaceutical company cannot be ruled out. Although the scientific credibility of false claims might be quickly demolished, dealing with the negative public impact of a concerted social media campaign could be costly and time consuming.
- The most appropriate means of analysing datasets arising from social media reports of adverse events requires evaluation although it does seem reasonable that the approaches based on disproportionality methods for spontaneous datasets described in this chapter have a role to play. However, whether such reports can be considered spontaneous for the purpose of analysis remains to be established and some means of separating the analysis of social media reports from true spontaneous reporting systems seems justified given the present state of knowledge surrounding their accuracy and overall contribution to pharmacovigilance.

Solutions to each of these challenges provide opportunities for innovations in computational sciences, including text analytics and natural language processing, record linkage, duplicate detection, pattern recognition (machine learning), forensic analytics and statistical analysis.

Having called out the specific challenges to the use of social media for safety signal detection, the benefits offered include: ease of access and speed of information availability and hence the potential to identify safety signals rapidly. However, perhaps the most compelling proposition is the ability to gain `soft intelligence' on how pharmaceutical products are used by patients in practice and use this information to trigger evaluation of potential new risk mitigation measures or communication.

REFERENCES

Almenoff, J., Tonning, J. M., Gould, A. L., Szarfman, A., Hauben, M., & Ouellet-Hellstrom, R. et al. (2005). Perspectives on the use of data mining in pharmacovigilance. *Drug Safety*, *28*(11), 981–1007. doi:10.2165/00002018-200528110-00002 PMID:16231953

Almenoff, J. S., DuMouchel, W., Kindman, L. A., Yang, X., & Fram, D. (2003). Disproportionality analysis using empirical bayes data mining: A tool for the evaluation of drug interactions in the post-marketing setting. *Pharmacoepidemiology and Drug Safety*, *12*(6), 517–521. doi:10.1002/pds.885 PMID:14513665

Baker, R.A., Pikalov, A., Tran, Q.V., Kremenets, T., Arani, R.B., & Doraiswamy, P.M. (2009). *Atypical antipsychotic drugs and diabetes mellitus in the US food and drug administration adverse event database: A systematic Bayesian signal detection analysis.* Psychopharmacol Bull., 42(1), 11-31.

Banks, D., Woo, E. J., Burwen, D. R., Perucci, P., Braun, M. M., & Ball, R. (2005). Comparing data mining methods on the VAERS database. *Pharmacoepidemiology and Drug Safety*, *14*(9), 601–609. doi:10.1002/pds.1107 PMID:15954077

Bate, A., Lindquist, M., Edwards, I. R., Olsson, S., Orre, R., Lansner, A., & De Freitas, R. M. (1998). A bayesian neural network method for adverse drug reaction signal generation. *European Journal of Clinical Pharmacology*, *54*(4), 315–321. doi:10.1007/s002280050466 PMID:9696956

Berry, S. M., & Berry, D. A. (2004). *Accounting for multiplicities in assessing drug safety: A three-level hierarchical mixture model*

Brown, J. S., Petronis, K., Bate, A., Zhang, F., Dashevsky, I., & Kulldorff, M. et al. (2011). Comparing two methods for detecting adverse event signals in observational data: Empirical bayes gamma poisson shrinker vs. tree-based scan statistic. *Pharmacoepidemiology and Drug Safety*, *20*, S144.

Candore, G., Juhlin, K., Manlik, K., Thakrar, B., Quarcoo, N., Seabroke, S., & Slattery, J. (2015). *Comparison of statistical signal detection methods within and across spontaneous reporting databases. Drug Safety, Casella, G., & Berger, R. L. (2001). Statistical inference* (2nd ed.). Cengage Learning.

Caster, O., Noren, G. N., Madigan, D., & Bate, A. (2013). *Logistic regression in signal detection: Another piece added to the puzzle*

CIOMS Working Group VIII. (2010). *Practical aspects of signal detection in pharmacovigilance.*

Cornelius, V. R., & Evans, S. J. W. (2009). The use of time to event models in signal detection. *Drug Safety*, *32*(10), 926.

DuMouchel, W. (1999). Bayesian data mining in large frequency tables, with an application to the FDA spontaneous reporting system. *The American Statistician*, *53*(3), 177–190.

Eu-Adr Group. (n. d.). Retrieved from http://euadr-project.org/

European Medicines Agency (EMA). (2014). EudraVigilance. Retrieved from https://eudravigilance.ema.europa.eu/human/index.asp

Evans, S. J. W., Waller, P. C., & Davis, S. (2001). Use of proportional reporting ratios (PRRs) for signal generation from spontaneous adverse drug reaction reports. *Pharmacoepidemiology and Drug Safety*, *10*(6), 483–486.

Gipson, G. (2012). A shrinkage-based comparative assessment of observed-to-expected disproportionality measures. *Pharmacoepidemiology and Drug Safety*, *21*(6), 589–596. doi:10.1002/pds.2349 PMID:22290739

Gipson, G., Schaaf, R., DuMouchel, W., Valentino, R., & Wisniewski, A. (2010). Impact of drug product litigation on safety signal detection in aers. *Pharmacoepidemiology and Drug Safety*, *19*, S224.

Gould, A. L. (2003). Practical pharmacovigilance analysis strategies. *Pharmacoepidemiology and Drug Safety*, *12*(7), 559–574.

Gould, A. L. (2008). Detecting potential safety issues in clinical trials by Bayesian screening. *Biometrical Journal. Biometrische Zeitschrift*, *50*(5), 837–851.

Heffernan, J. E., & Tawn, J. A. (2004). A conditional approach for multivariate extreme values. *Journal of the Royal Statistical Society. Series A, (Statistics in Society)*, *66*(3), 497–546. doi:10.1111/j.1467-9868.2004.02050.x

ICH MedDRA Maintenance and Support Services Organization (MSSO). (n. d.). MedDRA Retrieved from http://www.meddra.org/

Innovative Medicines Initiative PROTECT. (n. d.). Retrieved from http://www.imi-protect.eu/

International Conference on Harmonisation. (2004 November, 18). *ICH harmonised tripartite guideline pharmacovigilance planning E2E.* Retrieved from http://www.ich.org/fileadmin/Public_Web_Site/ICH_Products/Guidelines/Efficacy/E2E/Step4/E2E_Guideline.pdf

International Conference on Harmonisation. (1994). *ICH harmonised tripartite guideline - clinical safety data management: Definitions and standards for expedited reporting E2A.* ().ICH. Retrieved from http://www.ich.org/fileadmin/Public_Web_Site/ICH_Products/Guidelines/Efficacy/E2A/Step4/E2A_Guideline.pdf

International Conference on Harmonisation. (1996). *ICH harmonised tripartite guideline on good clinical practice E6(R1).* Retrieved from http://www.ich.org/fileadmin/Public_Web_Site/ICH_Products/Guidelines/Efficacy/E6_R1/Step4/E6_R1__Guideline.pdf

International Health Terminology Standards Development Organisation (IHTSDO). (n. d.). SNOMED CT - ihtsdo.org. Retrieved from http://www.ihtsdo.org/snomed-ct/

Juhlin, K., Ye, X., Star, K., & Noren, G. N. (2013). Outlier removal to uncover patterns in adverse drug reaction surveillance - a simple unmasking strategy. *Pharmacoepidemiology and Drug Safety*, *22*(10), 1119–1129. PMID:23832706

Karlin, S. (2014). Adverse events in social media: FDA expects signal detection “Revolution”. *The Pink Sheet.*

Kurz, X., Slattery, J., Addis, A., Durand, J., Segec, A., Skibicka, I., & Hidalgo-Simon, A. et al. (2010). The eudravigilance database of spontaneous adverse reactions as a tool for H1N1 vaccine safety monitoring. *Pharmacoepidemiology and Drug Safety*, *19*, S330–S331.

Lindquist, M., Edwards, I. R., Bate, A., Fucik, H., Nunes, A. M., & Stahl, M. (1999). From association to alert - A revised approach to international signal analysis. *Pharmacoepidemiology and Drug Safety*, *8*(SUPPL. 1), S15; S25.

Maignen, F., Hauben, M., Hung, E., Holle, L. V., & Dogne, J. (2013). A conceptual approach to the masking effect of measures of disproportionality. *Pharmacoepidemiology and Drug Safety*. PMID:24243699

Maignen, F., Hauben, M., Hung, E., Van Holle, L., & Dogne, J. (2014). *Assessing the extent and impact of the masking effect of disproportionality analyses on two spontaneous reporting systems databases*

Slattery, J., Candore, G., Tregunno, P., Wong, J., Seabroke, S., & Juhlin, K. et al. (2013). Comparison of disproportionality measures in eudravigilance. *Pharmacoepidemiology and Drug Safety*, *22*, 38–39.

Southworth, H. (2014). *Predicting potential liver toxicity from phase 2 data: A case study with ximelagatran.*

Southworth, H., & Heffernan, J. E. (2012a). Extreme value modelling of laboratory safety data from clinical studies. *Pharmaceutical Statistics*, *11*(5), 361–366. doi:10.1002/pst.1510 PMID:22684727

Southworth, H., & Heffernan, J. E. (2012b). Multivariate extreme value modelling of laboratory safety data from clinical studies. *Pharmaceutical Statistics*, *11*(5), 367–372. doi:10.1002/pst.1531 PMID:22888093

Szarfman, A., Machado, S. G., & O'Neill, R. T. (2002). Use of screening algorithms and computer systems to efficiently signal higher-than-expected combinations of drugs and events in the US FDA's spontaneous reports database. *Drug Safety*, *25*(6), 381–392. doi:10.2165/00002018-200225060-00001 PMID:12071774

aCRAN - package extRemes. extreme value analysis. (n. d.). Retrieved from http://cran.r-project.org/web/packages/extRemes/index.html

CRAN - package texmex. statistical modelling of extreme values. (n. d.). Retrieved from http://cran.r-project.org/web/packages/texmex/index.html

The Observational Medical Outcomes Partnership (OMOP). Observational medical outcomes partnership Retrieved from http://omop.org/

Tregunno, P. M., Fink, D. B., Fernandez, C., & Noren, N. G. (2013). Hit-miss model detects duplicates missed by rule-based screening of individual case safety reports. *Pharmacoepidemiology and Drug Safety*, *22*, 101–102.

Tregunno, P. M., Fink, D. B., Fernandez-Fernandez, C., Lazaro-Bengoa, E., & Noren, G. N. (2014). *Performance of probabilistic method to detect duplicate individual case safety reports.*

FDA adverse event reporting system (FAERS) (n. d.). U.S. Food and Drug Administration. Retrieved from http://www.fda.gov/Drugs/GuidanceComplianceRegulatoryInformation/Surveillance/AdverseDrugEffects/default.htm

U.S. Food and Drug Administration. (FDA). (2014). Sentinel initiative. Retrieved from http://www.fda.gov/Safety/FDASSentinelInitiative/default.htm

Uppsala Monitoring Centre (UMC). (n. d. a). VigiBase®. Retrieved from http://www.umc-products.com/DynPage.aspx?id=73590&mn1=1107&mn2=1132

Uppsala Monitoring Centre (UMC). (n. d. b). Welcome to WHO-ART. Retrieved from http://www.umc-products.com/DynPage.aspx?id=73589&mn1=1107&mn2=1664

Van Holle, L., & Bauchau, V. (2014). Signal detection on spontaneous reports of adverse events following immunisation: A comparison of the performance of a disproportionality-based algorithm and a time-to-onset-based algorithm. *Pharmacoepidemiology and Drug Safety*, *23*(2), 178–185. doi:10.1002/pds.3502 PMID:24038719

Van Puijenbroek, E. P., Bate, A., Leufkens, H. G. M., Lindquist, M., Orre, R., & Egberts, A. C. G. (2002). A comparison of measures of disproportionality for signal detection is spontaneous reporting systems for adverse drug reactions. *Pharmacoepidemiology and Drug Safety*, *11*(1), 3–10. doi:10.1002/pds.668 PMID:11998548

Wisniewski, A. F. Z., Juhlin, K., Laursen, M., Macia, M. M., Manlik, K., & Pinkston, V. K. et al. (2012). Characterisation of databases (DBS) used for signal detection (SD): Results of a survey of imi protect work package (WP) 3 participants. *Pharmacoepidemiology and Drug Safety*, *21*, 233–234. PMID:21786364

Chapter 6
Application of SMAC Technology

Manu Venugopal
Accenture, India

ABSTRACT

The current digital age is primarily driven by four technology forces namely, Social Media, Mobility, Analytics and Cloud computing. These technologies continue to evolve and shape the digital world, giving people and businesses newer experiences and opportunities that they were not exposed to in the past. Digital technology has the potential to change the world significantly which in turn has a disruptive impact in the world of business. Hence, 'digitizing' its business must be one of top priorities in the medium and long term of every business to ensure a successful future. This chapter begins with by defining each of the four technologies, its benefits and what it means to the key stakeholders in the healthcare business. It also covers many use cases of SMAC with a specific focus on clinical development and pharmacovigilance. The later part of the chapter lays the foundation for setting up a SMAC organization including key strategies, conceptual framework, technology and regulatory compliance considerations.

INTRODUCTION

In the past several decades, Information Technology has played a vital role in defining the growth of enterprises. From a support function, IT has quickly emerged as a function that enables enterprises to create a far more business value in the marketplace within a short timeframe that ever before. In the present era called the "digital age", technology that was once more confined within enterprises for back-end processing of information has crossed boundaries and has been transforming our day-to-day lives.

Social media, Mobility, Analytics and Cloud, widely known by its acronym –SMAC, are the four main sources of technology that is driving this revolutionary change across the business value chain, from customers to enterprises. It is very important for enterprises to be agile and ride on this new wave of digital shift to remain successful and relevant in the marketplace. Pharma industry has been a late adopter of technology when compared to industries such as financial services, automobile and telecom-

DOI: 10.4018/978-1-4666-8726-4.ch006

munications. However, with consumers (patients in the case of pharma and healthcare industry) having fully on boarded onto the SMAC bandwagon, it opens up new avenues for Pharma industry to understand and improve patient care and patient safety.

DEFINING S-M-A-C

This section introduces the basic concepts and types of each of the four technologies – social media, mobility, analytics and cloud computing. It will serve as the basis for the rest of the sections in this chapter which deals with application and implementation of the SMAC technology in the Life Sciences context.

Social Media

In the past few years, the term 'social media' has become ubiquitous in the world of digital media and internet. The major reason behind its success has been the rise of websites such as MySpace, Facebook, Twitter and so on. Many still consider social media to be confined with such networking sites. However, the term Social Media covers a much broader spectrum of which social networking sites are a part of.

Social Media is a virtual medium of communication through which users come together to access information, interact and connect with other users and share their information and opinions. Social media has enabled more interaction among people who are virtually connected based on their interests or background. For instance, people who have interests in wildlife photography get connected through sites such as wildlife photographic blog or naturephotohub.com to share their photos, experience and express their opinions about other posts. They may also comment and share photos or videos of wild animals shared through media sharing sites such as YouTube and Flickr. Or friends who have studied in the same school/ college or colleagues who have worked in the same organization network among one another through networking sites such as Facebook, LinkedIn. In the case of patients suffering from life threatening diseases such as cancer, AIDS, sites such as cancer.org, HIVAidsTribe.com provides them the much needed support, medical information to tackle various health related situations and opportunities to connect with other patients suffering from similar ailments. Social media has gained much popularity in the digital world since it is open to all and offers the freedom to express and promote themselves or their interests to a wider audience.

There are several tools and mediums under the realm of social media. This includes social networking sites, blogs, discussion forums, social news sites, media sharing sites, micro blogs, bookmarking sites and gamification. Table 1 provides a brief description of each type of social media tool and few popular examples existing today.

Mobility

The term 'Mobility' as per Oxford Dictionary means the ability to move or be moved freely and easily. In the technology context, mobility is the medium through which digital technology can be easily accessed on the move. Devices that enable technology to be portable and accessible are called mobile devices. Mobile devices are small lightweight handheld computing device which allows users to perform most of the activities that could be done with PCs or laptops. Mobile devices include smartphones, tablets and

Table 1. Types of Social Media

Tools	Use	Examples
Social Networking	Social networking services allow users of similar interests or background to engage with other users to connect and share information such as personal and professional details, provide status updates, express opinions and preferences, collaborate for specific events, market products and services	Facebook Google+ LinkedIn Pinterest
Media Sharing	Media sharing involves sharing of photos and videos. The sites allow users to upload pictures and videos as well as comment on other user submissions. Some sites enables user to maintain a personal profile, create a private group to view the posts, add preferences of media types	YouTube Vimeo Flickr Instagram
Blogs and Forums	Blogs and discussion forums are mediums through which users discuss on a specific topic or event. Blogs are generally used as online diaries where people jot down their opinion and thoughts. Comments can be provided on blog posts.	The Huffington Post The Verge Mashable!
Microblogs	Microblogs, as the name suggests, is a form of blogging that contain short and crisp messages or images posted by individuals or groups through their user accounts. These short messages are received by other users who follow (or subscribe) the senders.	Twitter Tumblr Google Buzz
News Feeds	Social news sites deliver the most talked about news articles, links or stories. It allows users to post news articles and request users to vote the articles. Based on the user member voting, the most popular and interesting articles are shown to users.	Digg Reddit
Wiki	Wiki is a piece of server software that allows users to freely create and edit Web page content using any Web browser. It allows users to freely edit site contents thus enabling collaboration in online content creation.	Wikipedia WikiMapia WikiTravel
Social Bookmarking	Bookmarking site is an online service that enables users to save, organize and share links to web pages. It also allows web links to be tagged using specific keywords which makes searching easier. Users can search bookmarked sites of other users as well.	Del.icio.us StumbleUpon Diigo
Gamification	Gamification is the use of game attributes to drive game-like player behavior in a non-game context (Wu, 2001). It works on using natural desires of humans when associated with a game such as competition, recognition, interaction, learning etc. for situations other than a game (marketing promotions, financial services, community health etc.)	Foursquare Nike+ Mint

e-readers. All these devices contain a mini display screen with touch functionality, powerful microprocessor, operating system, wireless access and large extensible storage capacity.

More than 2 billion mobile devices were shipped in the year 2013 and as per CCS Insights prediction, the total number of smartphones and tablets will exceed world's human population by year 2017 (CCS Insight, 2013). The huge success of smartphones and tablets can be attributed to the easy access of digital world at the fingertips at all times. Apps, short form of application software programs run of mobile devices, offer a multitude of uses such as web browsing, social networking, gaming, banking, shopping, location tracking, and health related services. What sets apart mobile devices from the rest of the technology platforms is its phenomenal adoption rate across various geographical regions. The staggering rate at which users have adopted mobile services is best explained with following comparison – It took 38 years for radio to reach 50 million users while TV and internet took around 13 and 4 years respectively to achieve the feat. However, 'Draw Something' App took just 50 days to reach 50 million users (Super Monitoring, 2013). Mobile usage statistics show that 50% of the average mobile web users use mobile as either their primary or exclusive means of going online. Also more than 80% of the mobile time is spent in Apps (Super Monitoring, 2013).

Mobile Platforms

From a gadget to make and receive phone calls and text messages, mobile phones have transformed into a multi-function computing device called smartphones. So is the case with tablets and iPads that are fast replacing traditional PCs and laptops. Mobile operating platforms grew in sophistication which determines what functions and features these devices should comprise and how effectively the phone resources need to be utilized. Developers create apps using these OS platforms. The popular mobile operating systems available in the market are iOS, Android, Windows Phone, BlackBerry OS, WebOS and Symbian. This together with BadaOS controlled about 97% of the total market share of worldwide sales in Q3, 2011 (Gartner, 2011).

Let us take a look at top 3 mobile operating systems available in the market today.

- *iOS* - This operating system, formerly known as iPhone OS, is owned by Apple and powers iPhone, iPad and iPad Touch. It was first released with iPhone in June 2007. iOS introduced the new way of user interaction using touch screen features such as tap, swipe, pinch on the phone's display screen. It is not open source and hence available on Apple products only. The latest version is iOS 7. Apple maintains an AppStore which is a collection of over 500,000 Apps built on iOS.
- *Android* – The fastest growing mobile operating system was originally created by Android Inc which was later acquired by Google in 2005. The core OS is open source and hence enables developers to build their own Andriod Apps and publish them. Known for its multi-tasking abilities and customizable home screens, Android OS continuously upgrades its versions with enhanced features. Latest version is called KitKat (4.4) and earlier versions were Jelly bean (4.1), Icecream Sandwich (4.0), Gingerbread (2.3) and Froyo (2.2). Android Apps are available in Google Play (formerly Android Market).
- *Windows Phone* – Owned by Microsoft, this mobile OS was launched in 2007 under the name Windows Phone 7. Windows Phone 7 introduced a new tile based interface called Metro that features interchangeable sections each designed for a specific function. The other feature is called Hub, an aggregator that collates content such as photos from different sources into a single location. The latest version is Windows Phone 8. Windows Apps are available in Windows Phone Apps+Games site.

Trends

Over the past several years, we have witnessed the phenomenal growth of mobile devices and its adoption as a means of communication and interaction. In the years to come, mobility will continue to transform the world around us and become a central hub for interaction in both personal and business environments. Following are a few trends in mobility that we see happening in the near future.

Smart Wearables

What is currently achieved using smartphones will soon be possible through wearable devices such as smart watches and smart glasses. Taking and receiving calls, connecting to the internet and social networking sites, transferring information with other devices will be possible using watches with wider screens, glasses or headsets. Though this area is in its nascent stages, few companies have started rolling out devices such as Samsung's Gear series (Gear 2 and Gear Neo), Nike's FuelBand, Huawei's TalkBand B1 etc.

World of Connected Objects

Connected objects are those that interact with one another virtually with mobile phones acting as the central hub for communication. Remotely operating your household appliances such as washing machine, television, gaming devices or your in-vehicle devices such as GPS navigation, sound system, AC controls via mobile phone apps will soon go mainstream giving users some exceptional experiences. Apple's iOS in the Car, Google led Open Automotive Alliance, Google's Nest acquisition are initiatives in this direction.

Mobile Advertising

With 4G connections, consumers can expect faster mobile speeds. This also opens opportunities for small advertising videos to be streamed in mobiles. This new trend in the advertising industry will gradually eliminate the 30 second online ads and give way for 5-10 second or 15 seconds media rich ads in mobile devices. It is likely that the duration and design will vary based on the device whether it is a smartphone, tablet or a smart wearable. New ad formats and purchasing models will emerge in the advertising industry to suit the mobile consumers.

Enterprise Mobility and BYOD

While mobile devices are becoming ubiquitous in the consumer world, enterprises have been slow in adopting mobility into their work environments. Employees are pushing for BYOD (Bring Your Own Device) culture so that they can use of their mobile devices running on iOS and Android for office related work. One of the main deterrents against this trend is the fear of data security. Enterprise infrastructure must beef up their security controls and make data as secure as possible at the source before letting it out to be accessed by any device. With an increasing number of a young generation in the employee mix, it is necessary that enterprises have acceptable BYOD policies in place.

Ultra HD 4K Mobile Video Recording

The new feature that most mobile manufacturers are gearing up is in the area of video recording using mobile devices. Many mobile phones available in the market today have a 1080p High Definition (HD) recording facility. 4K Ultra HD videos have 4 times the picture resolution when compared with a full HD. With new range of high power processors such as Snapdragan 801 series, Ultra HD recording feature in mobile phones will soon be seen in the marketplace.

Analytics

Business Analytics can broadly be defined as a means to drive decisions and actions for better business outcomes by employing statistical and quantitative analysis, explanatory and predictive modeling on the data at hand. Analytics is not a new discovery in the world of business. People have also been analyzing outcome of their actions and measuring performance against their competitors. Till quite recently this aspect was widely known as Business Intelligence (BI). Business intelligence can be considered as a branch of analytics that looks for trends in data and produces aggregated outputs with drill down and drill up capabilities. Business users get data in more consumable form as standard reports or dashboards which enable them to make decisions. Analytics encompasses not only this descriptive nature of analysis provided by BI but also allows users to forecast on what is likely to happen in future as well as offers possible course of actions and expected results for every action. This predictive and prescriptive capability brought a breath of fresh air for businesses looking to gain a competitive advantage over its peers in the marketplace. Apart from the accelerating pace of business and increased business complexity, another major driver for analytics to stage center stage is the huge volumes of data with various formats. Deriving useful insights from petabytes of data requires a combination of advanced technology and analytical minds leading organizations to form a separate analytics division.

Types of Analytics

Analytics covers a broader spectrum of services and can be divided into 3 main types namely, Descriptive Analytics, Predictive Analytics and Prescriptive Analytics. Figure 1 depicts the types of analytics from business value vs analytical complexity perspective.

Descriptive analytics focuses on questions such as "What happened to the business?" or "What is the impact on the business by an action taken in the past?" It helps business users with summarized data in a more consumable form and provides slicing and dicing of data using drill down and drill up features. Standard statistical aggregate functions such as mean, median, standard deviation, variance combined with data visualization techniques are employed to offer users business insights of what happened in the

Figure 1. Types of Analytics

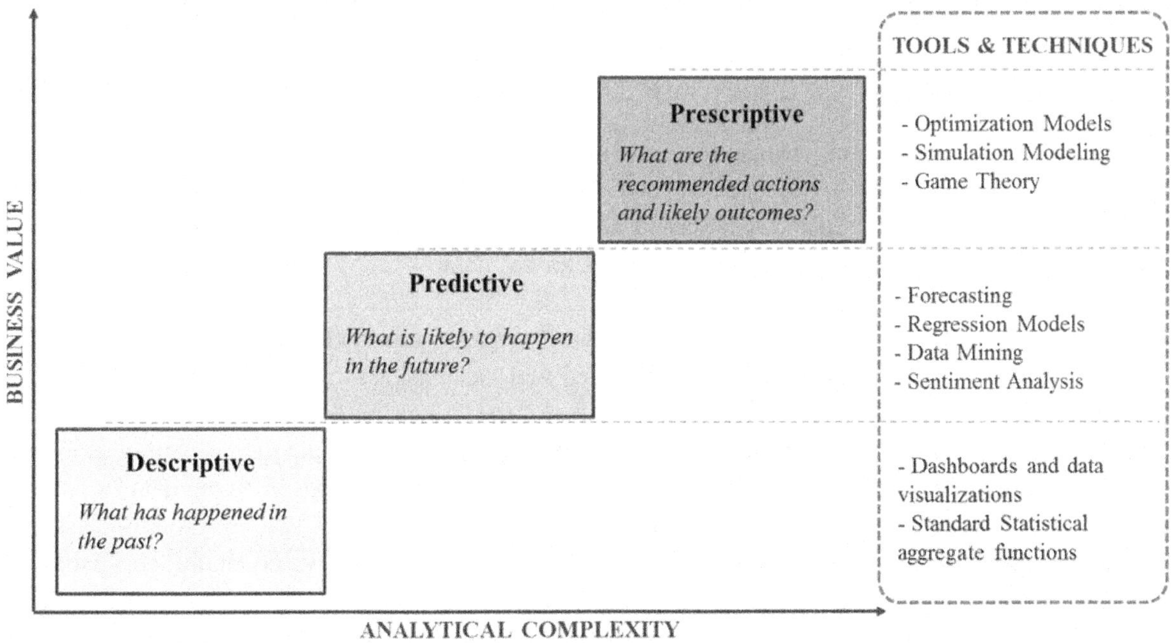

past. An example for descriptive analytics in the clinical trials domain is the measurement of progress and performance of sites during trial conduct. Plotting a trend on how the site is performing across various key indicators such as patient recruitment, data quality, document compliance and its comparison with the performance of other sites gives a clear indication whether it is a potential risky site. Using appropriate drill downs, users can get a view on what is going wrong in these low performing sites which will help them to take corrective measures.

Predictive Analytics is the next level of analytics that can forecast what might happen in the future based on the historical data. It must be noted that predictive analytics can only estimate what is likely to happen and cannot accurately tell what is going to happen in future. For prediction to be done, a model based on the available data must be built. This model can then be run for future time periods. The model will have to be fine-tuned as and when new data becomes available to improve the prediction accuracy. Extending the example of clinical trials, using predictive analytics one can forecast how many patients will be recruited by sites in the coming quarters or what will be the time delay and cost incurred if sites continue to show poor performance. Predictive analytics employs advanced statistical methods such as forecasting techniques, regression modeling, and sentiment analysis. This type of analytics can be both time based prediction (temporal) as well as non-time based prediction (non-temporal).

Prescriptive analytics is an advanced level of predictive analytics where the model recommends what are the possible courses of action that a decision maker can take and likely outcomes of each action. Since a prescriptive model predicts the outcome for multiple scenarios it can be viewed as combination of many predictive models. The model must have a feedback loop to adjust itself based on the action taken by the decision maker. An example in the case of clinical trials is patient recruitment optimization problem. During the clinical planning stage, based on the protocol sample population requirements and availability of site data, one can build a model to optimize the recruitment of patients in each site. During the course of the trial, to maintain the recruitment ratio, additional recruitment strategies have to be taken. Through a prescriptive model, one can predict what would be likely outcome on the cost and time when different strategies are employed. Optimization models, simulation modeling, canonical correlation analysis are some of the techniques used for prescriptive analytics.

Big Data and Analytics

Big data and analytics are interdependent because the value of big data is realized only through analytics and importance of analytics is enhanced by the rise of big data phenomenon. Big data is a term given to the high availability of digitized data that have the 3 characteristics namely, Volume, Velocity and Variety (Laney, 2001), popularly known as the 3V's for big data.

- *Volume:* Rising volumes of data in a digitized form from transactional systems, social media, instruments such as sensors and meters coupled with decreasing storage costs to store the exponentially growing data is one of the main factors for the rise of big data
- *Velocity:* The speed at which data is generated is phenomenal. Analyzing data streaming at exceptional speeds and taking real time decisions is yet another challenge posed by big data
- *Variety:* Data is pumped out in multiple formats by different sources that are broadly classified as structured and unstructured data. Unstructured data includes text, video, audio, geospatial information, outputs from electronic instruments/ equipment. Structured data adheres to some data

structure that is machine readable but the structure and data standards vary from system to system. Managing the complexity of multiple imprecise data types adds to a third dimension to the big data challenge.

It is clear from above characteristics that data of this nature cannot be managed with the available relational databases that were long used for data storage and analysis. New technologies that can hold and compute large volumes of multi-structured data in a relatively less time emerged in the market. For operational big data, NoSQL emerged which as faster and can scale more quickly and inexpensively than relational databases. For analytical big data, Hadoop MapReduce framework is considered very efficient since its new method of analyzing data complements the capabilities of SQL and can handle sophisticated algorithms and computations.

Deriving insights from big data to help make correct decisions is the need of the hour for modern organizations and hence companies are placing large bets on expanding their big data infrastructure and analytics capabilities.

Cloud Computing

National Institute of Standards and Technology (NIST) defines cloud computing as (Badger, Grance, Patt-Corner& Voas, 2011),

"Cloud computing is a model for enabling convenient, on-demand network access to a shared pool of configurable computing resources (e.g., networks, servers, storage, applications, and services) that can be rapidly provisioned and released with minimal management effort or service provider interaction."

In other words, cloud computing allows companies to access the computing resources and services including hardware and software over a network. Cloud services are scalable and configurable and due its pay-as-you-use pricing model, it offers minimal capital and operational expenditure when compared to traditional in-house IT implementations.

According to NIST, a cloud model is composed of five essential characteristics, three service models, and four deployment models.

Key Characteristics of the Cloud

- **Accessibility over network:** The main feature of setting up IT applications on the cloud is the ability to access the systems over a network, typically via the internet. It provides location independence such that regardless of where the users are located, applications and services can be accessed as if it was controlled from the company's own data center.
- **Flexible and Scalable:** To meet the demand, resources can be provisioned easily and the environment can be scaled up without any delays in approvals and high investment costs. It appears as if the capacities provided by the cloud are unlimited. Scaling down the existing capacity during a lean demand period can be accomplished hassle free.
- **On-demand service:** Cloud offers the flexibility to add, modify and remove the services in terms of provisioning, scaling up, monitoring based on user's preferences and business needs with minimal delays in implementation.

- **Resources sharing:** The computing resources such as servers, middleware, and monitoring and support activities are shared with multiple consumers leading to better utilization of resources and effective balance of demand. An environment shared by multiple users that operate similar set of tasks is called a multi- tenancy model.
- **Pay-as-you-use approach:** All services rendered by the cloud can be measured which enables a pay-as-you-use pricing model for companies who avail the services based on their needs. Resources usage can be monitored and reported thus providing transparency for the provider and the consumer.

Types of Cloud Models

There are four main types of cloud deployment models namely,

- **Private Cloud:** When the company operates a dedicated cloud environment for its own use, it is called a private cloud. The cloud can be set up within the company's own premises or off premises and can be setup, controlled and managed either by company's IT department or a cloud service provider. Since data is more secure in this form, life sciences companies prefer private clouds to public clouds.
- **Public Cloud:** When a third party service provider offers cloud services for use to multiple customers in a multi- tenant model, it is called a public cloud. It means that the same physical servers are utilized by several companies with data and services virtually segregated and secured from one another.
- **Hybrid Cloud:** As the name suggests, it is a combination of two cloud forms - private and public. It blends the benefits of both cloud systems such that confidential data is handled in the private cloud while the public cloud is utilized for accessing the wide range of applications and services deployed in it.
- **Community Cloud:** The cloud infrastructure is shared among various companies mainly from the same industry who have shared or common interests. The costs are shared among the partnering organizations. The cloud setup is either controlled by one of the organizations or by a third party service provider.

Service Models

There are three main layers or service models in cloud computing namely, Infrastructure as a Service (IaaS), Platform as a Service (PaaS) and Software as a Service (SaaS). Figure 2 (Czernicki, 2011) explains the differences among these three layers.

- **Infrastructure as a Service (IaaS):** In this model, the computing infrastructure such as the physical servers, data storage space, network cables, firewalls, IP addresses are provided by the service provider. This gives the flexibility for the company to setup their software stack and build their own applications without owning the hardware components. Examples are Amazon EC2, Raskspace, Google Compute Engine.
- **Platform as a Service (PaaS):** This is next level of abstraction in terms of service where along with the infrastructure, the underlying software including operating system, programming lan-

Figure 2. Layers of Cloud Computing

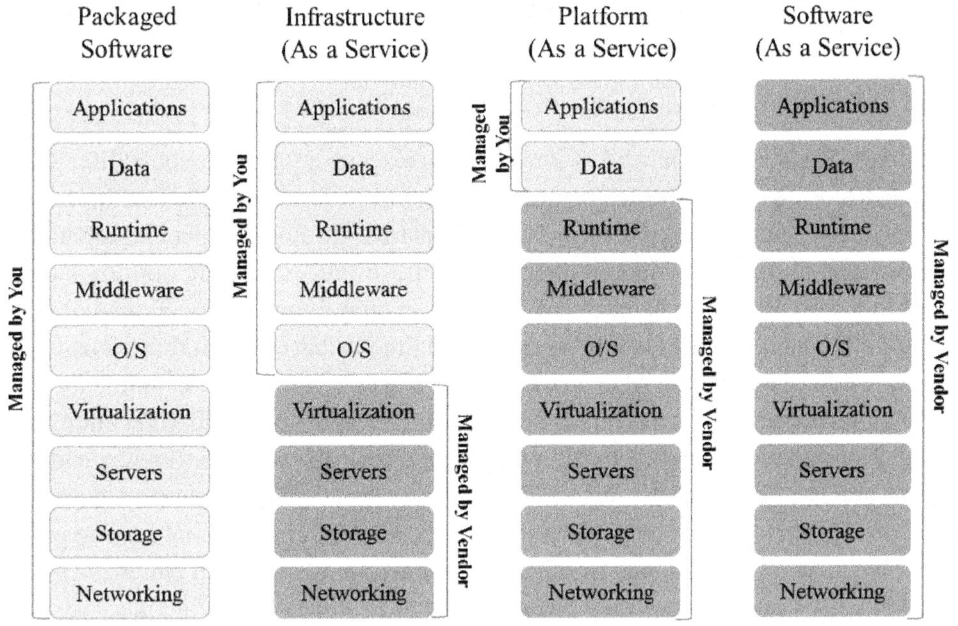

guage execution environment, databases are offered by the service provider. PaaS provides an integrated development and deployment environments, integration capabilities, APIs that will help developers to build their software applications and deploy them in the environment. Examples are Windows Azure, Force.com, Google App Engine.

- **Software as a Service (SaaS):** In this model, the software applications for the end users are provided over the network. The software is provided 'on-demand' giving the users the flexibility to access whenever they need. The service provider will have the ownership to maintain and upgrade the application, underlying software and hardware with minimal cost of ownership for the company that utilizes the service. Examples are Gmail, Google Apps, Microsoft Office 365, Dropbox.

PHARMA STAKEHOLDERS' VIEW ON SMAC TECHNOLOGY

The earlier section introduced the basic concepts of the four main technologies in the SMAC stack. This section deals with how do different stakeholders in the healthcare industry perceive the use and benefits of these technologies. If one looks at the world of medical products and services, there are four main types of stakeholders:

- **Patients:** This group, also called as the 'health' consumer, is the main stakeholder since all services and products are developed to improve the patient's health and well being
- **Pharma companies:** This group are the medicinal product manufacturers who ensure the best products that are safe and effective reach the patients
- **Regulators:** This group plays a vital role in ensuring the safety of the patient's health by continuously evaluating and monitoring the quality and safety of the products and services

- **Healthcare community:** The community includes the health professionals, hospitals, health insurance companies (or payers) who provides healthcare services to the patients

Patient's View

When a consumer uses social media and mobile devices extensively in their social life, it is very likely that as a patient, he/ she will look to such communication means for medical information as well. Social media including online patient communities such as PatientsLikeMe, RareConnect,AskaPatient, HealtheTreatment are excellent means to gain more insights into the disease, treatment options and other patient experiences. More than 40% of consumers say that information found via social media affects the way they deal with their health (Allied Health World, 2012).Studies have shown that patients with chronic illness, cancer, rare diseases or caregivers on their behalf actively participate on online forums for health related information. Sharing their experiences about various treatments and disease conditions with other patients, reading news articles related to the disease give patients more confidence and instills hope. Online patient communities have also impacted positively on patients in maternity, depression, weight loss programs etc. Another reason for patients to use social media is that the increasing participation of hospitals and physicians to disseminate health information and interact with patients. However, due to patient confidentiality issues, patient case discussions can be done only as a collective group without revealing the identity of patients.

Mobile device, from a patient's perspective, is an extension of the activity he/she does on the internet and social media. Since these devices are handy and accompany wherever one goes, patients go online using their mobile devices especially when waiting for the physician's appointment and during travel. Also, a slew of health apps are being developed that aid patients in terms of better disease management, improved drug adherence and enhanced wellness. B Braun's NuTRlscreen app that gives a quick nutrition screening of a patient, Philips Vital Signs app that measures heart & breath rate contact-free via an iPad 2 or iPhone 4 camera, Sanofi's iBGStar Diabetes Manager app that allows users to manage their diabetes information are some of the health apps targeted at patients and general public. While the scope for health mobile apps is overwhelming, it does not indicate that the scope is limited to smartphones only. Small Messaging System (SMS) technology has been put to tremendous use in the developing world to create health awareness, increase drug adherence and healthcare data capture. For example SIMpill, uses SMS messages to remind patients to adhere to their medication regime. A mobile phone chip embedded on the prescription bottle or blister pack sends reminders to patients according to their medication schedule and also warns if the dose taken is above the recommended dose or at incorrect times. Another organization Text to Change, piloted an initiative in Uganda in 2008 that focused on gathering public's knowledge on HIV/AIDS, improving their awareness about the disease and encouraging them to take up HIV testing via mobile phones (Text to Change, 2010).

Pharmaceutical Industry View

Many surveys conducted across the industries have shown that pharmaceutical industry is a late adopter of technology. It maintains the same stand in the case of SMAC or digital transformation initiatives as well.

Social media, the most popular among the new technologies, presents lot of benefits as well as risks to pharma companies. Patient centric websites, blogs and company sites to engage with patients and caregivers gives immense insights into their quality of life, their perception of the product, product's brand value in comparison to that of a competitor, product efficacy and safety profiles and so on. Establishing a two way dialogue between patient and the company in a social platform can improve the patient's confidence as well as eliminate any misconceptions. Getting involved with the patients and creating awareness about disease and treatment options, helps companies to increase brand image in the market. With benefits galore, drug makers are still 'social media shy' because of issues that can lead to breach of privacy, regulatory non-compliance, improper use of electronic communication platform etc. The apprehensions that drug makers have when it comes to use of social media are:

- Lack of proper regulatory guidelines by FDA and other regulatory bodies on use of social media
- Patient confidentiality issues since many social media platforms expose the patient's identity making two way interaction between patients and pharma companies liable for breach of privacy
- Lack of proper and effective way to deal with patient negative responses and potential liabilities arising from such issues
- Uncertainty over the return of investments on social media initiatives

Mobile technology has received far more acceptance in the pharma world and has been used across discovery, clinical development and sales and marketing. Using mobile devices to facilitate visually rich and interactive sales presentations to physicians, known as e-detailing, is a quite common and effective approach adopted by sales representatives in recent times. Mobile apps galore with most of the large pharma releasing iOS and Android apps for patients in the area of patient wellness, disease management, drug adherence, clinical trial recruitment, health news feeds and so on. With these mobile apps, companies get an opportunity to come closer to their patient community gaining their loyalty as well as understand their demographics, behavior and disease patterns. An interesting innovation using mobile technology was the invention of 'digital pill' or 'smart pill'. In the year 2012, FDA approved the ingestible sensor embedded in a pill (Proteus Digital Health, 2012) that can transmit electronic signals once the pill reaches the stomach of the patient. This signal is captured by a skin patch which transmits the signal to the patient's mobile device. Mobile device pass the information through the mobile network to healthcare providers thus ensuring the proper and timely use of the drug by patients.

The importance and benefits of analytics and big data is no different in pharma industry when compared with that in other industries. Large pharma companies have been investing a large share for enhancing their analytical capabilities within their R&D and sales and marketing divisions. Companies such as AstraZeneca has a Predictive Sciences group in their R&D organization which focuses on identifying the right target and right drug in its discovery phase using the large amounts of research and preclinical data available with them. It uses predictive modeling and simulation to help in deciding which projects have high success rates thereby spending their R&D funds judiciously. Optimization of clinical trial recruitment, early identification of high risk sites through risk based approach, better utilization of CRA resources are some of the examples in clinical development that companies employ heavy use of analytics. In the area of sales and marketing, companies sift through prescription data as well as social media data to perform customer segmentation analysis, sentimental analysis of their products and brand and sales force effectiveness. Apart from building the analytical skills, collecting and transforming data to make available and consumable for analytical techniques to run is a big challenge faced by many pharma companies. Incremental and proactive steps are taken to build an analytical environment and equip the workforce to bring analytics to life in their field of work.

In a volatile economic situation marred with steep patent cliffs, rising R&D cost pressures, increasing competition and stringent regulatory compliance, pharma companies are faced with a daunting challenge to contain costs and improve their shareholder value. Pharma companies have welcomed cloud computing since it offers big cost savings by significantly reducing the capital expenditure as well as has the potential to scale up and scale down the infrastructure based on demand. Several terabytes of data generated by data technologies such as next generation sequencing and gene profiling poses a challenge in terms of its storage and management. Cloud computing is ideally suited to address the burgeoning data needs of the pharma company. It also provides a platform for stronger collaboration and sharing large amounts of data among researchers, clinicians and business partners helping in drug research and development process. On the flip side of benefits, data security and privacy concerns of a cloud based environment have been a point of concern for many since it can lead to regulatory non-compliance or IP loss leading to criminal penalties and business interruption. Companies are started realizing the benefits with a more secure private cloud setup as opposed to a public cloud. In the R&D organization, systems such as Adverse event reporting system, clinical data warehousing platforms are finding their place in the cloud.

Regulator's View

Increasing use of social media and other new technologies by the industry players to reap the benefits of higher reach and better brand management of their products in a cost effective manner have prompted the regulatory agencies across the globe to act upon monitoring its use and minimize the safety risk on patients. FDA has stated that developing the guidelines for social media is one of its highest priorities. Since 2009, FDA has been seeking industry opinions, conducting hearings and releasing draft guidelines on the lines of online and social media use by life sciences players. At the time of writing this book, FDA has reportedly confirmed that a full social media guidance document will be released by July 2014. Similarly, European Medical Agency (EMA) and ICH have provided their viewpoints on specific topics such as adverse event reporting from a social media source. Below are some of the important guidelines released by major regulatory bodies on social media in the recent years:

1. *FDA draft guidance on post marketing safety reporting released in March 2001*:(Food and Drug Administration, 2001) It states that any adverse event experience submitted via the internet to the pharma company must be reported to FDA if it satisfies the four main elements for submitting an individual case safety report (identifiable patient, identifiable reporter, suspected drug information, adverse event information). It also states that companies should review any Internet sites sponsored by them for adverse experience information, but are not responsible for reviewing any Internet sites that they do not sponsor. However, if they become aware of an adverse experience on an Internet site that they do not sponsor, they should review the adverse experience and determine if it should be reported to FDA.
2. *ICH guidelines on adverse event information from Internet:*(ICH, 2003) On the similar lines of FDA, ICH recommends all MAHs to screen for adverse events in websites under their management and exclude external websites that do not come under the MAH purview. It also recommends MAH to consider utilizing their websites to facilitate adverse event data collection.
3. *FDA draft guidance on off-label use information available online in December 2011:*(Food and Drug Administration, 2011) FDA has been releasing multiple draft guidance each focusing on specific topics. In December 2011, the agency released its guidelines on unsolicited and solicited off label

use from social media. Unsolicited requests are those initiated by persons or entities completely independent of the relevant firm while solicited requests involve a prompt from the firm to publish off label uses of the medicinal product. Solicited request on off label use by pharma companies is against the law. As per the guidance, substantive responses of an unsolicited request must be provided only to private, one-on-one communication. For public requests, only the firm's contact details with a request to contact for more information must be provided in the public forums.

4. *EMA's GVP guidance citing social media use in February 2012*: (European Medicines Agency, 2012) Module VI of the Good Pharmavigilance Practices (GVP) released by EMA provides brief guidelines on adverse events reported in internet and digital media. The MAH must regularly screen for adverse reactions in digital media that they own, pay or control. It must also actively monitor special internet sites such as those of patients' support or special diseases groups in order to check if significant safety issues are discussed that may necessitate reporting. Frequency of screening must be such that valid ICSRs must comply with submission timelines based on the date the information was posted.
5. *FDA draft guidance on advertising promotional materials in social media in January 2014:*(Food and Drug Administration, 2014) FDA released its position on interactive promotional media, that is, tools and technologies that often allow for real-time communications and interactions. As per the guidelines, firm is responsible for promotional material communications on sites if it is owned, created, influenced or operated by the firm or on the behalf of the firm. On third party websites, firms are responsible only if they have editorial or review privileges on the content and not if there exists a financial support only. Firm is also responsible for the content generated by an employee or agent who is acting on behalf of the firm to promote the firm's product.

Mobile health apps have come under the scrutiny of FDA since some medical apps transform a mobile device into a medical device and provide diagnosis and treatment suggestions. It can increase the patient's health risk if the apps are not validated properly. In September 2013, FDA released a final guidance for regulating mobile apps (Food and Drug Administration, 2013).FDA will regulate a subset of mobile apps that satisfies the definition of a "device" which is intended to be used as an accessory to a regulated medical device, or transforms a mobile platform into a regulated medical device. If the intended use of a mobile app is for the diagnosis, cure, treatment or prevention of disease, the mobile app is a "device." Transforming a mobile device by attaching a glucose strip monitor to record blood sugar levels, displaying the ECG waveforms transmitted from remote medical equipment, suggesting treatment and dosage plans using patient specific parameters, remotely controlling medical equipment or implants are examples of mobile apps considered as devices. In such scenarios, FDA requires mobile medical app manufacturer to satisfy the regulatory requirements applicable to the specific device classification governing the app: class I (low risk), class II or class III (high risk). There are some apps that meet the definitions of a "device" but are considered lesser risk to patients such as app that coach patients to manage their disease, helps them to organize and track their health records, enabling them to communicate and collaborate with physicians and other patients. FDA will not enforce requirements for this category of apps. The third category of apps that has a medical focus but do not qualify as a medical device will not be regulated by FDA.

FDA, EMA and other regulatory bodies sit on a huge pile of clinical data and if analyzed can generate real world evidences on drug human interactions. Regulatory bodies are also actively considering the use of analytics in optimizing the various processes in the drug discovery, development and manufacturing phases. One such initiative is recommendation of risk based approach to monitoring (RBM) of clinical trials. Both FDA and EMA are actively pursuing the implementation of RBM by all pharma companies. FDA released draft guidance in Aug 2013(Food and Drug Administration, 2013) and EMA published a reflection paper in Nov, 2013 (European Medicines Agency, 2013) on RBM. As RBM guidelines, sponsors are required to develop monitoring plans that manage important risks to human subjects and data quality and address the challenges rather spent effort on time consuming and less important tasks at the site. One of the strategies in RBM is to enable remote monitoring where key risk indicators that measure the performance and quality of sites are monitored from a remote offsite location. Analytics has a big role to play in analyzing the risk indicators and provide alerts and warning signals of potential risky sites and help mitigate them by suggesting appropriate corrective actions.

None of the health authorities have issued a proper guidance on how to regulate a cloud computing technology. All initiatives that involve GxP applications to be hosted on the cloud will have to adhere to the existing regulations set for the traditional in house implementations. FDA's 21 CFR part 11 policies provide guidelines for computer system validation and infrastructure qualification for GxP applications. However, in the case of cloud computing, application of computer system validation (CSV) is not very straight forward. For example, IQ, or Installation Qualification is the first step to qualify the infrastructure environment where an application is hosted. It covers hardware, underlying software such as the OS and instructions for installation. It also requires specific information such as hardware serial number, system configuration, exact location of software, software versions which are very difficult to obtain in the case of a public cloud as opposed to private cloud. Also each server OS configuration in the cloud must be validated before it is used. This validation involves considerable time, effort and cost and diminishes cloud's advantage of faster scalability of infrastructure at a lesser cost. Lack of clarity in the guidelines for handling cloud setups has left companies to be conservative in their selection of cloud technology for new IT initiatives. An equally important aspect to keep in mind is the selection of a cloud service provider with sound knowledge of the regulatory framework for setup and management so that the risk of non-compliance is minimized.

Besides 21 CFR Part 11 policies, some of the important regulations and security standards that are applicable to cloud computing are listed below (Harjulampi, 2013):

1. *EU GMP Volume 4 Annex 11:* Annex 11 (EC 2011) describes how computerized systems which are used in conjunction with good manufacturing practices (GMP) regulated activities should be treated. The main principle stated in Annex 11 is "the application should be validated; IT infrastructure should be qualified"
2. *94/46/EC Data Protection Directive:* This regulates how personal information such as addresses and other personal records can be processed within the European Union. A new legal framework is being proposed to replace 94/46/EC which will be aimed against the privacy challenges from rapid technology evolvement including social media and cloud computing.
3. *US-EU SafeHarbor:* This has been setup by the US Department of Commerce to comply with the data privacy guidelines provided under 94/46/EC directive. Under this framework, a US company doing business in EU has sufficient privileges to process EU citizen personal data outside the EU.
4. *Health Insurance Portability and Accountability Act (HIPAA):* This requirement was brought into effect in US to protect the privacy of patient's medical information and restrict its access to selected individuals. The latest update on HIPAA stipulates that the healthcare industry must utilize elec-

tronic health and medical records (EHR and EMR) and migrate all health records to cloud. It also mandates that cloud service providers are as liable for HIPAA compliance as healthcare entities.

5. *ISO / IEC:* The ISO/IEC 27000 series focus on the information security management and there are a couple of new standards proposed for cloud computing in particular.
   a. *ISO/IEC 27001:2005:* The scope of the standard is to provide set of specifications of which organizations may use in seeking certification for their information security management systems. It has 8 sections that define Information Security Management System (ISMS) requirements, management responsibilities, internal ISMS audits, management review of the ISMS and continuously improvement of ISMS.
   b. *ISO/IEC 27002:2005 (formerly known as ISO/IEC 17799:2005):* This provides best practice recommendations on information security management for use by those responsible for initiating, implementing or maintaining ISMS. There are 15 sections which have total of 39 control objectives for information security management.
   c. *ISO/IEC 27017:* This is a draft standard currently under preparation which provides the guidelines on Information security controls for the use of cloud computing services based on ISO/IEC 27002.
   d. *ISO/IEC 27018:* This is another draft standard currently under preparation. The objective is to collect and organize security categories and their controls from current data protection regulations and help public cloud service providers to comply with their obligations and make this transparent to their customers.
   e. *ISO/IEC 27036-1 to ISO/IEC 27036-5:* This set of standards define how to evaluate and mitigate security risks when using IT services or information supplied by 3rd parties.
   f. *ISO/IEC 14971:2012:* This standard provides risk management tools for helping manufacturers to introduce medical devices on the market.

Healthcare Community View

Healthcare sector is not left behind in the adoption of newer technologies and have made a significant impact on improving the patient care and collaboration among physicians, patients and health insurance organizations. Many healthcare players have jumped onto the social media bandwagon to establish a better connect with the patients and enhance their experience and quality of care. Hospitals use Twitter to provide a live up-to-minute update on the proceedings in the operating room. Through online portals, hospitals and physicians engage with the patients to provide the awareness programs on disease management, wellness and good lifestyle tips. To ensure a hassle free visit, hospitals provide their patients with reminders of upcoming visits, latest information on waiting period on the day of the visit etc. Third party health websites such as HelloHealth and PatientsLikeMe facilitates an online communication between physicians and patients. Online physician forums such as Sermo provide a platform for physicians to interact and share knowledge among one another. Health insurance companies are also tapping into the social media such as Facebook, Twitter to market their insurance plans and provide expert advice on the health insurance related topics.

As pointed out in the earlier sections, mobile devices are transforming into medical devices with smart wearables measuring and monitoring health information of the user. In the healthcare industry, mobile devices are being widely used by physicians and nurses at the point of care. Getting the patient records on the handheld device during hospital rounds by the physicians is much more convenient than traditional on-paper recordings. Using mobile devices, physicians can continue to work on administrative and reporting activities even after office giving them more quality time to spend with the patients. Mobile apps such as AirStrip Cardiology app that gives doctors access to patients' heart readings and Epocrates that offers diagnostic lab test tools, medical dictionaries, drug-interaction checkers and treatment guides aids physicians and nurses to be more productive and effective.

In order to address the new regulatory requirements of switching to EMR and EHR for patient records as well as to lower down the healthcare costs, healthcare organizations are transforming their IT infrastructure to cloud computing. The benefits of cloud computing such as reduced capital expenditure, higher scalability, broader access and reduced operating costs address some major issues faced by the industry in terms of handling burgeoning patient records in digital form, rising healthcare costs and need for sharing data quickly and securely among regulators and other healthcare stakeholders. A cloud environment that transfers and stores data in an encrypted form can increase HIPAA compliance since most of the HIPAA breaches happen due to theft of storage devices and flash devices with confidential health information or otherwise referred to as "protected health information (PHI)". Cloud computing enables better collaboration among various stakeholders in the healthcare ecosystems. Many companies are creating what is called Healthcare Information Exchange (HIE) on the cloud where health systems, physicians, hospitals and other healthcare organizations can easily share pertinent information.

Healthcare data can be considered as a perfect playing ground for realizing the true potential of analytics. The enormity in patient data volumes, multi- structured formats of data, variety and variability in data types qualify healthcare data for a big data initiative. Healthcare data comprises data from EMR and EHR, individual genetic profiles, lab and imaging data, physician prescriptions, medical correspondence, claims data, demographic details, disease prevalence information and so on. Using this data, analytics can unlock interesting insights and patterns about the patient, sub population, disease incidence and prevalence, drug's efficacy and safety, patient care management, healthcare costs, claims and reimbursement patterns to name a few. Analytics can open up new avenues for healthcare organizations for enhancing patient care and improve the healthcare system to make it more sustainable and accessible.

SMAC AS THE CORNERSTONE FOR INNOVATIONS IN PHARMA R&D ECOSYSTEM

In the current situation where Pharma R&D is marred with multiple obstacles such as depleting product pipeline, rising clinical development costs and patent expiry, disruptive innovations across the R&D value chain are inevitable to ensure sustenance of growth. Use of SMAC technology within various R&D processes of drug discovery, clinical development and pharmacovigilance can enable increased success rate for identifying high potential compounds, reduced developmental costs, better collaboration among research and clinical communities, reduced risks for late stage failures and faster drug to market.

There are many social media, mobile, analytics and cloud computing use cases that can be incorporated in the R&D processes. In this section, we will discuss in detail four important use cases that have either already been adopted or have a high probability of getting adopted by the pharma industry in the near future. Each of these use cases exhibits the true benefit from the SMAC technology components, how different technology components work in unison in the R&D context and how it transforms the current processes to deliver significant business outcomes.

"Digital Pill" with Ingestible Sensors

Domain: Clinical Development
Sub Area: Drug Adherence and Clinical data monitoring
Technology Used: Mobility, Cloud Computing

This is one of the remarkable innovations in recent times seen in the pharma industry. Proteus Digital Health is the company behind this invention and it has captured the attention of the industry players as well as regulators with the European Union approving the device in 2010 followed by FDA approving it in 2012. Digital Pill or Smart Pill consists of a pin head sized ingestible sensor that must be consumed along with a medication pill. Once the sensor reaches the patient's stomach, the stomach fluids act as a medium to activate the sensor. The sensor transmits a small signal that gets captured by a transdermal patch placed on the patient's arm. The patch records the signal and also measures the patient's heart rate and other physiological responses. This information is transmitted via an inbuilt Bluetooth antenna to a mobile phone app which can interpret the information and send alerts and reminders to the patient and physician.

Let us explore what the major benefits this innovation brings to the table and how the technology is enabling it to happen. For long term medication ailments such as tuberculosis and diabetes or for life threatening diseases such as cancer, adhering to drug regime is very critical. It is noticed that patients undergoing long term medications tend to falter on their medication regime after a period of time leading to more complications such as multi drug resistance and other related disorders. Since there were no proper means to monitor drug adherence, one has to place trust on the patient that he/she takes the drug on time and in the appropriate quantity. With the use of an ingestible sensor, one can accurately measure the time when the body received the drug. It also helps patients and caregivers to remind in case the drug intake is missed as per the schedule. It can also send alerts in case there is an accidental overdose which is particularly common in elderly as they tend to forget the actual dose or if they had already taken the required dose. Drug adherence and its monitoring is one of the key aspects that determine the success of the clinical trials. Early alerts on subjects discontinuing the drug and providing them with the required assistance can reduce the number of subject discontinuations in the trial. Moreover, the data results of the clinical trial are purely based on how well the patients have followed the protocol. Inconsistency in the drug intake by subjects can result in misleading data analysis reports which can prove detrimental from the patient safety and clinical trial success standpoint.Another use of this device is the timely recording of patient's physiological responses and behavior. Any abnormal changes in the patient's health measures can trigger an alert to the physicians and care givers. Having access to the real time information can help physicians to administer the next course of action in a more effective manner.

From the above illustration of the use case, the use of mobile devices is evident in enabling the communication across various systems and users. It must be noticed that patients, care givers, physicians, pharmacists are spread across various geographical regions. The information transferred among these users is exchanged via a cloud environment which connects different systems and devices together. The flow of information can be better explained through an example. Let's take an elderly patient is participating in a clinical trial for Schizoprenia. The principal investigator after collecting his informed consent enrolls him to the trial and provides him with drug kit and smart pill device. The hospital configures the mobile app and Bluetooth connectivity to receive information into their centralized patient repository hosted in the cloud. The patient's close relatives/ caregivers are also enrolled into the system to receive alerts and reminders. The principal investigator and lead nurse staff will have access to all patient records and the information exchanged by the digital pill device present over the cloud via their mobile devices. As the schedule, the patient starts his treatment and along with the drug, the ingestible sensor is also consumed. Once the sensor reaches the body, it sends signals to the wearable patch. The patch records the time of intake, dosage amount and also measures the patient's heart rate, temperature, body activity and sends it to the patient's mobile app. The mobile app determines the intake time and dosage are within the acceptance range and transmits the patient's parameters via a mobile connection to the cloud. In the cloud business rules engine, the patient body parameters are checked for abnormality and depending on the change the investigator is alerted in his mobile device. Investigator will monitor the previous day's readings and readings on the next few days and if required, arrange an unscheduled visit to the hospital for the patient. In the event that the patient fails to take the drug on time, the mobile app will remind the patient at the appropriate time. Continuous failures in drug intake will be tracked and an alert will be send to the investigator and hospital staff to contact the patient.

Risk Based Monitoring

Domain: Clinical Development
Sub Area: Trial Monitoring
Technology Used: Analytics, Cloud Computing

Ever since FDA and the European Medical Agency (EMA) started recommending on adopting risk based approaches to clinical trial monitoring, the players in the pharma industry have made its implementation one of its top priorities in the R&D space. FDA and EMA have brought out guidelines and released reflection papers on the RBM initiative to help companies in the design and implementation of RBM. Also, industry consortium like Transcelerate is working together to build a common framework for the industry to follow in their future trials. Traditionally, monitoring of trial sites and their progress is done by Clinical Research Associates (CRA) on-site (meaning at the trial site). These routine visits involved doing 100% source data verification (SDV) and review of source documents. Bulk of CRA's visit time is spent on less critical tasks and administrative activities. With the RBM initiative, regulatory authorities have turned their focus from mandatory 100% SDV to identifying risks associated to critical data elements and processes necessary to achieve the study objectives and ensure patient protection and overall study quality. Based on the risk level of the sites, the monitoring activities on the specific risk areas can be increased or decreased thereby reducing the workload of the CRA, providing quality time on issue resolution and minimizing the overall study costs. Another related area that RBM initiative advocates is to use of centralized monitoring concept wherein all the sites are assessed and monitored on a set of Key Risk Indicators (KRI) from an offsite location. When the Key Risk Indicators cross the acceptable threshold values, appropriate alerts/ triggers are sent to the monitoring team to act upon. It is noted that centralized monitoring identifies not only a great extent of on-site monitoring findings but also helps in performing comparisons between sites across a range of categories such as patient safety, site performance, data quality, study compliance and so on.

For a pharma company, full-fledged implementation of RBM to realize its true benefits requires a great deal of collaboration with technology. Analytics is the main underlying technology that plays a great role in the identification of risks and providing appropriate triggers for further action by the monitoring team. Analytics comprises of both data interpretation and data visualization. Representing the KRI data in a more visual and insightful manner is key to identifying the underlying risks involved and the reasons for the risks. Showing trends in data, comparing KRIs among sites and regions, calculating statistical functions such as mean, standard deviation at the project, country and site levels, identifying fraud in data entered at sites, providing drill down features, generating ad-hoc reports can be realized if appropriate visualization tools are employed. Another important factor to consider is the data since analytics can give insights only when the source data is well defined and is of high quality. For RBM to be successful, data from various external and internal sources must first be consolidated and standardized. This comprises clinical and non-clinical data from internal systems such as Clinical Trial Management System (CTMS), Clinical Data Management System (CDMS), Electronic Data Capture (EDC), Interactive Voice Recognition System (IVRS), Financial System, Clinical Trial Supplies System, Adverse Event Reporting System (AERS) and external sources such as CROs, central labs and business partners. Consolidating these large multi formatted data from different sources and running analytics engine require an environment with great scalability and compute power. Cloud computing provides the best choice for this kind of a technology requirement. Using a combination of cloud and analytics platform, pharma companies will be in a comfortable position to implement RBM for their clinical trials. It must be noted that RBM is not a pure technology initiative and lot of changes must happen at the business process level as well. However, above mentioned technologies must be established and well aligned to support the new business processes.

Smart Patient Management System

Domain: Clinical Development
Sub Area: Patient Recruitment and Patient Retention
Technology Used: Social media, Mobility

Patient recruitment is still one of the top reasons contributing to the delay of clinical trials. Some studies show that almost 80% of trials fail to meet their initial enrollment quotas on time (Earls, 2012). Despite various strategies to improve recruitment numbers through better interaction with patients and investigators and connecting with people using impactful advertisements through traditional and digital media, companies have not been able to increase participation and recruitment of patients. A number of issues contribute to this complex problem. Issues include increasing complexity in protocol designs leading to stringent inclusion/ exclusion criteria, stiff competition among peers to tap into the same patient pool for their trials, lack of awareness among people on the needs and benefits of clinical trials, unwillingness by physicians to become principal investigators due to huge workloads and large responsibilities bestowed on them during the course of the trial. Patient retention in the trial is another challenge that directly impacts the patient recruitment planning. Increased number of discontinuations can be the result of pressures on the patients from frequent hospital visits, time-consuming and elaborate medical assessments, low co-operation and engagement by the site staff, lack of knowledge on trial progress and no proper interaction with fellow trial participants and the trial investigator.

The number of patients and caregivers using internet and social media for health related information is increasing steadily across all age groups. Online health communities and health discussion forums bring together patients and well-wishers to discuss about specific diseases or drugs, share health tips and provide moral support and encouragement for the fellow patients. Online communities such as Dr Susan Love Foundation not only invite patients but also healthy people who would want to know about the disease and be part of the success movement that fights the disease. This scenario presents a great opportunity for pharma companies since the members of these communities can be potential trial subjects for their trials in the specific therapeutic areas. Online messages highlighting the trial objectives, benefits of the drug, trial processes involved and experiences of volunteers from previous trials when shared with the communities of interest can help in two ways – one, increase the awareness of clinical trial participation among the public and second, motivate the target population to enroll in the trial. Effective and continuous communication via webcasts and videos of key medical personnel and opinion leaders talking about clinical trial benefits and drugs can improve the participation. Patient engagement by providing disease management and lifestyle tips, arranging online chat sessions with key medical personnel to talk about diseases and treatments can improve the trust in the sponsor company and increase the patient's rate of acceptance to be part of new trial initiatives happening in the concerned field. Companies through active participation in these communities can gain a wealth of knowledge on patient influences and real world outcomes that can prove useful in trial designs and recruitment planning strategies. CROs who conduct clinical trials for multiple life sciences companies across therapeutic areas are also finding ways to recruit patents online. ClinicalResearch.com by Quintiles is one example where patients are given with search features to select the most suited trial based on their disease condition, location and trial regime.

Social media play a very important role in the patient retention programs as well. Trial sites can start online communities or a Facebook page and request all patients to get registered to the service. This medium can form the basic communication channel for sharing study information, study progress, timely inputs and tips from investigator, creating and scheduling visit calendars for each patient and sending reminders on upcoming visits. It also enhances the engagement levels among patients as well as the site staff and investigator. A flipside to having social media channels is addition of more effort and resources to maintain the site and address queries in the best possible manner. Another challenge is the greater risk of data contamination. In blinded studies, patients in study and control groups may have different experiences and sharing this information can influence other patients introducing skewness in the data collected. Hence, moderation of the comments and information must be done diligently to remove any response that lead to bias in the test results.

Where engagement with people is a key requirement, mobile devices can bring a huge successful impact in a short span of time. Hence, the use of mobile devices becomes an important strategy in patient recruitment and retention in clinical trials. Exco InTouch, a leading provider in patient engagement and data capture solutions in clinical research is one appropriate example. Exco InTouch uses SMS (Short Messaging System) technology to send trial related information to all the members from a client database with an option to respond back as a short code (eg: Text 'TRIAL' to 1234). Two way dialogues with the screening center can be enabled over the phone and determined if the patient is suitable for enrolment. If not, this patient information is stored in the 'potential patient pool' and further engagement activities are performed for consideration in future studies. Mobile devices can be a very useful tool for continuous patient engagement during the course of trial. Sending mobile reminders for upcoming visits, scheduling daily alerts according to the drug regime, sharing health related articles and sending wishes on birthday and other important occasions can establish a better connect with patients. Mobile apps can be employed not only to exchange information among patients, sites and sponsor companies but also as a self-monitoring tool to track progress and activity completion. The data emerging from apps or SMSes of the patient groups can be consolidated and analysed to derive meaningful insights on patient outcomes, treatment compliance, visit variations and overall patient engagement.

Signal Detection and Analysis Using External Data Sources

Domain: Pharmacovigilance
Sub Area: Safety Signal Detection
Technology Used: Analytics, Cloud Computing

Signal Detection is a one of the critical activities in the field of drug safety surveillance. Council for International Organizations of Medical Sciences (CIOMS) defines a safety signal as "information that arises from one or multiple sources, which suggests a new, potentially causal association, or a new aspect of a known association between an intervention [e.g., administration of a medicine] and an event or set of related events, either adverse or beneficial, that is judged to be of sufficient likelihood to justify verificatory action." Safety signals can be detected from a number of adverse event sources such as spontaneous reports, clinical data and post marketing study data. Since it is difficult to uncover the significance of an association between a drug and an event with manual analysis from thousands of cases, statistical techniques play an important role in signal detection. Over the years, number of techniques has evolved to identify safety signals. Commonly used techniques are the classic disproportionality measures such as Proportional Reporting Ratio (PRR), Reporting Odds Ratio (ROR), Relative Reporting Ratio (RRR) and newer techniques such as Multi Gamma Poisson Shrinker (MGPS) used by FDA and Bayesian Confidence Propagation Neural Network (BCPNN) used by the UK MHRA. It is mandatory that every market authorization holder responsible for adverse event reporting must perform signal detection and report them to the authorities for further active surveillance. The signals are then prioritized and evaluated on the basis of causaility, frequency and clinical implications. Depending on the severity of the event and drug's potential to cause large safety risks in patients, appropriate action is taken. Some of actions taken include updating the product label in appropriately, sending notices to physicians on potential adverse events and their circumstances, conducting further clinical trials to ascertain the event occurrence or withdrawing the drug from the market to prevent further safety hazards.

Pharma companies conduct safety signal detection using the data available to them in the internal adverse event reporting system. FDA performs signal detection activities on the FDA AERS database which contains more than a decade worth of data on all safety cases reported in the US. Similarly EMA and other regulatory authorities perform the same activity on their respective databases. Since the databases used by companies and regulatory authorities differ, there is a high probability that the results of signal detection will vary significantly even for the same statistical measures used. The reason is statistical measures attempt to identify signals in comparison with the other adverse events present in the source database. Many a time pharma companies are at the receiving end when regulatory authorities send their signal findings of the company's drugs which were not captured during company's signal detection efforts. For better signal detection, the source must contain fairly large amounts of cleaned and verified data.

Apart from the company's own safety repository, companies must consolidate data other sources available externally. Public safety data registries such as FDA AERS, WHO Vigibase, European EVDBMS, safety data from EHR/EMR records and information from health insurance claims data must be consolidated into single standardized format for performing signal detection. Since both internal and external data sources are distributed with large amounts of multi formatted data, a cloud based integrated safety repository will benefit the pharma company. The other benefits that cloud based solution offers such as scalability, great computational power, reliability are equally important in this scenario. Transforming the data into a well-defined safety data model is a real challenge and requires a robust Meta data repository consisting of data transformation rules, calculations and derivations. Once the data is standardized and cleaned, the data gets fed into an analytics engine where standardized data is manipulated and data mining algorithms are run to produce the desired statistical outputs. The analytics engine feeds the manipulated data as well as the outputs to the data visualization component to produce visual indications of the signals across drugs with a varying level of significance. The visualizations help in comparing signals across products, across data sources and across various statistical measures. It also provides drill down and filtering options that aids in signal prioritization and signal evaluation processes.

This cloud based signal detection platform offers lot of flexibility in plug and play integration of many visualization and reporting tools and as well change sources and data mining algorithms with minimal effort and system changes. Employing a robust platform will help companies to proactively assess safety signals and gain the reputation and trust of the regulatory bodies and patient community.

Other Use Cases

Apart from the main use cases described in detail above, one can think of a host of other use cases that the SMAC components can be put to good use in the clinical development and pharmacovigilance space. Following is a list of additional use cases that can be adopted by pharma companies to improve their drug development cycles.

- Competitive Trial intelligence using analytics on the publicly available data on clinical trials conducted by peers and competitors across specific therapeutic areas
- Better collaboration and relationship among investigators in clinical trials using social media and mobile apps driven communities
- Investigator selection and profiling via thorough evaluation of their compliance to regulations, adherence to clinical trial processes, patient recruitment success rates using data from internal and external data sources
- Drug Repositioning by analyzing the characteristics and PK/PD parameters of hundreds of compounds that failed during drug discovery stage and comparing them with other successful compounds to discover new uses of compounds
- Adverse Event data capture through mobile devices by investigators and patients for quicker medical response and on-time submission compliance
- Conduct of virtual clinical trials by recruiting, conducting and engaging the patients over the internet with minimal face to face interaction

IMPLEMENTATION OF SMAC IN LIFE SCIENCES BUSINESSES

SMAC Strategies

One of the first aspects when thinking of implementing SMAC is the development of a clear and concise strategy that aligns well with the organization's business and growth strategy. SMAC essentially might be a set of digital technologies; however it has far reaching implications on driving business value, transforming ways of working and improving the brand equity of the firm and its products or services. It is therefore, important that SMAC is regarded as business game changer and not just a new addition to the organization's IT strategy.

Below are the main few approaches that are highly recommended when implementing SMAC:

Consider SMAC as an Integrated Stack and Not in Isolation

Organizations will realize the true benefits when social media, mobility, analytics combined with cloud computing is considered as a single ecosystem instead of viewing each of the technologies in silos. Take the example of Netflix, a major online entertainment company. It uses advanced analytics and modeling to understand the customers' movie preferences and viewing patterns and recommends other movies that match their preferences. The company hosts their entire movie library on the cloud which can be accessed via multiple channels – television, laptop, mobile devices. There is a social interaction platform that allows users to share their experiences, reviews and opinions.

Get the Objectives Correct

The organization must have clear objectives on how it is going to utilize the services of these new technologies and more importantly, whether it aligns with the vision, business strategy and model. A good approach will be to list out the business priorities set for the future and map how these priorities can be achieved using SMAC stack. A risk-benefit analysis must be done to compare between traditional methods of achieving the priorities and SMAC approach. Well defined objectives set the boundary and areas to operate and provide clarity in creating a better implementation plan.

Some of the objectives that Life Sciences can achieve using SMAC technologies are around improving patient outcomes, optimizing drug development cycles, enhancing patient care and patient experiences, increasing collaboration among peers and healthcare community and reducing overall healthcare costs.

Keep a Watchful Eye on the Regulatory Requirements and its Implications

Uncertainty and lack of maturity in regulatory legislation must not act as a deterrent to SMAC adoption. It is also important to note that major regulatory authorities only provide broad guidelines on a technology adoption and implications of violating them. This provides companies ample flexibility in designing and implementing the technology across the pharma value chain. The same holds good regarding SMAC adoption as well. Life sciences industry players must keep themselves updated on this changing regulatory landscape and built solutions that are easily adaptable and scalable to meet the regulatory requirements.

Accountability Must Start from the Top

Introducing new technologies that have the potential to drastically change the current ways of working and doing business require unconditional support and uninterrupted focus of the company's top management. To ensure success from the SMAC initiatives, it is necessary to appoint C-Level executive in the firm to head the initiatives. This will also help gain an enterprise wide reach to such initiatives and will not get restricted to certain departments or divisions that initiated the change. A task force having senior representatives from various business functions, IT, suppliers and vendors must be formed who can bring their expertise and knowledge in devising the strategy and roadmap for SMAC implementation.

Start Small and Grow Big

SMAC technology is relatively new and is undergoing an incredibly fast evolution. In an environment where technologies are becoming obsolete and replaced by other disruptive innovations, it is difficult to go for a "Big Bang" implementation approach spanning across multiple years. Simply because, it is challenging to predict that benefits of implementation and there is no guarantee that the solution used will stay relevant during the several years it takes to implement. It is therefore recommended to start pilot initiatives in a specific business division, understand the problems and shortcomings of the solution, get feedback from the users and built the solution such that it is scalable, flexible and technology agnostic.

Create Robust Policies and Best Practices

This is particularly important for social media initiatives since any information shared in social media can be viewed by public. Most of the organizations and institutions actively participating in social media have set ground rules clearly that revolve around the approach of "using sound judgment and common sense" before posting any information. Employees must realize that they represent the company they work for and therefore must share only publicly available information and be cautious of social engineering activities. Clear information security policies must also be laid out on the use of mobile devices in the workplace and exchange of information on the cloud.

Embrace both Internal and External Focused Initiatives

While initiatives involving the customers and suppliers such as setting up social media platform for physicians and patients to collaborate or mobility solutions focusing on patient wellness and care can bring direct business value, organizations must not lose focus on their internal customers – the employees. Virtual collaboration systems for teams and divisions, social media such as Yammer to express their opinions and share ideas, apps to access internal systems via mobile devices, internal cloud based storage to share information among one another can help create a vibrant and productive work environment.

Prep Up Your Internal and External Stakeholders/ Users

It goes without saying that the best of the technologies will fail to create any business impact if not used properly. Many of the SMAC initiatives have the potential to challenge the current ways of working and resistance to change is an obvious phenomenon. Targeted training focusing on the benefits such as better productivity and quality, effective balance of workload, improved transparency at work will help users to embrace the change. It is also essential that employees undergo training on SMAC policies and sign the acknowledgement that they will abide by the terms and conditions. Patient or health professional focused initiatives will also require extensive training through workshops, product launches, webinars and online tutorials to increase its uptake and usage.

Invest in the Right Technologies and Technology Partners

SMAC, after all, is a combination of four technologies which makes technology and IT services an essential part of the overall strategy. These technologies are in different stages of enterprise adoption in the life sciences industries with mobility and cloud computing gaining more prominence when compared with big data and social media. Companies must initiate programs based on its adoption levels of these technologies and gradually move towards enterprise architecture that converges all four technology elements. Also special care must be taken in selecting the vendors who have deep understanding of the industry and its regulatory aspects. Technology partners play an important role as consultants guiding the company in the evaluation and selection of suitable technologies.

Monitoring and Measuring is Key

Whether a social media initiative or a cloud computing platform is meeting its business objectives can be assessed only through proper monitoring and measuring its key performance indicators. Defining KPIs for each initiative and setting up a monitoring team is hence essential for the success of the initiative. Another key aspect of monitoring is with respect to social media where all employee and customer responses must be routinely monitored for adverse events, negative comments, non-compliance and appropriate actions must be taken to address them.

SMAC Architecture

To create a competitive edge over its peers, majority of the life sciences companies considers the adoption of SMAC as one of their main strategies. It is quite challenging for the IT departments of these enterprises to scale up their existing architecture and make it compatible with the new technologies. To implement these technologies successfully, IT architects have to collectively work with their business functions, vendors as well 3rd party service providers. It must be noted that while SMAC as an integrated ecosystem bring in synergistic benefits, from an IT architectural point of view there is no 'one size fits all' architecture that enterprises can implement. Each of the technologies is distinct and requires different technology components in the IT stack. When these technology components come together and integrate among one another to enable seamless exchange of information, the true value of SMAC stack is realized.

Figure 3. SMAC Technologies and its users in Pharma R&D

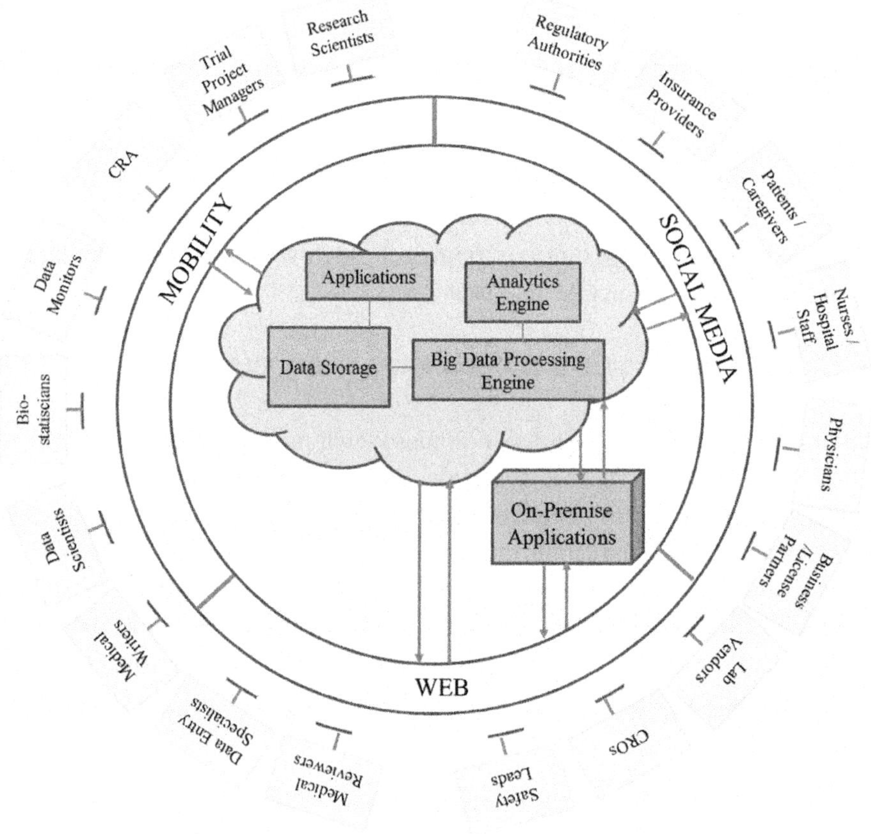

Figure 3 depicts a conceptual view of each of four technologies are connected to one another and exchange information with different stakeholders in the Life Sciences R&D space.

As depicted in the figure, cloud infrastructure forms the platform on which mobile, social media and analytics solution components co-exist. It does not mean that the new technologies will wipe away the existing IT landscape. Especially in the Life Sciences R&D space, there will be lots of software applications such in the research and discovery areas that the business would prefer to be run on dedicated on-premise environments. However, On-premise private cloud setup is a good alternative to dedicated data center environment. Since cloud environments provide easy scalability of computational power which is required to process and analyze large volumes of high complexity data, big data and analytics engines are deployed on the cloud. Mobile platforms and social media platforms essentially act as delivery systems that capture, collect and transfer data to Analytics and transactional systems and sent the processed output back to the consumers. The cloud environment facilitates data exchange through Application Programming Interfaces (API) that integrates with social media and mobile systems.

Figure 4 drills down one step deeper into the SMAC architecture for an R&D organization. It depicts the high level architecture that pieces together solution building blocks of social media, mobile, analytics and cloud computing technologies. This diagram highlights various layers in the IT stack and what each of these layers comprises of.

Figure 4. SMAC Architecture

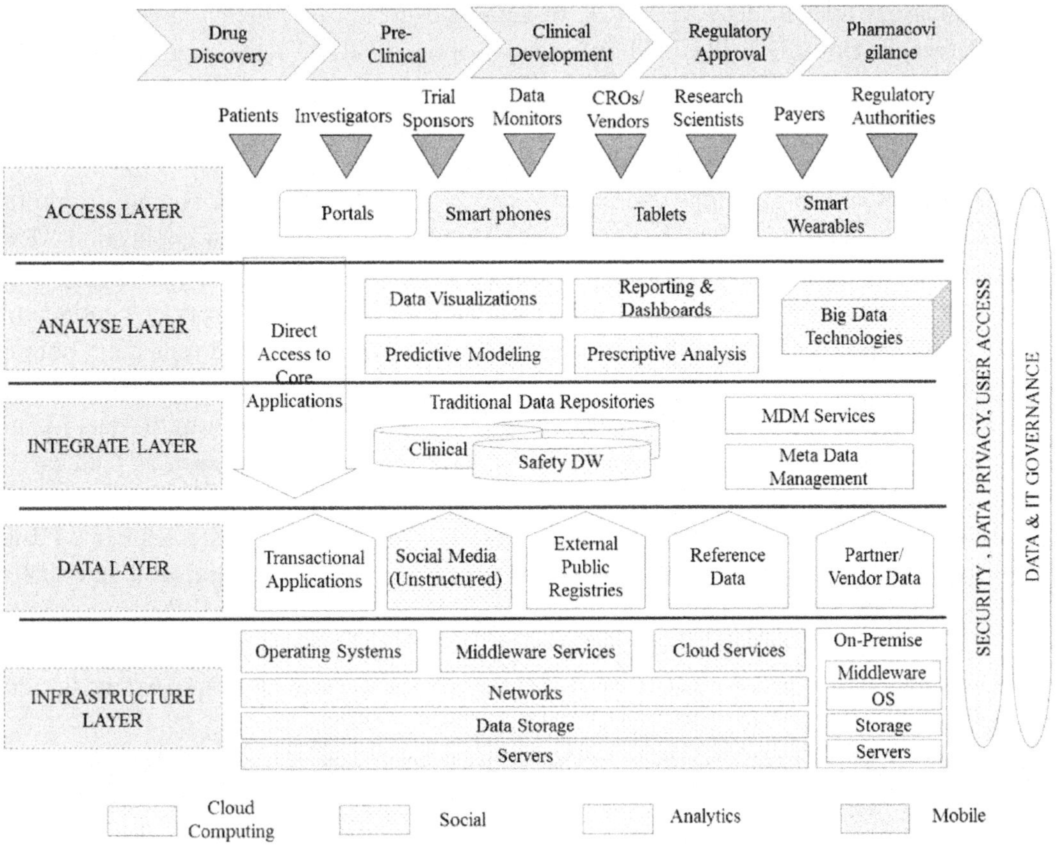

Let us go through each of the architectural layers in further detail.

- **Infrastructure Layer:** It comprises the IT hardware and software components required to operate the hardware and communicate with the other software applications. It forms the platform on which the other layers consisting of data, software applications, user access mechanisms reside. Cloud infrastructure provides easy scalability of its services and provides API interfaces for integrating with other solution components. There are different deployment models such as private cloud, public cloud or a hydrid variety combining the two. Cloud configurations can either be setup and monitored by an experienced 3rd party service provider or by the company's own IT department. On the other hand, on-premise infrastructure is deployed in company owned data center and any modifications to the infrastructure will be the responsibility of the company.
- **Data Layer:** This layer consists of all the data requirements of the R&D organizations. Data flows into the company's data pool through various channels:
  - **Transactional systems:** Data from core systems such as EDC, IVRS, CTMS, AERS, LIMS, chromatographic instruments, genome sequencing systems. Most of the data are structured with the exception of data coming from research lab instruments.
  - **Social Media:** Data from external sites such as facebook, Youtube, health communities as well as internal collaboration sites such as yammer, community portals. Data is mostly unstructured in nature.

- **External public registries:** Data from public data sources such as FDA AERS, Who Vigibase, Clinicaltrials.gov, EMR/EHR data repositories
- **Reference Data:** Data that holds information about the R&D data entities such as GenBank, MedDRA dictionary, WHO Drug dictionary
- **Partner/ Vendor Data:** Data arising from various business partners or CROs such as clinical trial information, Adverse Event cases, laboratory and Image data

- **Integrate Layer:** The main purpose of this layer is to consolidate data coming from different channels into a standardized format that can be consumed by upstream applications. Existing data repositories such as Clinical Data Repository (CDR), Clinical operations data warehouse, safety data warehouse/data marts with a robust meta-data management service continue to play a role in providing a clean standardized data for reporting, analysis and regulatory submission purposes. To handle big data files, NoSQL databases such as MongoDB, Cassandra, HBaseand Hadoop MapReduce, Hive, Hama frameworks can be employed to transform the data for analytical use. Master Data Management platform will have separate MDMs such as Clinical MDM, Safety MDM that provides a single point of reference for the data entities used in R&D.
- **Analyse Layer:** This layer sits on top of the integrate layer and utilizes the aggregated data provided by the integrate layer to develop analytical solutions. This layer comprises data visualization applications such as Spotfire, Tableau, statistical analysis tools such as SAS, R, Matlab to perform data manipulation, regression modeling, simulation, decision trees, clustering tasks and host of other applications to perform various business analytics such web analytics, social media analytics and text analytics.
- **Access Layer:** This layer acts as the interface between users and technology enabling the users to utilize the technology services through various access channels. Access mediums consist of traditional web access through portals as well as newer mediums such as mobile devices and social media websites. Data from the applications and tools are formatted to be compatible with both web and mobile access mediums.

The common services that run across these architecture layers focus on data security, data privacy user access, data governance, workflow management and IT monitoring services. These services are an essential component that enables the IT platform to adhere to its architectural principles of security, reliability, availability and scalability.

KEY CONSIDERATIONS WHEN IMPLEMENTING SMAC IN LIFE SCIENCES BUSINESSES

Architectural Considerations

When gearing up for transforming IT architecture to include the S-M-A-C components, companies must keep the following principles in mind:

Architecture Must be Flexible and Dynamic to Easily Adapt with the Changing Business Needs

Since the digital technologies are fast evolving and businesses are fast incorporating the changing needs in the marketplace, IT architecture must be designed such that applications, services and resources can be easily deployed or modified depending on the business demand. (Vitalari, Strain & Shaughnessy, 2012)

Security Inbuilt for Cloud and Externalization

With applications and services on the cloud that might be hosted off premises with external and internal users connected to the cloud, it is important to build a secured environment with well-defined user access management system. With inbuilt security mechanisms, applications and services can be externalized without comprising on unauthorized access to confidential data and critical systems.

Employ SOA Principles and Technologies for Better Integration and Collaboration

Social media and mobile devices exist outside the boundaries of the organization and hence one must employ agile and loosely coupled service architectures to collaborate with these external services. The type of service oriented architecture to be employed depends on how the services are utilized. A traditional SOA, which uses a synchronous request/response pattern, requires the subscriber of the service to know all information of the service to complete its execution. On the other hand, an event driven SOA is based on an asynchronous pattern where event publisher pushes the events to the subscribers and do not wait for any response. In the current SMAC scenario, an API-based integration model is best suited since businesses are allowing 3rd party developers to build functionalities by exposing their business functionalities through APIs. Through APIs, enterprises can easily integrate relevant applications and reduce the overall development effort to a great extent. RESTful APIs are preferred to SOAP APIs since it is simple, consumes less development effort, flexible to use XML or JSON models, lighter and exhibit better performance. REST is a better choice for both mobile and Web environments. (Banerjee, 2013; Cloud Standards Customer Council, 2013)

Build Application with Access for Multiple Clients

Software applications must be developed such that they can be accessed by both web and mobile clients. This flexibility in access can increase the usage and bring more productivity in the workplace.

IT System and Processes Must Follow All Regulatory Guidelines and Data Privacy Norms

This is perhaps the most important and relevant for Life Sciences industry since violation of any regulatory requirement and data privacy norms will be detrimental to the company's reputation and existence. Where the data resides, how it is stored, how it is consumed and transformed, who has access to the data and so on must be considered while designing an IT solution.

Analytics Inbuilt Than Afterthought

Success of analytics is largely dependent on the availability of high quality data in a consumable form. To achieve this, analytics needs must be considered during the conception stage and data design, data consolidation and its access must be clearly thought through that aids analytics initiatives in future.

Regulatory Compliance Considerations

The progress on embracing SMAC technologies by Life Sciences companies relies on the clarity and confidence provided by the regulatory authorities. Fear of regulatory non-compliance is cited as one of the important reasons of delayed and cautious uptake of these technologies by Life Sciences players. Hence, when incorporating SMAC components in their IT and business functions, proper assessment of the architecture, solution and business process must be conducted from a regulatory perspective.

All the important guidelines set by various authorities have been briefly explained in the earlier section on "Regulatory View". The key considerations to be taken care from a regulatory perspective are summarized below.

Take Responsibility of the Content of Sites That You Own or Control

As per FDA and ICH guidelines, any content posted on the internet sites created, owned or controlled by companies will come under the company's responsibility. Hence adverse events, promotional materials and company's request and responses will come under the regulatory scanner.

Keep a Watchful Eye on any Adverse Event Posts on External Websites that You Monitor

Apart from the company owned online sites, companies must also assess the adverse events posted on other sites that they regularly monitor for any purpose. The regulation states that if any company official becomes aware of adverse events, it qualifies for regulatory reporting.

Adhere to the Guidelines When Requesting and Responding on Social Media

Special care must be taken care when companies either request users (solicited requests) or respond to their queries. Policies on dealing with off label requests in social media are clearly laid out by FDA. Also, when dealing with adverse events, it is prudent for companies to collect the information online and then follow up through traditional communication mediums unless a robust one-on one secure communication channel is established between the patient/ reporter and the company.

Patient Confidentiality and Privacy are Paramount

Interactions through social media, mobile apps and data transfer over a cloud environment have a higher probability of data leakage and unauthorized access. Secured communication channels employing robust encryption technologies and firewalls and differentiated user role access must be in place so that no patient data and Intellectual Property (IP) data is compromised.

Comply with Data Portability Guidelines when Transferring Data Across Regions and Mediums

Regulatory policies such as HIPAA and Safe-Harbour pose restrictions in the storage and transfer of data across geographical regions. Every country has their own data privacy laws that must be strictly abided when dealing with data originated from the respective countries. These laws and policies must be taken into consideration when setting up cloud environments as well as dealing with data emanating from social media sites and mobile apps.

Get Sufficient Regulatory Clearance before launching a Medical Mobile App

Any mobile app and associated platform acting like a medical device will have to undergo product approvals from FDA. In the event of creating apps that resembles any medical device in functionality, it is recommended to seek guidance from FDA and understand the regulatory implications to avoid non-compliance issues and financial loss after launch.

CSV Principles Remain the Same Irrespective of How and Where the Application is Deployed

In the R&D space, all data dealing with clinical and safety data must be validated and must comply with 21 CFR Part 11 policies. Computer System Validation (CSV) must be performed on such applications with appropriate documentation of IQ, OQ and PQ similar to that followed in conventional software application implementations.

Organization's Culture and Change Management

Creating a well-defined strategy and managing a successful implementation of SMAC initiatives within the organization is only half the distance covered to reach the goal. Any new initiative, be it in a business or IT function, is deemed successful only when it has achieved the objective and realized the true benefits. Same is the case with SMAC initiatives also. Since SMAC extends beyond the boundaries of an organization, a robust change management drive must be executed involving both the internal and external stakeholders. Employees will be willing to use social media and mobile phones in their personal lives. But when it comes to work related activities, there will be a big resistance to change the existing ways of working. This resistance to change is obvious and hence planning on better SMAC adoption early on is essential. By empowering employees on the need to adopt disruptive technologies in the changing business climate and inculcating a culture of innovation in the workplace can help in gaining acceptance for change. Providing upfront and open communication from leadership on the newer ways of working highlighting the benefits both for the employees and business, requesting everyone's active participation in the new initiatives and setting up interactive and collaborative forums to discuss and express opinions and viewpoints will prove beneficial. Extensive training on changing rules of the game at the workplace must be planned and executed well in advance of the deployment to minimize any degradation in the quality of work.

Bringing external stakeholders into the fold of the change is equally important. For example a health mobile app will be successful only when large number of patients download and use it in their daily life. A collaboration platform between physicians and patients is considered useful only when sufficient number of physicians and patients interact with one another on a very frequent basis and patients realize the benefits through this interaction. To increase uptake of these new technologies, companies must consider organizing grand product launches, running promotions over various social media channels and roadshows in major physician/ patient networking events, incentivizing members for effective use and conducting workshops to introduce the functionalities and benefits of the initiative. Since health related initiatives are used by people across age groups having varying exposure to technology, it is important to have an efficient helpline/helpdesk service support to address people issues, concerns and feedback. A dedicated service management team comprising of business and IT professionals must be formed to listen to patient/physician queries and experiences and prompt action/response must be delivered adhering to well-defined SLA agreements. External stakeholders will also include business partners and 3rd party vendors and introduction of SMAC technologies will potentially require a change at their end so that the integration of data and services remain undisrupted. Hence, assessing the impact on all external interfaces and invoking appropriate change management controls with vendors and partners must be taken into consideration during the planning stages itself.

SUMMARY AND CONCLUSION

Life Sciences companies are currently faced with many economic, regulatory and research challenges leading to the decline in Pharma R&D productivity. However, the four evolving forces of technology – Social Media, Mobility, Analytics and Cloud Computing, have opened new avenues for transformation which will be driven by better collaboration, improved decision support systems and cost effective operating models. Collaboration among researchers, patients, healthcare professionals, insurance payers and regulators that social media offers for achieving better patient outcomes will revive the much needed growth in the industry. Convenience and agility put forth by mobile devices can help gather the momentum towards better communication and collaboration within the enterprise and beyond. With burgeoning digital data, analytics can uncover the rich insights locked in the data helping companies to discover more successful compounds, minimize risks during clinical trial planning and conduct and perform proactive drug safety surveillance. With rising costs to run the business, cloud computing offers huge opportunity for companies to bring down the IT capital expenditure and operating costs by moving into a "pay-as-you-use" resource model. While benefits galore, SMAC also presents few big challenges for implementation within the Life Sciences enterprise. Regulations around SMAC technology are very nascent which hinders a quick adoption by companies. Fear of data privacy and intellectual property loss in a collaborative environment set by SMAC is a still looming large on the Life Sciences companies.

Over next few years, the four forces will continue to evolve rapidly disrupting the status quo and paving new avenues for consumers and businesses alike. Life Sciences industry among with the regulatory authorities must address the current challenges quickly and become part of the exciting journey laid out by SMAC innovations.

REFERENCES

Allied Health World. (2012). A tweet a day keeps the doctor away. Retrieved from http://www.mediabistro.com/alltwitter/files/2012/12/social-media-healthcare.png

Badger, L.,Grance, T.,Patt-Corner, R., & Voas, J. (2011). Cloud computing synopsis and recommendations. NIST special publication, 800, 146

Banerjee, S. (2013). SMAC: Social, mobile, analytics, and cloud. *Cutter IT Journal*, *26*(2), 27–30. Retrieved http://www.cutter.com/content-and-analysis/journals-and-reports/cutter-it-journal/sample/itj1302/itj1302.pdf

Cloud Standards Customer Council. (2013). Convergence of social, mobile and cloud: 7 steps to ensure success. Retrieved from http://www.cloudstandardscustomercouncil.org/Convergence_of_Cloud_Social%20_Mobile_Final.pdf

Czernicki, B. (2011, February 7). IaaS, PaaS and SaaS terms clearly explained and defined. Retrieved from http://www.silverlighthack.com/post/2011/02/27/iaas-paas-and-saas-terms-explained-and-defined.aspx

Earls, E. (2012, July 19). Clinical trial delays: America's patient recruitment dilemma. Retrieved from http://www.drugdevelopment-technology.com/features/featureclinical-trial-patient-recruitment

European Medicines Agency. (2012). Guideline on good pharmacovigilance practices- Module VI. Retrieved from http://www.ema.europa.eu/docs/en_GB/document_library/Scientific_guideline/2012/02/WC500123203.pdf

European Medicines Agency. (2013). Reflection paper on risk based quality management in clinical trials. Retrieved from http://www.ema.europa.eu/docs/en_GB/document_library/Scientific_guideline/2011/08/WC500110059.pdf

Food and Drug Administration. (2001). Guidance for industry: post marketing safety reporting for human drug and biological products including vaccines. Retrieved from http://www.fda.gov/downloads/BiologicsBloodVaccines/GuidanceComplianceRegulatoryInformation/Guidances/Vaccines/ucm092257.pdf

Food and Drug Administration. (2011). Guidance for industry responding to unsolicited requests for off-label information about prescription drugs and medical devices. Retrieved from http://www.fda.gov/downloads/Drugs/GuidanceComplianceRegulatoryInformation/Guidances/UCM285145.pdf

Food and Drug Administration. (2013). Mobile medical applications-guidance for industry and food and drug administration staff. Retrieved from http://www.fda.gov/downloads/MedicalDevices/.../UCM263366.pdf

Food and Drug Administration. (2013). Oversight of clinical investigations — a risk based approach to monitoring. Retrieved from http://www.fda.gov/downloads/Drugs/.../Guidances/UCM269919.pdf

Food and Drug Administration. (2014). Guidance for industry fulfilling regulatory requirements for postmarketing submissions of interactive promotional media for prescription human and animal drugs and biologics. Retrieved from http://www.fda.gov/downloads/Drugs/GuidanceComplianceRegulatoryInformation/Guidances/UCM381352.pdf

Gartner (2011, November 15). Gartner says sales of mobile devices grew 5.6 percent in third quarter of 2011; smartphone sales increased 42 percent. Retrieved from http://www.gartner.com/newsroom/id/1848514

Harjulampi, V. (2013). *Adopting cloud computing and hosted services in pharmaceutical industry*. Jyväskylä, Finland: Jamk University of Applied Sciences.

Insight, C. C. S. (2013, June 10). Mobile phone sales will hit 1.86 billion in 2013 as strong smartphone growth continues. Retrieved February 15, 2014 from http://www.ccsinsight.com/press/company-news/1655-mobile-phone-sales-will-hit-186-billion-in-2013-as-strong-smartphone-growth-continues

Laney, D. (2001). 3D data management: controlling data volume, velocity and variety. Retrieved from http://blogs.gartner.com/doug-laney/files/2012/01/ad949-3D-Data-Management-Controlling-Data-Volume-Velocity-and-Variety.pdf

Monitoring, S. (2013, September 23). State of mobile 2013. Retrieved from http://www.supermonitoring.com/blog/2013/09/23/state-of-mobile-2013-infographic

Post approval safety data management: definitions and standards for expedited reporting. (2003). Proceedings of ICH-International Conference on Harmonisation of Technical Requirements for Registration of Pharmaceuticals for Human Use. Retrieved from http://www.ich.org/fileadmin/Public_Web_Site/ICH_Products/Guidelines/Efficacy/E2D/Step4/E2D_Guideline.pdf

Proteus Digital Health. (2012, July 30). Proteus digital health announces FDA clearance of ingestible sensor. Retrieved from http://www.proteus.com/proteus-digital-health-announces-fda-clearance-of-ingestible-sensor

We created an SMS campaign to increase HIV/AIDS awareness in Uganda. (2010). Text to Change. Retrieved from http://www.simpill.com/index.html

Vitalari, N., Strain, W., & Shaughnessy, H. (2012). Creating elastic digital architectures. Retrieved from http://www.cognizant.com/InsightsWhitepapers/Creating-Elastic-Digital-Architectures.pdf

Wu, M. (2011). What is gamification, really. Retrieved from http://community.lithium.com/t5/Science-of-Social-blog/What-is-Gamification-Really/ba-p/30447

APPENDIX

Table 2.

Term	Description
21 CFR Part 11	Title 21 CFR Part 11 of the Code of Federal Regulations deals with the FDA guidelines on electronic records and electronic signatures (ERES). Part 11, as it is commonly called, defines the criteria under which electronic records and electronic signatures are considered to be trustworthy, reliable and equivalent to paper records
4G	4G is the fourth generation of mobile telecommunications technology, succeeding 3G. In addition to the usual voice and other services of 3G, 4G provides mobile ultra-broadband Internet access, to laptops with USB wireless modems, to smartphones, and to other mobile devices
API	API or Application Programming Interface specifies how software components must interact with each other
Big Data	A blanket term for any collection of data sets so large and complex that it becomes difficult to process using on-hand database management tools or traditional data processing applications
BYOD	Acronym for Bring Your Own Device, an initiative that enables employees to bring their personal electronic devices to the workplace and perform work related activities
CRA	CRA or Clinical Research Associate, also called as Clinical Monitor, is the person who monitors the progress of clinical trials in various investigative sites coming under his/her geographical area
CRO	Contract Research Organization undertakes outsourcing work by Pharmaceutical, Biotechnology and medical devices companies in the area of research and clinical development
CSV	Computer System Validation is a documentation process assuring that a computer system does exactly what it is designed to do in a consistent and reproducible manner
EMA	European Medical Agency is the Health Authority for European Union region
EMR / EHR	Electronic Medical Record (EMR) or Electronic Health Record (EHR) is an electronic copy of the patient's medical files
FDA	Food and Drug Administration is the Health Authority of USA
Hadoop MapReduce	MapReduce is the heart of Hadoop. It is a programming model composed of two separate and distinct tasks that Hadoop programs perform. The first is the map job, which takes a set of data and converts it into another set of data, where individual elements are broken down into tuples (key/value pairs). The reduce job takes the output from a map as input and combines those data tuples into a smaller set of tuples
ICH	ICH or International Conference on Harmonisation of Technical Requirements for Registration of Pharmaceuticals for Human Use is an international body that brings together FDA, EMA and Japan health authority with an aim to achieve greater harmonisation in the interpretation and application of technical guidelines and requirements used in the research and development of new medicines.
ICSR	Individual Case Safety Report is a safety report submitted to the health authorities comprising the details of an adverse event occurrence in a patient
JSON	JavaScript Object Notation, is an open source data interchange format that uses human-readable text to transmit data objects between a server and web application
MATLAB	MATLAB® is a high-level language and interactive environment for numerical computation, visualization, and programming. Using MATLAB, you can analyze data, develop algorithms, and create models and applications
MDM	MDM or Master Data Management comprises the processes, governance, policies, standards and tools that consistently define and manage the critical data of an organization to provide a single point of reference
NoSQL	NoSQL or Not Only SQL database provides a mechanism for storage and retrieval of data that is modeled in means other than the tabular relations used in relational databases

continued on following page

Table 2. Continued

Term	Description
PK/PD	PK stands for Pharmacokinetics and PD stands for pharmacodynamics. Pharmacokinetics is defined as the study of the time course of drug absorption, distribution, metabolism and excretion in the body. Pharmacodynamics refers to the relationship between drug concentration at the site of action and the resulting effect, including the time course and intensity of therapeutic and adverse effects.
R	R is an open source statistical analysis tool
R&D	Research and Development
RBM	Risk Based Monitoring is a new risk based approach to monitor the quality and safety aspects of a clinical trial
REST	REST stands for Representational State Transfer. It relies on a stateless, client-server, cacheable communications protocol such as the HTTP protocol. REST is an architecture style for designing networked applications. The idea is that, rather than using complex mechanisms such as CORBA, RPC or SOAP, simple HTTP can be used to make calls between machines
SAS	SAS or Statistical Analysis System is a software suite developed by SAS Institute for advanced analytics, business intelligence, data management, and predictive analytics
SDV	Source Data Verification is a review process performed during clinical trial monitoring to ensure that the patient data has been accurately captured from the source document into the clinical database.
SMAC	Acronym for Social Media, Mobility, Analytics and Cloud Computing
SMS	Simple Messaging System is a mobile technology to transfer short text messages
SOA	Service Oriented Architecture is an architectural design pattern that supports communications between services. SOA defines how two computing entities, such as programs, interact in such a way as to enable one entity to perform a unit of work on behalf of another entity. It requires a service provider, mediation, and service requestor with a service description.
SOAP	Simple Object Access Protocol is a protocol specification for exchanging structured information in the implementation of web services in computer networks. SOAP based Web services are one of the most common implementation of SOA.
XML	XML or Extensible Markup Language is a markup language that defines a set of rules for encoding documents in a format that is both human-readable and machine-readable. It is one of most preferred means to transport and store data because of its simple, easy to understand structure.

Chapter 7
Architecture of an Integrated Collaboration Portal for Clinical Trial
A Case Study

Partha Chakraborty
Cognizant Technology Solutions, India

ABSTRACT

Collaboration is defined as the actions for individuals and teams to work together for a common goal. There are several bottlenecks to an efficient and effective collaborative model of clinical trial including: the lack of a centralized, consistent, globally accessible platform to manage and store essential study related documentation; inconsistent or incomplete work assignments; inefficient notification of key events requiring follow-on action; and incomplete, missing, expired, or redundant documentation and training activities and need to maintain multiple credential to access various system, Removing these barriers is an important part of establishing an environment that fosters collaboration among all constituencies involved in managing clinical trial keeping them connected, informed, and on task by providing access to everyone at any time, from anywhere.The case study below introduces need of an integrated clinical collaboration platform, addressing key functionality of such an platform and describes the architecture & design consideration to industrialize such a platform. The intended audience of this case study is the architects & designers of similar systems. The clinical trial activity for a drug in research is approximately 70% of the overall drug development cost. It is estimated that 4% of the cost of a trial is in 'rework' involving communication, regulatory issues, patient enrollment, document review and replacement of patients. The integrated clinical collaboration platform has potential to eliminate significant amount of cost of re-work, which is in order of $3.5M per trial.

1. INTRODUCTION

In the Clinical trial process an Investigator plays a pivotal role in the documentation, compliance and procedures involved in a trial. To run a successful trial an investigator is anticipated to:

DOI: 10.4018/978-1-4666-8726-4.ch007

- Collaborate with various stakeholders and team in effective and efficient manner .
- Access multiple discrete applications like clinical trial management system (CTMS), procurement systems, Enterprise resource planning (ERP)systems, Electronic data capture (EDC) systems, learning management systems, patient recruitment information and Document management system (DMS)- to track, trace and update information
- Fill battery of time consuming forms like FDA forms 1572 and equivalent, financial disclosure forms, investigator agreement form, confidential disclosure form and facilities description form- to be compliant
- Communicate and coordinate to numerous stakeholders like Clinical Research associate, multiple stakeholders within sponsor organization like business sponsor of the trial, Regulatory affairs personnel, physicians, IT personnel etc.
- Manage grants and payments to site co-coordinators, patients etc. The process is time intensive and involves lots of stakeholders for auditing, receipt acknowledgement, records management etc. Keep a tab of trial master file (TMF) - contains every piece of essential information associated with a trial under Good Clinical Practices (GcP). Management of a paper TMF is resource intensive; documents are handled by multiple people from collection at the investigational site to placement in regulatory binders. TMF documents are tracked manually using spreadsheets or checklists that provide little visibility, often, causing duplication of effort. This causes decreased operational efficiency, higher costs, and the risk of non-compliance, and possibly approval delays
- Keep oneself updated and complaint with the trainings and certifications as well as the profiles and resumes

Due to wide array of activities and diverse spectrum of work, often plagued redundant or missing information causing rework, escalating the time and cost of the Clinical trial. Research found that large and mid size Pharmaceutical companies may have more than 15,000 investigators engaged at a given point in time.With the activities listed above and the associated challenges, the reduction in rework cost per trial is a good enough business case for Pharmaceutical companies to implement an integrated solution. Furthermore, shrinking pipelines, pharmaceutical companies are increasingly recognizing the value of timely communication, simplified process and strong and lasting relationships with investigators and site personnel to improve clinical trial execution, reduce rework cost and to ensure the ongoing success of clinical programs.

2. INTEGRATED CLINICAL COLLABORATION PLATFORM

An integrated clinical collaboration platform is one of the key solutions to combat the challenges/costs listed above. Such a portal solution must have following components:

- *Self registration* of the expternal collaborators and access to multiple applications through a single sign on platform are one of the key features. In past one decade, globalization, specialization, and outsourcing have changed the way clinical trials are conducted. In order to support this transformation, corporate IT is being asked to provision collaboration with individuals outside the organization. Automating the user account provisioning process eliminates the need for corporate

Figure 1. Functions of the integrated clinical collaboration platform

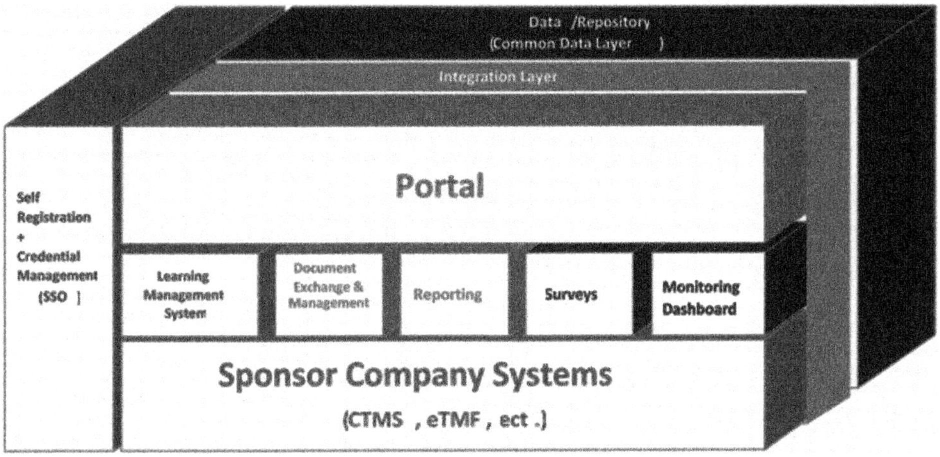

IT intervention and improves the efficiency of opening the clinical collaboration environment to extranal users. This automation is accomplished through a workflow that routes extranet user self-registration requests to the Clinical Program Manager for review, credential verification, and approval. Once the request is approved, the workflow sends an e-mail to the registrant containing a link to the collaboration workspace, a system generated username, and a temporary password. When the extranet user logs in for the first time, they are re-directed to an area of the workspace where they must first complete the prerequisite tasks and change their password before gaining complete access. Through automation, the administrative burden typically associated with workspace creation and external user account provisioning by corporate IT is eliminated and another barrier to effective clinical trial collaboration is removed.

- *Document exchange & management* and information repository: Many sponsor organizations adopts method for storing and exchaning trial documentation using shared drives and e-mail. This method cannot scale to meet the complexity of large global clinical trials. As the volume of clinical trial documentation is growing, resulting in increased management costs and the need to disseminate important information both quickly and efficiently. Apart from helping exchange and storage of documents, the integrated clinical collaboration platform can provide online completion and submission of various forms, avoiding laborious paper documentation. This also providies recruitment and screening tools, automating scheduling, managing finances, and providing the reporting and metrics needed for business oversight and growth.
- Provisioning of *clinical collaboration workspace* to different users: The clinical collaboration workspace simplifies the management of essential trial documentation through implementation of workflow by effectively integrating document management processes with the way people work on a daily basis. Users can collect and collate information in one place, avoiding duplicates with version control. . A key aspect of electronic workflow is the inclusion of *automated alerts and notifications*. Both alerts and notifications are used to keep those involved with the collaborative process aware of what is happening on the trial. Alerts however, require the recipient to complete an action while notifications are usually informational in nature. Both can leverage e-mail or SMS technology to communicate with the recipient.

- An *electronic trial master file* (eTMF), which will collect essential trial documents electronically in a central location and makes them available to disparate constituencies, via the Internet, from any location at any time.
- Provide a common *training and certification* rendering services
- Provide updated and timely information globally equipping all stakeholder with informed knowledge through *dashboard* and have all stake holders on a single page by effectively keeping a tab on trial progression, oversight facilities for the sponsors
- Online *Grants management system* enabling streamlined, tracking and distribution. Keeping the fund allocation within budget and on time.

3. BUILDING ARCHITECTURE OF INTEGRATED CLINICAL COLLABORATION PLATFORM

This chapter addresses elements of building architecture of intehrated clinical collaboration platform and key design consideration associated with it.

The chapter is intended for various types of audiences with diverse priorities, objectives and technical perspectives including business analysts, IT architects, IT designers and developers. Given this diversity, this chapter seeks to provide a high level and focused view on the architecture capabilities and definitions..

The key sections of this chapter are:-

- Section 5 highlights the key architectural guiding principles that clinical collaboration platform should adhere to, throughout the design and implementation in order to create a robust, scalable and extensible solution.
- Section 6 elaborates an illustration of use of custom off the shelf product in the architecture. It is important to note that use of specific product is for illustration only. Author does not intend promote any specific product for this purpose,
- Section 7 & 8 elaborate the enterprise view and technical view of the platform. The architecture views will provide for a comprehensive and layered architecture with all the required functional and system components that will enable a robust, scalable and extensible architecture for the platform. In particular the architecture will capture the key architectural elements for integration, modularity, service reuse and security for the integrated clinical collaboration platform. These section provide in detail the various core components of the Product Architecture – Portal, Security, and Integration Architecture views. For each of the views the respective sections provide elaboration for all the key components of the architecture and its interactions, dependencies. Also the architecture views will demonstrate how the functional scenarios/use cases (and system use cases) are realized through the core components (and its interaction with other components) that take up the architecture.
- Section 9 elaborates consideration of deployment of the integrated clinical collaboration platform.
- Appendix A elaborates in detail the key architecture and design decisions, motivation, implications and dependencies. The architecture and design decisions has been rationalized based on the specific functional and non-functional requirements of the integrated clinical collaboration platform.

4. ARCHITECTURE GOALS AND CONSTRAINTS OF INTEGRATED CLINICAL COLLABORATION PLATFORM

Following section describe the key architecture requirements and constraints those have a significant bearing on architecture, design and building the integrated clinical collaboration platform (Bass L., Clements P., & Kazman R. (2005) *Software Architecture in Practice: Second Edition)*

5. USE OF OPEN SOURCE AND COMMERCIAL CUSTOMIZED OFF THE SHELF PRODUCT / SOLUTION IN THE ARCHITECTURE

Industry has witnessed new & innovative products and solution in the emerging areas of technology in form of social media, collaboration platform, security solution, analytics and SaaS / cloud solution. The new integrated clinical collaboration platform needs to utilize these innovative solutions. As it is eveident from description of functional & architectural goal of the platform, specific solution is required in the area of self registration and security, portal technology, middleware, learning management, databse technology, document management and cloud technology. The case study mentions specific solution, availalble in this area, It does not intend to say those specific solutions are the best in the area and it

Table 1. Architecure goals of the integrated clinical collaboration platform

Category	Goal/Constraint
Service-Oriented	Architecture is to be built on the fundamental principles of service orientation promoting ability to independently build, assemble, deploy and consume specific business services for varied stakeholder (sponsor users, Investigators and later on patients) needs. The principles of service orientation thereby ensures an extensible architecture that is loosely coupled allowing high levels of service autonomy and composition
Extensibility	The platform needs to provide for an extensible architecture allowing seamless integration with new solutions coming in the industry and other sponsor systems as well as enabling newer services and components to be seamlessly plugged into the platform.
Scalability	The platform will provide for scalability at the various layers – web, app and persistence layers using a combination of mechanisms that ensure scalability and elasticity such as Load Balancing and Clustering services, virtualization.
Security	The platform architecture will enable a federated security model ensuring message confidentiality and message integrity during message conversation between two or more end points. For example as part of realizing the stated functionality of a sponsor user or an Investigator during clinical planning functions, the message conversations will involve more than two end points.. It is expected that message conversations (transmissions) will occur over Internet as secured message transmissions. The architecture is to comply with SAML 2.0 authentication scheme enabling web-based authentication and authorization scenarios for sponsor users and Investigators as well as cross-domain single sign-on (SSO), which will help reducing the administrative overhead of distributing multiple authentication tokens to the user. The architecture is to comply with industry specific regulatory requirements and compliance such as CFR – Part 11
High Availability	Integrated clinical collaboration platform and other solution in its eco-system is to ensure high availability of its resources (service instances, data) as the services will be deployed and delivered for global users. A integrated platform must be able to handle up to 200,000 global users..
Response Time	The architecture needs to ensure reasonable response times for regular transactions when Investigators, sponsor users perform clinical study planning, and document exchange.
Build vs. Buy	Integrated clinical collaboration platform uses product / solution (COTS) already available in the industry.

does not intend to promote these solutions. Solutions with similar capabilities can certainly be used to create similar architecture and commission an industrialized solution.

Use of open source technology can assist to bring the overall cost of ownership to bring down. This case study has referenced to certain set of open source technology. Sepcfic steps need to be adopted to ensure that platform remains compliant with regulatorary guideline

The table as described below provides for the various software products, platforms, technology stacks and frameworks that will be used to realize the integrated clinical collaboration architecture.

Security, self registration form a significant part of the integrated clinical colloaboration solution. This can be accomplished by using commercial off the shelf solution. The table below outlines different components of such solution.

Table 2. **COTS products** *refered in the architecure of integrated clinical collaboration platform*

Component	COTS Product / Technology	License (commercial / open source)	Description
OS	Red Hat Enterprise Linux	Commercial Open Source License	Operating system for the environment.
Portal	Liferay Portal	Commercial Open Source License	Liferay portal enterprise edition for version 6.2.1 with service pack. Used to build the integrated clinical collaboration platform
Language / VM	Java Runtime Environment	Open Source	This is the runtime environment of java, which would be used for creating the java virtual machine (JVM).
Database	Oracle	Commercial	Database to store the information
Application Server Platform	JBoss EAP	Commercial Open Source	JBoss Enterprise Application Platform 6.1.0 comes with JBoss application server 7.2. This application server acts as J2EE web container, where Liferay components will be deployed.
Web Server	Apache HTTP Server	Commercial	Apache HTTP server acts as web server, where first the https requests arrive and then the request is forwarded to the Liferay. Apache HTTP server would act as both static content storage (images, JavaScript, CSS files etc.) and as HTTP request load balancer.
Routing & Mediation engine	Apache Camel	Commercial	For out-of-box adapter component support, support for all relevant integration patterns and to serve as the core integration framework
Message Queue	JBoss A MQ	Commercial Open Source	One of the most optimized messaging providers today. Will be uses as the internal messaging system for ensuring message reliability and scalability
Search	Lucene	Commercial	Search engine
Self registration and security	CA solution	Commercial	Single Sign On, Self Registration, Credential Management, Level 1-2-3 security.
Learning Management Solution	SumTotal	Commercial	Learning management & training

Table 3. Components with in security solution

Component	COTS Product / Technology	License (commercial / open source)	Description
Access Manager/Federation	CA SiteMinder	Commercial	CA SiteMinder Secure SSO & Flexible Access Management can provide enterprise-class secure single sign-on (SSO) and flexible identity access management so that integrated clinical colloaboration platform can authenticate investigators, sponsor users, other site users and control access to applications and services deployed on member companies' trusted network. CA SiteMinder with Federation capability will be used for securing access to Portal and LMS
Policy Store	CA Directory	Commercial	The SiteMinder policy store (policy store) is an entitlement store that resides in an LDAP directory server. The purpose of this component is to store all policy-related objects, including the: ▪ Resources SiteMinder is protecting ▪ Methods used to protect those resources ▪ Users or groups that can or cannot access those resources ▪ Actions that must take place when users are granted or denied access to protected resources
Web agent +Web agent option pack	CA Secure Proxy Server	Commercial	This component is used as web agent and also as proxy engine for Reverse Proxy
IDM	CA Identity Minder	Commercial	IDM is used for provisioning the users and to maintain work flows
Provisioning	CA Provisioning Server	Commercial	Provisioning server will be used by IDM for provisioning the users to end points.
Provisioning Directory	CA Directory	Commercial	Provisioning Directory will be used for storing the users
Reports	CA Reports Server	Commercial	Used by IDM and SiteMinder to generate reports

6. HARMONIZATION OF CLINICAL COLLABORATION: PROPOSAL FOR A CONSORTIUM BASED APPROACH

The Clinical Data Interchange Standard Consortium (CDISC) started in 1997 as an open, multidisciplinary, neutral,501(c)(3), non-profit standards developing organization (SDO) that has been working to develop global standards and innovations to streamline medical research and ensure a link with healthcare. This has created open standard for clinical data storage, exchange, submission and display through SDTM, ODM and CDASH standard. This has helped esbalishing better method for exchaning information among mutiple technology in eClinical framework. The simailar approach and method can be used to to harmonize clinical collaraboration with in the industry. That can lead to creating efficient interaction among all the stakeholders in the collaboration process and results in to complete digitilization of the process.

At its core the integrated clinical collaboration platform is a system of engagement for different stakeholders participating in clinical study planning and monitoring functions enabling the stakeholders to connect with the platform from anywhere any time. The collaboration platform can provide for a shared, cross-industry, web-based investigator portal specifically designed to improve communication among users. Through the development of this common interface, Idea can be to streamline and standardize investigator and clinical trial site access, while enhancing communications through the harmonized delivery of content and services. Additionally, in terms of economic benefit, a shared platform is expected to result in substantial cost savings to individual companies by reducing the need for companies to run their own portals (which currently have different looks and feel), and require site-specific training for users.

7. WHAT IS AN ARCHITECTURE PATTEN?

An architectural pattern is a general, reusable solution to a commonly occurring problem in software architecture within a given context. Architectural patterns are similar to software design patterns but have a broader scope. The architectural patterns address various issues in software engineering, such as computer hardware performance limitations, high availability and minimization of a business risk. Some architectural patterns have been implemented within software frameworks. Even though an architectural pattern conveys an image of a system, it is not an architecture. An architectural pattern is a concept that solves and delineates some essential cohesive elements of a software architecture. Countless different architectures may implement the same pattern and share the related characteristics. Patterns are often defined as "strictly described and commonly available". For example, the layered architecture is a call-and-return style because it defines an overall style to interact. When it is strictly described and commonly available, it is a patternAnonymous(2015) Architectural Pattern,Retrieved from Architectural Pattern Wiki:https://en.wikipedia.org/wiki/Architectural_pattern

Integrated clinical collaboration platform requires combination of architectural patterns & design patterns. Both functional as well as the architectural goal need consideration before elaborating in to specific patterns, those are required to commission a platform like this. A detailed description of design consideration are described in Appendix A. A design cosnideration is described with the specific topic, description, assumptions & constraints, motivation, implications, dependencies, alternatives & recommendation.

8. KEY ARCHITECTURE PATTERNS USED FOR CLINICAL COLLABORATION PLATFORM

The platform architecture is composed of the following key architecture tenets and patterns:

- The platform architecture applies the principles of service orientation to model components and services that will directly enable the business processes and functional features of the integrated clinical collaboration platform. Given the integrated clinical collaboration platform to be used as "the system of engagement" for different stakeholders such as investigators, sponsor users and later on patients, the principles of service orientation enables composing reusable business services for clinical study planning, site collaboration and thereby ensuring an extensible architecture.
- The platform architecture applies the principles for loosely coupling, message orientation to enable service based integration to integrate clinical collaboration platform with member systems (such as CTMS).

- The platform architecture provides for orthogonal services for unified security and governance for the integrated clinical collaboration platform

8.1 Portal Services

The primary components of the portal are addressed in sections below:

User Access Layer

The user access layer defines the different roles in the clinical collaboration platform will specific access control levels. Following are sample roles.

- Trial Administrator
- Trial Research Coordinator
- Primary Investigator
- Clinical Research Associate
- Portal Administrator

Presentation Layer

The portal can be implemented using multiple sections. These portal sections would serve the following:-

- Displaying multiple web contents.
- Custom or out of the box portlets, which would actually perform the business implementation of different functional requirements for platform.
- Built in UI libraries, themes and layout services.
- Plugin for customizing Liferay server behavior.
- Document and Media library portlet for managing static documents and media files.
- Report and graph portlets depicting data in graphical and textual reports and charts or graphs.
- Search engine for searching content within the portal using keywords. Lucene search engine is out of the box feature of liferay portal.
- Search implementation for study related, site related and member company related data searching.
- Access control, Login and User Profile for management of users' information.

Content Management System and Document Library

Portal provides for a rich content management system (CMS) and document library and media. This is required as platform document library would be extensively used for implementation of "Document Dropbox". CMS would be leveraged for displaying static web contents. Following are the few key features of document library and CMS that can be used. -

- Document Storage provides for a scalable storage option and enables documents to be stored according to specific hierarchical location.

- Taxonomy is tagging of documents, help to provide tag names against documents. This helps to search document against keywords.
- Categorization helps to organize documents within the storage, as well as to help during searching against keywords.
- Workflow is Liferay's out of the box feature, which enables reviewing of documents, web contents.
- Authoring is associated with web content, where business user can create web content through rich text editor.
- Versioning of web contents and documents help to keep the old documents within the portal in an organized manner.
- Publishing of web contents takes place after reviewing the written content by reviewer.
- Remote staging is another feature by which administrator can publish contents from authoring system (environment) to the production environment. This mechanism helps a lot for reviewing, testing, validating of look and feel of contents before publishing.

Portal Custom Frameworks

Following are a few custom java based frameworks that would be built which are common for all portlets that will built and deployed for the integrated clinical collaboration platform -

- Task framework - Framework which enables easy creation of tasks, management of tasks at any time. All business objects can access the task interfaces to search, create, update, delete individual task and can retrieve list of tasks.
- Logging framework enables different kinds of logging throughout the portal application.
- Auditing framework enables to capture user events at any point of the state of portal application.
- Portal Scheduler framework enables to manage scheduling batch jobs within the portal application.
- Exception handling framework is another important framework which takes care of exception handling at any level of portal application.
- Security framework components will provide for overall portal security related aspects like cross site scripting, request forgery etc.
- Message listener framework to get transactional push messages from ESB layer. In this scenario member company system sends information to ESB layer and in ESB layer puts these messages to message queues, which is accessible from portal.

8.2 Integration Services

The Integration Architecture view provides a comprehensive integration approach and mechanism to integrate the integrated clinical collaboration platform with member company systems such as CTMS. The integration architecture will provide for a standardized API (as RESTful services) enabling sponsor companies to consume the interfaces to integrate the same with their CTMS systems

Integration Architecture will serve to provide the necessary abstraction (in terms of business services/components and data services) for the the integrated clinical collaboration portal layer components while executing business functions such as clinical site study planning, site collaboration, document exchange

The Integration Architecture is realized by using ESB layer and framework which provides for both design time and run time APIs for generation and management of RESTful services

8.3 Secuirty Services

In this case study, the security system architecture view for the integrated clinical collaboration platform is realized via enterprise security products/components to deliver the security elements as indicated below.

The case study provisions architectural scenario where sponsor users are authicated via in-house security system (example: CA) and site users, who are external users, are authicated via externized security system (example: Exostar)

The following list of steps depicts security interaction for site users, external to the sponsor organization:

1. Site User tries to access the clinical collaboration portal.
2. SPS displays page to user with list of IDP's
3. Site Users selects Exostar from list of IDP's.
4. SPS will redirect the user to the Exostar.
5. Exostar will display the login page to the site user to enter his/her credentials.
6. Site user enters his/her credentials on the login page and submits.
7. Exostar authenticates the site user.
8. Exostar generates SAML and post the SAML to assertion consumer Service URL of Policy Server.
9. SPS forwards the SAML to policy server.
10. Policy server validates SAML and generates SM Session for user.
11. Policy sever authorizes the site user against Portal DB.
12. Policy server sends the user profile attributes required by the platform in header.
13. SPS will redirect the user to the platform along with the header.

The following list of steps depicts security interaction for sponsor users, internal to the sponsor organization

1. Sponsor User tries to access the the clinical collaboration portal.
2. SPS displays a page with List of IDP's.
3. Spnsor User selects Sponsor IDP from the list of IDP's
4. SPS redirects the user to Sponsor IDP.
5. Sponsor IDP will display the login page to the sponsor user to enter his/her credentials.
6. Sponsor user enters his/her credentials on the login page and submits.
7. Sponsor IDP authenticates the sponsor user.
8. Sponsor IDP generates the SAML and post the SAML to assertion consumer URL of Policy Server.
9. SPS forwards the SAML to policy server.
10. Policy server validates SAML and generates SM SESSION for user.
11. Policy server will redirect the user to the platform.

9. ARCHITECTURAL CONSIDERATION FOR DEPLOYMENT OF INTEGRATED CLINICAL COLLABORATION PLATFORM

The integrated clinical collaboration platform architecture is divided into multiple layers. The request from the browser will be redirected to a load balancer. The load balancer will redirect the traffic to one of the Apache web server cluster nodes. The request will be processed by the web server, and then will be directed to another load balancer which will route to one of the application servers based on current load. Load balancer and clustering of web/application servers would be done to ensure high degree of scalability, performance and availability of the system. The deployment architecture of the integrated clinical collaboration platform needs to support all the functional and architectural goals stated earlier including high availability and scalability.

In order to make specific architectural provisioning for high availability and disaster recovery, uptime, recovery point objective (RPO) and recovery time objective (RTO) need to be set. RPO is the maximum tolerable period in which data might be lost due to an incident. RTO is the duration of time and service level within which the system must be restored after a disaster. Combination of RPO, RTO, availability goal and anticipated load in the system will determine exact number and size of the servers and equiptment.

Clustering and load balancing are the features of the deployment architecture.They ensures architectural goal of high availability and disaster recovery.

A cluster is a set of nodes (servers) that communicate with each other and work toward a common goal. In the above depiction of deployment of the clinical collaboration platform

- The web server cluster consists of multiple Apache httpd servers
- The app server cluster consists of multiple JBoss Application servers for life ray portal & ESB layer and
- The persistence layer has an active-passive Oracle server cluster. .In case, the active primary server goes down, the passive secondary server will be promoted as a primary until the primary instance comes alive.

The integrated clinical collaboration platform implements a load balancer to moderate the requests hitting the web and app servers. No one web or app server will be fully utilized when the other is available for utilization. The web browser sends in requests and receives responses directly over the wire using the HTTP protocol. A load balancer is required to process all requests and dispatch them to server nodes in the cluster.

State replication is directly handled by JBoss. When JBoss is run in the *all* configuration, session state replication is enabled by default and is replicated across all JBoss instances in the cluster. F5 load balancer is recommended to use for the platform.

Following consideration needs to be made for the load balancer:

Oracle Data Guard ensures high availability, data protection, and disaster recovery for enterprise data. Data Guard provides a comprehensive set of services that create, maintain, manage, and monitor one or more standby databases to enable production Oracle databases to survive disasters and data corruptions. Data Guard maintains these standby databases as transactionally consistent copies of the production database. Then, if the production database becomes unavailable because of a planned or an unplanned outage, Data Guard can switch any standby database to the production role, minimizing the downtime associated with the outage. Data Guard can be used with traditional backup, restoration, and

Table 4. Load Balancer Architecture Considerations

Security	A firewall is deployed in front of the Load Balancer
Availability	Use of at-least two load balancers in a HA configuration to reduce single point of failure.
Session Persistence	Session persistence should be enabled using SSL
Algorithm	Least Connections, Round-Robin is generally acceptable
Web application Firewall	Can be used to apply URL restrictions on the vCloud load balancer access to admin based on source address
Advanced configurations	Can be used to allow certain types of client traffic to dedicated nodes

cluster techniques to provide a high level of data protection and data availability.With Data Guard, administrators can optionally improve production database performance by offloading resource-intensive backup and reporting operations to standby systems.

10. CONCLUSION

It is evident that the cost of clinical trials continues to escalate. Collaboration among site, sponsor and related stakeholders can increase operational efficiency, which is a key component of cost control / reduction and speed of the trials. Advent of better digital media, collaboration and cloud technology and acceptance of externalization of IT infrastructure can make process of collaboration real. Industry wide effort to make the process harmonized shall make adoption of technology much master.

REFERENCES

AMR Clinical Metrics Study. (2008). BearingPoint.

Anonymous(2015), Oracle Help Centre-Oracle Database Online Documentation, 10g release 2 (10.2) / Administration - Data Guard Concepts and Administration(p. 9).Retrieved from: http://docs.oracle.com/cd/B19306_01/server.102/b14239/concepts.htm

Anonymous,(2015).501© Organization,Retrieved from 501© Organization Wiki: https://en.wikipedia.org/wiki/501(c)_organization

Anonymous,(2015).Standards Organization,Retrieved from Standards Organization Wiki: https://en.wikipedia.org/wiki/Standards_organization

Architectural Pattern. (2015). Wikipedia. Retrieved from https://en.wikipedia.org/wiki/Architectural_pattern

Avgeriou, P., & Zdun, U. (2005, July). Architectural patterns revisited:a pattern language. Proceedings of the 10th European Conference on Pattern Languages of Programs (EuroPlop 2005). Irsee, Germany.

Bass, L., Clements, P., & Kazman, R. (2005). *Software Architecture in Practice* (2nd ed.).

Buschmann, Meunier, Rohnert, Sommerlad & Stal (1996). Pattern-Oriented Software Architecture: A System of Patterns.

Oracle Help Centre-Oracle Database Online Documentation (p. 9). (n. d.). Oracle. Retrieved from http://docs.oracle.com/cd/B19306_01/server.102/b14239/concepts.htm

KEY TERMS AND DEFINITIONS

Clinical Collaboration Workspace: The clinical collaboration workspace is a workflow management tool that eases the management of essential trial documentation by effectively integrating document management processes with the way people work on a daily basis. This electronic workflow also enables the inclusion of automated alerts and notifications

CTMS: A clinical trial management system (CTMS) is a software system used by biotechnology and pharmaceutical industries to manage clinical trials in clinical research. The system maintains and manages planning, performing and reporting functions, along with participant contact information, tracking deadlines and milestones.

EDC: An Electronic Data Capture (EDC) system is a computerized system designed for the collection of clinical data in electronic format for use mainly in human clinical trials. EDC replaces the traditional paper-based data collection methodology to streamline data collection and expedite the time to market for drugs and medical devices.

Electronic Trial Master File (eTMF): An electronic trial master file is a formalized means of organizing and storing documents, images and other digital content for pharmaceutical clinical trials that may be required for compliance with government regulatory agencies.

Integration: Integration in the chapter referes to the exchange of data and services of the Clinical collaboration portal with other corporate systems and system of records

Portal: A *Web portal* or *public portal* refers to a Web site or service that offers a broad array of resources and services, such as e-mail, forums, search engines, collaboration and document exchange. An *enterprise portal* is a Web-based interface for users of enterprise applications. Enterprise portals also provide access to enterprise information such as corporate databases, applications (including Web applications), and systems.

Site Monitoring Visits: Study sites are monitored to ensure oversight of the clinical research study by the sponsor. Regular site monitor visits can be broken down into four types: pre-study visits, initiation visits, periodic monitoring visits, and close-out visits. Study sites may also be monitored or audited by the FDA, Clinical Research Organizations (CROs), IRBs and sponsors

APPENDIX

Key design considerations of architectural components for the integrated clinical collaboration platform This appendix describes various design considerations and illustrates a structure on how these considerations can be documented, mentioning topic, description, assumption & dependencies, motivation of the choice, alternatives and implications. These considerations are in relation to the case study mentioned in this chapter. Practioners need to make similar considerations, which are apt to their architectural situation. Buschmann, Meunier, Rohnert,Sommerlad & Stal(1996). Pattern-Oriented Software Architecture: A System of Patterns.

Table 5. Acronyms and Definitions

Acronym	Meaning
CA	Computer Associate's
CDL	Common Data Layer
CTMS	Clinical Trial Management System
DMZ	Demilitarized Zone
IDM	Identity Minder
IDP	Identity Provider- A SAML authority that generates an assertion for use by a relying party (Service Provider). Identity Provider creates, maintains, and manages identity information for users and provides user authentication.
OTP	One Time Password
PAP	Policy Administration Point
REST	Representational state transfer
SAM	Secure Access Manager
SOAP	Simple Object Access Protocol
SAML	Security Assertion Markup Language
SP	Service Provider- A SAML entity that uses information from Identity Provider to provide access to services. The Service Provider uses SAML assertions from an IDP to authenticate and authorize the user.
SPS	Secure Proxy Server
SSO	Single Sign On
XML	Extensible Markup Language
CCD	Common Cookie Domain

Table 6. Design Considerations: Service Orientation

Area	Design & Construction
Topic	**Service Oriented Approach in designing and building the integrated clinical collaboration platform**
Description	The integrated clinical collaboration platform architecture will be built on the core principles of service orientation enabling an extensible architecture that enables reusable business and entity services to be independently built, assembled, deployed and consumed by stakeholders in varied business contexts and service delivery models. This will provide the following benefits to clinical collaboration stakeholders – - Service Autonomy - The integrated clinical collaboration platform architecture will enable reusable business and entity services such as Facility Site Study Planning function, which will be delivered and consumed as autonomous services (with absolutely no dependencies with other services/functions in the architecture) - Service Composition and Federation - The integrated clinical collaboration platform architecture will ensure business and entity services such as for Facility Site Study Planning function to be composed and re-composed (re-assembled) based on varying needs/requirements that could emerge in the future. For example it is possible that a sponsor may want to use some of the on premise applications/services for Facility Study Planning function and at the same time maximize the benefits of the services that are available in integrated clinical collaboration platform. Service Composition as a principle will allow integrated clinical collaboration platform at runtime to recompose newer Facility Site Study Planning services aggregating parts of study planning functionality that is available as sponsor on-premise services along with study planning services part of integrated clinical collaboration platform. This approach provides extensibility to the integrated clinical collaboration platform architecture, maximize reuse and integration, thereby resulting in tangible business benefits for all participant organizations - Service Interoperability and Standardization – The integrated clinical collaboration platform architecture will ensure interoperability by complying with the principles of REST messaging protocol, XML driven message formats for message exchange.
Assumptions and Constraints	An SOA approach is adopted in implementing the solution. This approach will assist in promoting extensibility, reusable business and entity service as well provide for service integration with member company systems.
Motivation	One of the fundamental themes of the SOA approach in achieving business flexibility is to achieve a level of decoupling between the consumer and provider. This enables the provider implementation to change and evolve without affecting the requestor. This level of decoupling can be achieved in many ways with varying levels of decoupling and service levels.
Alternatives	Direct Connection: Using this communication architecture a requestor directly connects with the provider using the message format and the protocol specified in the service description from the provider. Message and Event Driven: Messaging - Using this communication architecture, the service provider(s) are decoupled from the service requestor(s) by a messaging component. The requestors send messages to the intermediate messaging component which in turn sends the message, with possible additional mediation and transformation to the receiving service provider. The flow of communication among the requestor(s) and provider(s) is determined by the predefined message flows loaded in the messaging component (as Message Exchange Patterns [MEPs]). Event Driven: Using this communicating architecture, the service provider(s) and service requestor(s) are unaware of each other and communicate by sending predefined events. The event processing component serves as an intermediary mediating the event messages among multiple requestor(s) and provider(s). The flow of communication among the requestors) and provider(s) is determined by the subscriptions of the services.
Implications	Significant custom coding may be required to achieve certain service requirements, such as assured delivery, 'guaranteed once' delivery for direct communication and event driven architecture.
Recommendation	Service Oriented Design and Development using both Message and Event driven architectures.
Dependencies	-

Table 7. Design Considerations: Multi-Channel Integration

Area	Design & Construction
Topic	Design for RESTful interfaces rather than SOAP interfaces
Description	One of the key architectural goals of the integrated clinical collaboration platform is extensibility and flexibility to support data formats that can be delivered to multiple channels (web, mobility) without relinquishing performance for its site users. RESTFUL interfaces/services provide for flexible data representation, enabling serialization of data in either XML or JSON format. RESTful APIs are easier to understand because they add an element of using standardized URIs and gives importance to HTTP verb used (i.e. GET, POST, PUT and DELETE). RESTful services are also lightweight, that is they don't have a lot of extra xml markup. All that is required is a browser or HTTP stack and pretty much every device or machine connected to a network has that.
Assumptions and Constraints	REST interfaces enables only point to point communication on top of HTTP transport while promoting flexibility in producing and consuming data formats for different content delivery channels . This imposes restrictions on integrated clinical collaboration platform to serialize message over other arbitrary transports (such as binary). Also REST interfaces does not provide rich support for RPC type semantics and message exchange patterns that one will find in SOAP interfaces
Motivation	4. Better suited for point-to-point communication 5. No formal description standards, hence easy to use 6. Built in error handling
Alternatives	NA
Implications	NA
Recommendation	NA
Dependencies	NA

Table 8. Design Considerations: Service Component Interface Definitions

Area	Design & Construction
Topic	Message Format to be serialized between service end points in the integrated clinical collaboration platform
Description	Provide for standardized message formats for the integrated clinical collaboration platform optimized and can be serialized over HTTP transport. The messages need to be consumed over varied content delivery channels such as web, mobility. A canonical model can be built.
Assumptions and Constraints	RESTful services will generate and serialize XML/JSON objects over HTTP transport. The schema and structural representation of the data will obey the rules of JSON and XML and will be encoded as part of HTML. The consumers of the REST services are expected to decipher and process the data as appropriate
Motivation	Both XML and JSON provide flexible and optimized message formats that can be easily serialized over HTTP. JSON objects provide the best option for delivering content/message to mobility
Alternatives	NA
Implications	NA
Recommendation	NA
Related Recommendation	See: Design consideration of multi channel integration through RESTful interfaces

Table 9. Design Considerations: Metrics Services

Area	Design & Construction
Topic	Service Metrics, SLRs and SLAs
Description	In order to govern a SOA infrastructure such as any middleware / ESB, it is important to add, enforce and audit 'Policies' and 'Service Level Agreements (SLAs)' across multiple security and identity domains.
Assumptions and Constraints	Middleware / ESB provides some level of metering and metrics
Motivation	Clinical colloaboration portal needs to be adaptable in meeting the service levels and business objectives. An important element of achieving the service levels and business objectives is monitoring, reporting, and metering of the services based on performance indicators. Utilize policy driven orchestration to monitor and manage the services resources and maintain SLAs based on feedback mechanisms will provide for close-looped metrics. Using vendor tools to not only report but automatically manage and orchestrate helps ensure most flexibility and ability to react to rapidly changing conditions. This is definitely most complex to implement but provides maximum functionally.
Alternatives	Embed resource consumption measurement within the services themselves by coding and instrumenting the services individually. This approach should be used if there are minimal services to be measured and reported on and if tools available do not provide the desired metering.
Implications	Expensive to implement and maintain.
Recommendation	Use Registries to enforce policies and service level data to ensure the metrics are available when the services are invoked.

Table 10. Design Considerations: User Interface

Area	Design &Construction
Topic	The web applications must be compatible with clinical trial users, including assistive technology. HTML and XML must be validated to ensure they adhere to formal W3C specifications.
Description	The usability and extensibility of the UI or front end layer provides rich, smooth, seamless experience to the end user. This characteristic ensures the quality of the web platform.
Assumptions and Constraints	Presentation / UI: Using open and W3C compliant options such as • XHTML/JS/CSS3/JQuery for standard web site • All the business and technical logics would be addressed through business service.
Motivation	Following are the different key features of chosen tool/technologies. JQuery: • Cross-browser JavaScript library designed to simplify client-side scripting of HTML. • Simplifies HTML document traversing, event handling, animating, and Ajax interactions for rapid web development • JQuery is used for front end development. CSS3: • Cascading Style Sheets level 3 is the most recent iteration of CSS. • Device compatible CSS and themes can be applied quite easily and once loaded it is consistent over the entire website. Presentation and content modeling for standard website should be seamless and based on flexible information architecture.
Alternatives	Not Applicable
Implications	The views would be generated through Liferay theme, layout which is combination of JSP files, CSS, JavaScript and JQuery modules.
Recommendation	Proper organization of different source CSS and JavaScript files, minify of JQuery libraries are also recommended for better performance.

Table 11. Design Considerations: Caching

Area	Design & Construction
Topic	Should Caching be used? If yes, at which layer in the integrated clinical collaboration platform
Description	the integrated clinical collaboration platform requires effective caching techniques to allow investigators, sponsor users and other site users to have faster access to data that is frequently used while performing clinical study planning, site collaboration and document exchange functions.
Assumptions and Constraints	Only specific business scenarios foster usage of cache. Caching should be introduced (when applicable) only when it can provide significant benefits to reduce the load on services and backend systems in the integrated clinical collaboration platform.
Motivation	Increase the integrated clinical collaboration platform performance, decrease load on external systems. Every layer implements caching. The data access layer uses Hibernate which in-turn implements 2 levels of caching. Level 1 within the current database session. Level 2 is across database sessions. Apart from these Entity and Query cache is also used. EHCache is used by the Liferay portal for boosting performance.
Alternatives	Flexible and in-memory databases. No-SQL Databases
Implications	Cached objects have a temporary lifecycle of their own. the integrated clinical collaboration platform components/services that use cached objects must be prudent of this fact and ensure information it uses from the cache do not undergo significant changes during the lifetime of the object.
Recommendation	Cached objects should be used only for read-only domain objects whose content do not change. Memory cache without replication should be used; each component that will provide caching feature should be responsible to manage its cached objects.

Table 12. Design Considerations: Managing State

Area	Design & Construction
Topic	State management at services level
Description	Should a service provider or related components hold any state, or obtain all required processing information from the request message? If it does hold state, how often should enterprise resource data is synchronized with the resource layer (assuming concurrent access from multiple services and applications)?
Assumptions and Constraints	NA
Motivation	The services are implemented using REST which is stateless
Alternatives	NA
Implications	NA
Recommendation	For all practical purposes, state should be managed at the orchestration level and not at the services level. Services should be stateless in nature.

Table 13. Design Considerations: Logging and Audit Capabilities

Area	Design & Construction
Topic	Design Considerations for Logging and Audit
Description	Logging framework in the integrated clinical collaboration platform
Assumptions and Constraints	Logging: There will not be any centralized logging framework, which would be implemented and used by each layer of the the integrated clinical collaboration platform. Auditing: As auditing is a user initiated event capture mechanism, the primary role is to capture user actions within the the integrated clinical collaboration platform. Auditing would be taken care by Liferay portal layer. Rest of auditing requires during the processing logic would be implemented as per requirements.
Motivation	Each layer (portal, integration, security etc.) would implement its own logging mechanism, and this approach of logging eliminates the inter dependency during construction and environmental setup. Audit trail would be implemented by portal and integration layers. Portal would capture the user activity and pass them to the ESB layer. ESB layer would insert the audit trail record in common data layer (CDL).
Alternatives	Not Applicable
Implications	
Recommendation	Proper designing approach as mentioned above sections is required to be taken, so audit trail services can be invoked from anywhere of the portal.
Related Recommendation	Construction – Service Oriented Approach

Table 14. Design Considerations: Integration with ESB

Area	Integration
Topic	Use of ESB technology for integrating heterogeneous platforms through high performance, reliable and guaranteed communication
Description	The discussion is around how ESB technology is suited to integrate or communicate between the integrated clinical collaboration platform services, other applications within the platform and member company systems.
Assumptions and Constraints	Integrations between different systems will be a mix of synchronous and asynchronous messaging based or service-based communication. Some of the communications may be statically and / or dynamically invoked.
Motivation	Multi-tenancy • Multi-tenancy will be applicable since there will be scenarios where the same service will be invoked by multiple consumers (a member company or a member company system). There will also be scenarios where there will be dedicated service for a particular consumer (a member company or a member company system). Each service and API exposed will be secured by an access token that will be acquired by the consumers. The consumers will be restricted to access the services that they are not authorized for. Platform Synergy • The portal solution chosen is Liferay that runs on JBoss. If ESB technology also runs on JBoss and it uses some of the state-of-the-art features offered by JBoss. So it will work seamlessly with the portal solution Time to Market • Out-of-the-box support for multi-tenancy resulting in enablement of new tenant for the integration platform • Out-of-the-box support for many different protocols resulting in quick integration with the supported protocol of the on-premise or cloud based tenant systems • Out-of-the-box support for many different formats resulting in quick integration with the supported format of the on-premise or cloud based tenant systems • Out-of-the-box support for monitoring and management of the artifacts of the different tenants resulting in highly secure integration environment • Pre-built code libraries and modules available for reuse across different components including services, integration flows, etc. • Out-of-the-box support for all relevant integration patterns Platform Flexibility • The entire platform is built as a set of cohesive services collaborating with one another using well-defined interfaces and standard protocols (HTTP, JMS, native, etc.) • Support for Spring (arguably the best dependency injection framework today) as the de-facto component development and wiring framework • Additional capability introduced inside Spring to support advanced features such as sharing beans across different applications, inheriting application contexts from other code modules, swapping implementations seamlessly, to name a few High Performance and Scalability • Failover and load-balancing through active-active as well as active-passive clustering of JBoss AS and HornetQ • Modular class-loading architecture using JBoss Modules for fast runtime resolution of application binaries and their dependencies. Sophisticated techniques to ensure that one class only gets loaded once by the runtime virtual machine • Rapid application provisioning through very low deployment footprint. Typical monolithic mega-bytes of application binaries replaced with lightweight kilo-bytes of modular bundles Patterns • The prevalent ESB technology supports most of the industry standard Enterprise Integration Patterns. Pipes and Filters, Message Router, Message Translator, Content Based Router, Recipient List are just to name a few.
Alternatives	ESB solutions available in the market
Implications	Most of the ESB solutions are heavy footprint and they do not cater to the unique requirements of cloud integration scenarios.
Recommendation	Small footprint ESB technology
Related Recommendation	Construction – Service Oriented Approach

Table 15. Design Considerations: Use of Portal Technology

Area	Presentation
Topic	Use of portal technology for accessing web content and resources (data, docs, apps, integrations, notifications etc.) securely, providing the smooth and seamless web experience to the end user by interacting views.
Description	Liferay has the capability to provide the the integrated clinical collaboration platform on easy and on demand accessibility to the resources, provide for custom web experience based on user's identity, and would also serve as an enterprise desktop
Assumptions and Constraints	NA
Motivation	• Liferay is the Most Popular Java CMS in Water & Stone's 2010 and 2011 Open Source CMS Market Share Report. • Has been named a Leader in Gartner's Magic Quadrant for Horizontal Portal Products in 2011 & 2012. • The mostly widely deployed portal system in the world with more than 250,000 deployments. • Proven real world performance with Fortune 500 client websites across industries. • A very strong and contributively community with roughly 3.5 million downloads. • Guaranteed SLAs up to 24/7/1 with regular service packs and a five-year EOSL policy. • Liferay supports rapid innovation with customer contributed development and new enterprise releases every eighth month. • Liferay has the lowest Total Cost of Ownership (TCO) compared to its competitors starting with its licensing and getting it up and running through development costs, operational costs, and training/ support costs (from the perspective of infrastructure, developers, administrators and end users).
Alternatives	Other popular portal solutions
Implications	NA
Recommendation	Liferay Enterprise Portal or other popular portal solution

Table 16. Design Considerations: Use of Life-ray portal for Multi-tenancy

Area	Support for multi-tenancy
Topic	Use of Liferay Portal for multi-tenancy support.
Description	Multi-tenancy defines the use of the integrated clinical collaboration platform service in a shared basis by multiple member organizations, without compromising on data integrity and data security. Also the product architecture and design of data model should be in such a way that, usage for individual member organizations and investigators and other site users should be retrieved easily.
Assumptions and Constraints	NA
Motivation	• Liferay provides a very versatile architecture for any portal, where multiple organizations can be created and maintained. • The information within the portal stored in a well architected database schema, where data are segregated organization wise. • Within the same portal instance multiple organizations can be created and all the organizations can share the same portal application.
Alternatives	Within other popular portal solution very few supports this feature.
Implications	NA
Recommendation	Liferay Enterprise Portal or any other standard portal

Table 17. Design Considerations: Use of Hibernate for ORM

Area	Object Relational Mapping (ORM) framework
Topic	Use of Hibernate for object relational mapping (ORM) framework
Description	How much Hibernate framework capable enough to provide ORM for database
Assumptions and Constraints	No specific assumption or constraint.
Motivation	• Hibernate is an industry proven framework used as ORM framework. • Hibernate provides first level cache, which enables proper object management, which is a default behavior. • Second level cache supported by Ehcache framework, can be configured without changing any source code. • Hibernate provides easy development API, where creation of SQL code is not required for CRUD operations on the databases • Hibernate is a java based framework and hence integration with java based module, applications are seamless.
Alternatives	Other ORM frameworks like JPA (Java Persistence API, OpenJPA, HibernateJPA)
Implications	NA
Recommendation	Hibernate

DESCRIPTION OF COMPONENTS IN THE INTEGRATION ARCHITECTURE

API Gateway

The API Gateway supports the following features:

- Support for hypermedia controls and state-of-art media types
- Ability to render partial as well as embedded resource responses based on consumer queries [this can be used to compose multiple API calls into one to reduce chattiness as well as to limit data returned to exactly what is required by a client]
- An intuitive API browser for browsing through the API resources
- APIs can be rate limited using various configurable rules based on time, location, client application, results per request, user quota definitions, etc.
- Custom API clients can be registered and centralized access control policies can be defined by the API resource owners which can be uniformly applied across multiple APIs if required
- The API Implementation is responsible only for populating the state of a resource from backend applications. Hypermedia composition is done by the API Execution Framework itself and relayed to the API consumer using standard media types
- The JAX-RS 2.0 specification along with custom extensions provides the foundation for the API execution runtime
- Apache Camel with its rich support for integration patterns and system connectors can be used as one of the implementation types for an API resource

Integration Framework

The Integration Framework consists of a host of technology connectors. The technology connectors can be used by individual integration flows to connect to source and target systems through different types of protocols. The Integration Framework connects to a Metadata Repository that holds a set of database tables required to run the different types of integration flows and services.

Technology Connector

This component is responsible for connecting to various transports in a uniform way. This is a key abstraction which allows flows to consume and produce data from and to disparate mediums in a standard manner without altering the APIs used. The following key Camel constructs together realize the Technology Connector component of the integration framework. Currently supported connectors include File, FTP, JMS, HTTP, Servlet, CXF, Direct, VM, Seda, SMTP and Quartz. Other connectors can be added on demand based on project requirements

Integration Flow

Integration flows are the individual routing definitions which are processed and executed at runtime by the integration framework. They are defined by stringing together technology connectors and processors to realize a concrete integration scenario.

Integration Patterns

The Integration Framework supports most of the commonly used integration patterns such as Splitter, Aggregator, Resequencer, Wire Tap, Router, to name a few. The framework translates the pattern language into concrete implementation constructs through specialized processors which when used in flows can solve many of the typical integration challenges.

Type Converters and Data Formats

Type Converters

These allow automatic conversion between well-known types thus reducing the need for boiler plate code to manually handle this in flows. Type Converters are implemented using the Apache Camel Type Converter constructs. Apache Camel comes with more than 150 type converters out-of-the-box for handling frequently used.

Data Formats

Data Formats are pluggable transformers that can transform messages from one form to another and vice versa. They are implemented using the Apache Camel Data Format constructs. Data Formats expose operations for marshalling and unmarshalling raw data to Java object trees. Apache Camel comes bundled with more than 18 data formats out of the box.

Expression Evaluators

These components are used to evaluate arbitrary expressions on data objects. Expressions add a layer of flexibility and dynamism to components. Expressions are used widely in integration flows to define routing conditions, variable extraction rules, dynamic computations, etc. hence the Express Evaluator components find wide usage here. Expression Evaluators are implemented using the Camel Language constructs for Expressions and Predicates.

Camel Languages

To support the flexible and powerful Enterprise Integration Patterns, Apache Camel supports various Languages to create an Expression or Predicate within either the Routing Domain Specific Language or the Xml Configuration. Both expressions and predicates can be combined to form more complex expressions.

Lifecycle Management

The lifecycle of the components developed on the integration framework needs to be synchronized with the lifecycle management capabilities provided by the application development framework. Within application development, the Spring framework is used to provide lifecycle management capabilities for application startup and shutdown. Apache Camel as the implementation of the integration framework hence needs to be able to hook its lifecycle management with that provided by Spring.

Camel Context

The Camel Context is the runtime container for Camel which holds components, routes, endpoints, processors, type converters, data formats and languages and executes and manages them at runtime allocating necessary resources and instances as needed. It is the Camel Context which starts when the application starts and stops when the application stops

Service Provider Framework

Transport Receiver

This component receives raw services requests over various transport protocols and constructs a protocol agnostic representation of the service request before handing it downstream for further processing. Currently, two such transport receivers are available, one which is implemented as an HTTP Servlet for receiving requests over HTTP and the other for receiving in-memory requests (Local transport) from other components and services within the same Java virtual machine. The transport receiver looks up the service configuration for the requested service, creates an instance of the Message Context based on the incoming payload and forwards to the Server Message Processor for actual processing.

Message Context

This is an interface through which all framework components get access to the message that is being processed. The message context is comprised of two parts. One is the message itself. Message is an abstraction of the incoming / outgoing data and it can be accessed either raw or serialized. The second part is the context, which contains state information pertaining to the current message processing. Each of the framework components are free to add additional state (name, value) as they process the message. This context should however not be used as a dumping ground to store arbitrary information, thereby increasing its size unnecessarily.

The overall state that is maintained in the Message Context includes the following -

- References to Service and Operation Descriptions
- References to Request and Response messages (inbound and outbound messages)
- References to other runtime objects, such as Protocol Processor and Transport
- Various information properties, such as authenticated user id, request id and GUID
- Flexible map of user properties (name-value pairs)
- Flexible map of system properties (name-value pairs), not modifiable by user
- List of errors collected during processing

Service Configuration

The service delivery framework uses service configuration files to extract information such as supported transports, supported data bindings, pipeline configurations, header mapping options, monitoring configuration, etc. The service configuration is stored in the form of XML files and cached by the service delivery framework. Configurations can be done both at a global level and at a per service level. Service level configurations override similar definitions in the global one. This component allows service behavior and configuration to be externalized so that they can be managed and altered independently.

Data Binding Factories and Configuration

The service delivery framework provides pluggable data binding strategies so that various data formats can be supported for services. All raw data formats are translated into the same Java content tree by using the data binding abstractions. Currently, JAXB classes are used as the Java content tree though the framework is not restricted to the usage of JAXB and arbitrary Java content trees can be used as desired through configuration changes. The supported data bindings for a service are configured through the Service Configuration.

The currently supported data bindings include:

- Plain old XML - This data binding uses the Woodstox parser to marshal and un-marshal XML encoded data into Java objects. In case of SOAP, the SOAP protocol processor takes care of the wrapping and unwrapping to and from Envelopes.

- Fast Infoset - This data binding is for handling data encoded using the Fast Infoset format. Fast Infoset is a binary XML representation which is efficient in terms of size and parsing times.
- JSON - This data binding uses the Jackson parser to marshal and un-marshal JSON encoded data into Java objects.
- Binary JSON (SMILE) - This data binding uses the Jackson parser to marshal and un-marshal binary JSON encoded data into Java objects.
- Protostuff - Based on the high performing Google Protocol Buffers, this is an ultra-fast data format which can be used for rapidly exchanging data within and outside the platform. This data binding marshals and un-marshals Protostuff encoded data into Java objects.
- NV (Name - Value) - This data binding can be used to perform REST style service invocations. It is meant to be used in conjunction with HTTP GET requests where the payload is passed as name-value pairs in the HTTP query string.

Server Message Processor

This is the overall orchestrator and coordinator within the service delivery framework. It is responsible for invoking protocol processors, pipeline handlers and request / response dispatchers. The message processor has four stages of processing -

- Running the message through an input pipeline
- Giving it to the Request Dispatcher
- Running the message through an output pipeline
- Giving the response to a Response Dispatcher

A message processor is also responsible for invoking the Protocol Processor to perform protocol specific tasks. For example, it plugs in the Axis2 message handling chains using a SOAP Protocol Processor while processing SOAP requests.

The message processor also records monitoring events which are collected by the Monitoring Frameworks to track service usage and response times.

Protocol Processor

The Protocol Processor is invoked at various times during the pipeline processing. It is initialized by the framework once at startup. It defines actions that need to be taken for a particular protocol at various stages, before entering the request pipeline, before request dispatch, before entering the response pipeline and before the response is dispatched. Two protocol processors are at present defined. The SOAP Protocol Processor uses the Axis2 libraries to unwrap the payload from the SOAP envelope before the request pipeline is executed and adds the SOAP envelope back around the raw XML message before the response is dispatched. The Null Protocol Process is a no-op processor which does no manipulation to the message at all. This is used to keep the interfaces and programming model consistent across the framework.

Pipeline Handlers

Handlers apply cross-cutting concerns in a generic manner to services during request and response processing. They are initialized by the framework once and during message processing they are passed a handle to the Message Context instance, hence giving them access to the entire state at that instant. Handlers can be used for various purposes such as enforcing security, storing state, validating request data or modifying requests and responses in a generic manner.

Pipelines are the drivers for running a message in sequence through the configured handlers for a service. They have little message manipulation logic of their own and focus on invoking the handlers and handling any exceptions that potentially arise during this invocation. They are created on a per-service basis, and shared among multiple invoking threads.

Request Dispatcher

The Request Dispatcher is responsible for actually invoking the service method by passing the necessary Java arguments. A base dispatcher takes care of most of the pre-dispatching and post-dispatching logic. However, if the dispatcher was to be a common class, Java reflection would need to be used to dispatch to correct service implementation which is relatively expensive. To avoid this, service specific request dispatchers are generated during code generation time. These classes are strictly typed and hence circumvent the need of reflection for service method invocation. The request dispatcher also sets the Message Context to a Thread Local before dispatch so that it can be accessed anywhere by downstream classes in a thread safe manner. It unsets the Message Context after the dispatch is complete.

Response Dispatcher

This dispatches the response for a request to the caller. It normally delegates to the underlying Transport abstraction for dispatching the response.

Service Implementation

This is the class which actually implements the business logic for the service interface. Based on the category of service, this may have dependencies with various platform components such as the Integration Framework, Utility Libraries, Application Development Framework, etc. Instances for these classes are created by the service delivery framework and pooled for subsequent reuse.

Authentication Service

This service accepts varied credentials (can represent an end user, organization, IP address, etc.), chooses the right authentication scheme to use for the resource being accessed and routes to the appropriate scheme specific provider implementation. The service follows a provider - implementation architecture where multiple authentication providers can be plugged in and discovered using appropriate schemes. The service clients pass the correct credential set and the service chooses the correct provider implementation based on the received set.

The Default Access Token provider is available for directly authenticating user security credentials against the local user store.

The LDAP Access Token provider is available for authenticating users against an enterprise LDAP directory.

The API Access Token provider is available as the default provider implementation while accessing services. This can authenticate access tokens against a local token store and return the subject and subject group information once successfully authenticated. The API Access Keys (Access Token and Refresh Token) are issued at the time of user registration. On expiry of the access token, the Security Token Service is used to renew access and obtain a fresh token.

Authorization Service

This service accepts the subject information, optionally his subject groups, the resource being accessed and optionally the action on the resource that has been requested. It uses the Policy Service to retrieve the necessary XACML authorization policies and evaluates them. Based on the outcome, the subject is granted access to the resource with a Boolean success result.

The authorization policies are cached by the service for ease of access while handling subsequent requests. The cache is reloaded at configurable intervals from the persistent policy store.

Rate Limiter Service

This service can be used to rate limit consumers based on quotas, daily limits, data limits, subscription levels, etc. This service is not provided out-of-the-box as it is mainly relevant for cloud-based deployment scenarios. Details of implementation are out of scope of the current document.

Policy Service

This service manages subjects, subject groups, resources, resource actions and the various kinds of policies that can be applied to the resources and subjects. The policy definitions are based on the XACML standard. The policy service exposes all operations for creating and modifying resources, creating policies, adding subjects to existing policies, removing policies, updating policies and above all retrieving various kinds of policies such as authentication, authorization, rate limiting, whitelist and blacklist.

This service also uses a provider - implementation style of architecture and pluggable providers can be introduced on a case-to-case basis. The Policy Service interacts with the Data Persistence components to access the Policy Store for managing policies.

Event Notification Components

These components provide the ability to send alerts and notifications to various endpoints through multiple channels such as E-mail. Notifications are in principle processed asynchronously using the messaging infrastructure as these require reliability but at the same time seldom obstruct the main application flow. The service delivery framework is used to expose a generic service component which exposes functions for sending notifications using a canonical notification structure. This allows calling components to be agnostic of channel specific technical complexities.

For connecting to the appropriate channels, the notification sender makes use of the connectors provided by the integration framework. The notification components also comprise of a metadata store where intended recipients, channel preferences and rules are stored along with pre-defined alert templates which are populated with data dynamically at runtime. This streamlines content rendering by centralizing it, thereby forcing other components to adhere to standards rather than creating ad-hoc content on a case to case basis.

Monitoring Dashboard

This component allows users and administrators to analyze service usage and access metrics using intuitive graphs and charts. The component accepts search criteria (service id, time units, time ranges, etc.), passes them on to the Metrics Query Service and relays the information back to be displayed appropriately using graphs and charts. Normal users are limited to access only their individual access details. Administrators can access all metrics and further filter using the consumer id. Graphs include line charts for trend analysis and bar charts for comparative analysis (call counts, response times, error count, etc.)

Application Management

This component allows administrators to manage the lifecycle of applications deployed on the platform. It allows administrators to view a list of applications deployed, the resources that they contain (services and flows) and the resources that they reference using a hierarchical tree style representation.

Administrators can start and stop the applications as required. For integration flows which are deployed within the applications, the component allows administrators to suspend and resume flows temporarily at runtime for handling ad-hoc crisis situations.

Monitoring Components

Service Monitoring

Service Monitoring components collect key metrics during service access and record them to a persistent store on a scheduled basis. They also include components which expose APIs for querying these metrics based on multiple dimensions. (BearingPoint / AMR Clinical Metrics Study 2008)

Metric Definitions

Metric Definitions will be at a global level but the metrics will be reported at a tenant level. These are standard structures for representing metrics. They are represented in the application layer as Java beans. Values are recorded against these definitions and persisted in the metric store. A metric definition consists of the following elements:

- Metric ID: It is a unique identifier for the metric and is generated by combining the Metric Name (CallTime, TotalTime, PipelineTime, AuthenticationTime, etc.) with the Service UUID and Operation Name. So a Metric ID would look like *'SomeService.someOperation.CallTime'*.

- Monitoring Level: This defines the level of a metric in a way similar to the log level defined in Java logging. A metric level can be specified in global service configurations as well as service specific configurations, to enable logging of the metrics. It consists of three possible values, 'NORMAL', 'FINE' and 'FINEST'.
- Metric Category: This categorized the metric into logically related groups. It consists of three possible values, 'TIMING', 'ERROR' and 'OTHER'.
- Metric Value Factory: A factory for creating metric value types which compute the value for a metric at runtime. Different value types compute data differently. For example a timing metric might calculate millisecond differences while a normal metric might increment counters.

Metrics Collector

This is the runtime manager for all metric values collected. It is initialized a singleton and accumulates metrics periodically. Based on the snapshot interval configuration, it uses the Metrics Storage Provider to persist the snapshot data collected in that cycle. The metrics collection will be asynchronous.

Metrics Aggregator

This component collects metrics for a given metric value. Metric classifiers (combination of consumer, source data center, target data center) are used to identify the data collection point. The update methods here based on the metric classifiers to find the right data point to aggregate.

Metrics Storage Provider

This provides the interface for defining storage mechanism for the metrics. The interface provides methods to store registered metrics and the aggregated metric values. Two providers are available, a File based provider which stores metric data onto the local file system and a Database based provider which stores the metric data to a relational database. The active storage providers to be used are configured at service level or at global level through the standard service configuration sections for monitoring.

Metric Query Service

This is implemented as service on the service delivery framework. It exposes operations for getting access to metrics data based on various search criteria such as time ranges, time units, consumer id, service id, operation id, data centre, etc. Operations exist for data retrieval both at a summary level and at a detailed level. The service uses a provider - implementation style architecture where multiple query providers can be plugged in and the service delegates calls to the appropriate provider based on configuration. Currently, the File based and Database based query providers are available which access data from files and databases respectively. For databases, the provider uses the Data Access Objects from the data persistence layer.

Error Handling Components

Service Exceptions

Exceptions occurring during service invocation are caught and translated into the common error structure above by the Request Dispatcher component of the service delivery framework. The exception is then sent to the exception handling queue for asynchronous handling. If the service operation is a request-response one, the Common Error Data is serialized into the appropriate data format (SOAP Fault, XML, JSON, etc.) and relayed back to the caller.

Flow Exceptions

Exceptions occurring in integration flows are captured by standard Camel error handling constructs. Custom error relaying flows are implemented which can retry the interaction for a specified number of times at a specified interval, post which they relay the exception using custom processor components which translate the exception into the common error structure and post it to the exception handling queue.

Exception Handling Flow

This is implemented as an Integration Flow on the integration framework. It listens to a generic JMS queue for receiving messages in the common error structure. Once it receives messages, it records the same using the Error Logging Provider. Based on the severity level of the exception, it sends notifications to the system administrator or other concerned personnel using their preferred channel of communication using the Event Notification Service.

Error Logging Provider

This logs errors to a persistent store. It follows a provider-implementation architectural style. Currently, a Database logging provider is implemented which used the data access objects in the data persistence layer to record the error in the error store.

Flow Monitoring

Monitoring of integration flows is realized through the inherent capabilities provided by the Apache Camel framework monitoring constructs.

JMX Console and Notifications

The Java provided JConsole user interface can be used to monitor the integration flows which are registered as MBeans. Almost all relevant information such as number of messages processed, errors, payload tracking, in flight messages, thread usage, etc. can be analyzed easily through the various attributes exposes by these MBeans.

Apart from this, there is also a JMX component which can be used inside integration flows to receive JMX event notifications. This provides the ability to subscribe to these events and take necessary actions such as integration with existing enterprise monitoring data stores, etc.

DESCRIPTION OF COMPONENTS IN THE SECURITY ARCHITECTURE

The primary components of the security architecture are addressed below:

Secure Proxy Server (SPS)

Secure Proxy Server (SPS) provides a proxy-based solution for access control. Like traditional proxy server, CA Site Minder SPS does not provide resource caching. SPS acts as a single gateway for access to enterprise resources. Based on the configuration of proxy rules SPS determines how to manage a user request.

In integrated clinical collaboration platform, CA SiteMinder SPS can act as proxy to portal application server and also the Learning Management System (LMS) Application server. SPS contains in-built Apache webserver and Web agent. Web Agent will act as policy enforcement point which will protect the Portal and LMS by polling the Policy Server.

For implementation of Cross Domain SSO (Federation) the users accessing integrated clinical collaboration platform need to be authenticated against multiple identity stores across multiple domains. Federation Gateway service of SPS will be leveraged for Cross Domain SSO (Federation).

SPS will intercept all requests which are coming to the integrated clinical collaboration platform and displays page with list of IDP's. The user will select the identity provider where he/she wants to be authenticated. After authentication IDP will generate the SAML and redirect the user to Service Provider (SiteMinder).The SAML thus generated will contain multiple attributes as per the need of the SP and pass the same to SiteMinder. SiteMinder will validate the SAML and pass the user attributes received to SPS in header which in turn will pass the header to portal. Portal parses the header and allows the user to access integrated clinical collaboration platform.

SiteMinder

CA SiteMinder provides enterprise-class secure single sign-on (SSO) and flexible identity access management that authenticates users and controls access to Web applications and portals. Across Internet, intranet and cloud applications, it helps enable the secure delivery of essential information and applications to users via secure single sign-on.

CA SiteMinder Federation capabilities will be leveraged in the integrated clinical collaboration platform to redirect the users to IDP for Authentication and also for validating the SAML sent by IDP and in turn providing the users access to portal and LMS systems. A SiteMinder environment typically includes following components.

Policy Server

A SiteMinder Policy Server (Policy Server) acts as the Policy Decision Point (PDP). The purpose of the Policy Server is to evaluate and enforce access control policies, which it communicates to a Web Agent. A Policy Server provides the following:

- Policy-based user management
- Authentication services
- Authorization services
- Password services
- Session management
- Auditing services

Policy Store

The SiteMinder policy store is an entitlement store that resides in an LDAP directory server. In the integrated clinical collaboration platform CA Directory Server (version R12 Sp12) is being used as policy store. The purpose of policy store is to store all policy-related objects such as

- Resources SiteMinder is protecting
- Methods used to protect those resources
- Users or groups that can or cannot access those resources
- Actions which are triggered by access grant or denial of protected resources to users

User Store

A SiteMinder user store is an user directory or database in your enterprise network. The purpose of the user store connection is to make user data available to the Policy Server, which includes the following:

- Organizational information
- User and group attributes
- User credentials, such as passwords
- User attributes, such as first and last name

In the integrated clinical collaboration platform, user store will be used to store SPS Admin and Policy Server Admin credentials. SPS Admin trying to access SPS admin user interface will be authenticated against the user store configured. Similarly Policy Server Admin trying to login to SiteMinder admin ser interface will be authenticated against this user directory.

Identity Minder (IDM)

CA Identity Minder delivers a unified approach for managing users' identities throughout their entire lifecycle and providing them with timely, appropriate access to applications and data.

Identity Minder will sync users from multiple user bases and as per the need of the downstream applications (Portal, LMS) will provision users with minimal attributes to Message Queue. ESB Layer will fetch the user profile data from Message Queue and provision the data to downstream applications (Portal, CDL and LMS). The following are the components of CA Identity Minder leveraged in integrated clinical collaboration platform

- Identity Minder Server
- Identity Repository (User Store)

User Store

CA Directory has been used as User Store in the platform. This serves as the Identity Repository for IdM. User profile data which will be provisioned to the downstream applications will store in this user store.

Identity Minder Server

Identity manager server is where Identity manager solution gets hosted. It can be hosted on Jboss, Web Sphere and Web Logic. In integrated clinical collaboration platform, CA IDM solution will be hosted on JBoss. CA IDM will perform following functionalities:-

- Recieves the Site User Profile data from Exostar
- Sending Site User Profiles to ESB layer using Message Queue

IDM will push the Site user profile details into Messaging Queue and thereafter, ESB layer will pass on the data to portal database and all other down stream applications. If any changes to Site User profile occurs in Common data layer then it will be first moved to ESB layer, which will push in the changes to CA IdM using Message Queue. In this way there will be a synch between IdM records and records maintained in Common data layer.

Exostar

The Secure Access Manager (SAM) is Exostar's comprehensive portal solution for registration, account management, and authentication as a means of providing controlled access to Exostar hosted applications and applications located within other enterprise domains.

SAM allows enterprises to enable controlled access to information and applications. A standard deployment approach ensures that a single sign-on (SSO) federation infrastructure can be seamlessly extended to support integration with Exostar. Exostar's SAM solution provides the following capabilities:

- Authentication (SAML)
- User Self-Registration
- Password Management
- One Time Password

All the External users (Site users and Investigators) accessing the integrated clinical collaboration platform will self-register in Exostar. Exostar will act as an Identity Provider for Site users. Password management and Authentication for the Site Users will be also be handled by Exostar. Exostar will authenticate the user on SAML request from SiteMinder (Service Provider) and will pass a SAML containing the required attributes back to SiteMinder after successful authentication. Exostar will be used as an Identity provider which has two basic responsibilities,

- User Registration
- Password Management

An investigator receives a registration mail along with the registration link. The link will navigate the investigator to the registration page. Upon successful registration the user details will be sent to IDM using Exostar user provisioning capability. Admin will define standard password policies in Exostar and manage the same.

Level of Assurance (LOA)

The level of assurance is needed based on the consequence of authentication errors and/or misuse of credentials. As the level of risk is increasing in accessing the online enterprise resources, therefore assurance should also increase. Informal or low value requests will require less stringent assurance. Higher value or legally significant requests will require more stringent assurance. LOA level 3 has been considered as optimum level of assurance. LOA 3 basically consists of User ID and Password authentication along with One Time Password.

SAM User Provisioning Mechanism

SAM manages users by their affiliation with an organization. When user accounts are created in SAM provisioning, records will be generated and sent to the CA IdM. Changes to the organization or user attributes in SAM will generate a provisioning record that will be sent to IdM.

The provisioning specifications for propagating updates from SAM to the application systems are:

1. To generate xml encoded files for file-transfer between Exostar and a Service Provider.
2. To provide the schema section of the provisioning web-service WSDL. This WSDL defines a SOAP interface between Exostar and a Service Provider.

Chapter 8
Architecture of an Integrated Regulatory Information Management Platform for Clinical Trials:
A Case Study in Regulatory Information Management System Implementation

Ayan Choudhury
Cognizant Technology Solutions, India

ABSTRACT

The pharmaceutical and medical manufacturing sectors have entered a period of disruptive transformation in the way regulatory affairs are conducted globally. The global clinical and regulatory landscape is evolving more quickly in this decade than ever before. The advent of adaptive trial designs, rolling submissions for indications, as well as the impact of regulatory policies in emerging markets, are influencing Pharma's ability to secure approvals efficiently and effectively and with required emphasis on safety and compliance. The impact of these changes on Regulatory Information Management can be significant over the next 5-7 years. Companies are rightfully asking what the transformation in business processes and technology might look like and what types of innovations they can adopt now to prepare them for the future state. The case study below introduces the need for an integrated Regulatory Information Management (RIM) platform, addressing key functionality of such an environment and describes the architecture & design consideration to industrialize such a platform.

INTRODUCTION

According to industry sources and publications, the pharmaceutical and medical manufacturing sectors have entered a period of disruptive transformation in the way regulatory affairs are conducted globally and the global clinical and regulatory landscape is evolving more quickly in this decade than ever before. The advent of adaptive trial designs, rolling submissions for indications including oncology and HIV/

DOI: 10.4018/978-1-4666-8726-4.ch008

HCV, as well as the impact of regulatory policies in emerging markets and the BRIC countries, are influencing Pharma's ability to secure approvals efficiently and effectively and with the required emphasis on safety and compliance. The impact of these changes on Regulatory Information Management can be significant now and over the next 5-7 years. Companies are rightfully asking what the transformation in business processes and technology might look like and what types of innovations they can adopt now to prepare them for the future state.

The case study below introduces the need for an integrated Regulatory Information Management (RIM) platform, addressing key functionality of such an environment and describes the architecture & design consideration to industrialize such a platform. The intended audiences of this case study are the architects & designers of similar systems

The desired transformation objectives of a Regulatory Information Management (RIM) transformation initiative are:

- Move from a present state architecture which is highly distributed (with a slew of point solutions) to a fully integrated Regulatory system
- Optimize and drive business transformation by leveraging the best practices and approaches that will create a strong regulatory affairs value proposition among key stakeholders - Pharma, CRO and health authorities
- Demonstrate effective methods for incorporating health authority and industry standards
- Deliver a high quality, timely and reliable business integrated environment
- Develop an approach and solution for Pharma, device and diagnostic combined and separately
- Gain advantage through cloud to synergize the processes across geographies and increase collaboration internally and externally

In the Drug development life cycle multiple Regulatory roles/users play a pivotal role in planning and tracking information related to one or more submissions, authoring primary labeling document and other regulatory activities associated with a product.

In this case study we will first review the various *business drivers* that drive the need for an integrated regulatory platform to enable and orchestrate the activities of the different regulatory roles and subsequently look at the possible architecture options to realize the following business requirement scenarios.

- Transform Product and Life Cycle Registration Management
    - Consolidate multitude of central systems and hundreds of local systems
    - End to end product lifecycle management
    - Regulatory information shared throughout cycle
- Augment Publishing Capability
    - Replace end-of-life eCTD system in short term
    - Long term solution to include comprehensive publishing capabilities: via eCTD with seamless exchange of documents and workflows
- Improve Regulatory Data Quality
    - Data standards and data governance
    - Regulatory Information Foundation establishment: Master Data
    - Build and enforce standardized policies across LOCs and systems with clear roles and responsibilities

Figure 1. Business Drivers of the integrated regulatory information management platform

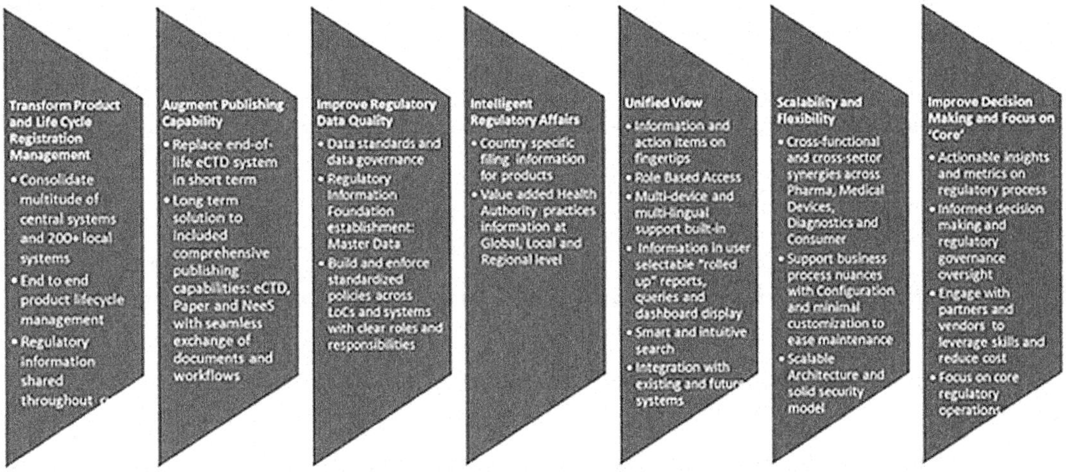

- Intelligent Regulatory Affairs
  - Country specific filing information for products
  - Value added Health Authority practices information at Global, Local and Regional level
- Unified View
  - Information and action items on fingertips
  - Role Based Access
  - Multi-device and multi-lingual support built-in
  - Information in user selectable "rolled up" reports, queries and dashboard display
  - Smart and intuitive search
  - Integration with existing and future systems
- Scalability and Flexibility
  - Cross-functional and cross-sector synergies across Pharma, Medical Devices, Diagnostics and Consumer
  - Support business process nuances with Configuration and minimal customization to ease maintenance
  - Scalable Architecture and solid security model
- Improve Decision Making and Focus on 'Core'
  - Actionable insights and metrics on regulatory process
  - Informed decision making and regulatory governance oversight
  - Engage with partners and vendors to leverage skills and reduce cost
  - Focus on core regulatory operations

Detailed Business Outcome Scenarios

The *functional components* of the desired regulatory information management platform are depicted Figure 2.

Herein we list some of the key business realization scenarios required of the different key regulatory roles, while working on an integrated platform:

Figure 2. Functions of the integrated regulatory information management platform

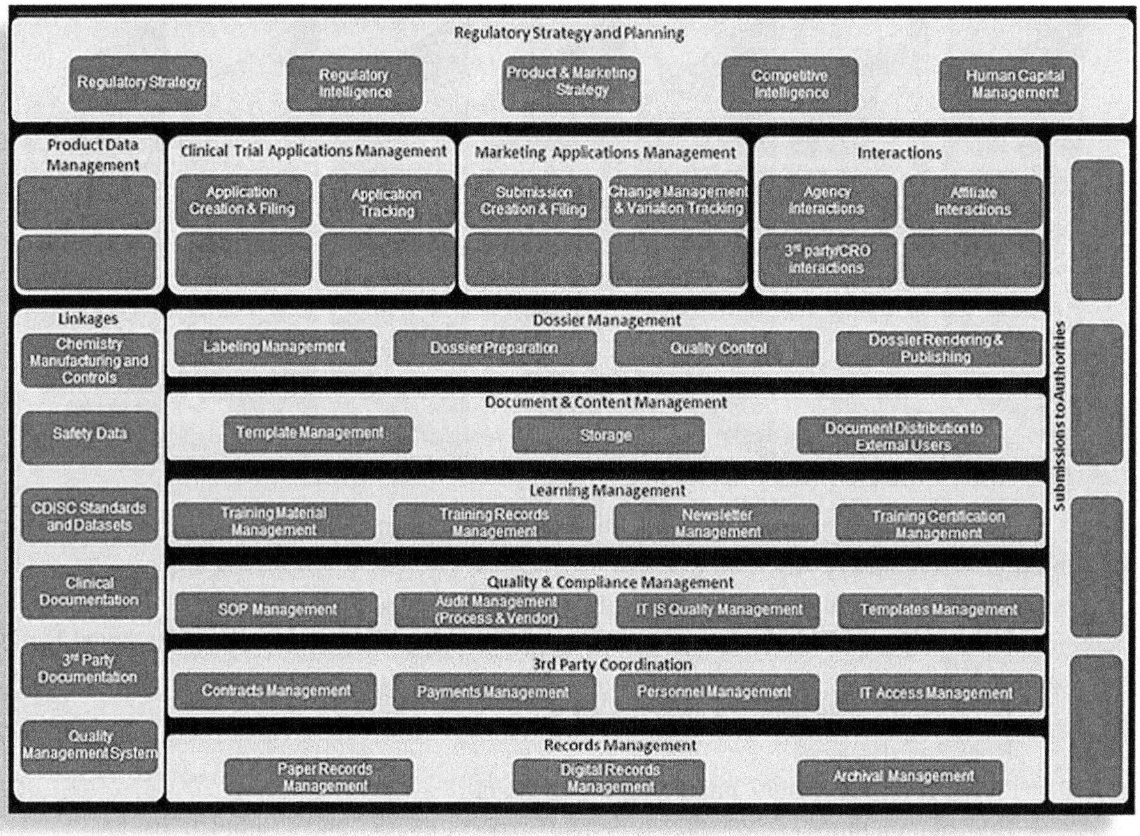

View Real-Time Updates on Product Life Cycle Activities and Reports

- A regulatory user can view a timeline of planned and actual activities (e.g. by quarter, region/country; submission type; pre/post approval, renewal) and can prepare a metrics report for selected fields. A regulatory user can view planned major submissions that they subscribed to, see the status of the team's portfolio of products. The user can click through any views to the underlying data or document in its native system without logging in again. Other linked systems have common interfaces (e.g. like MS Office).
- A regulatory user can drill down one level to see the products within their responsibility. When the user clicks on a product, he can see the global regulatory life cycle of the product; a summary of ongoing and planned activities, open questions from HA (Health Authority), submissions by status, planned submissions, delays (per step in the process by country), change controls (waiting for assessments or documents, pending implementation), the manufacturing or R&D team members, Regulatory team members, upcoming team meetings (linked to Outlook to exchange information). Information is linked to the current approved submissions for the product.
- A regulatory manager can see upcoming deadlines or missed deadlines for his group (submissions, commitments, HA questions) and can tailor information for his team. The manager can see review tasks with deadlines.

Planning

- A user can develop a submission plan, create dependencies within a project, assign resources to tasks, track progress, manage the budget, and analyze workloads.
- From an overall submission management process, a user can link submission types to cost per submission type in order to manage the submission budget.
- A user can view the portfolio of submissions and analyze resource and sequences across projects and set priorities.
- A user plans a submission and tracks the progress of the submission by milestone dates. This information is communicated to the LOCs (Local Operating Company) and other technical stakeholders (who can access the system to see the progress) in an interactive way so that notifications of upcoming activities and completion of tasks are transparent and dynamic. Milestone dates include projected and actual; the user is given the ability to analyze the information.
- When a submission is planned, the LOC is notified. When the LOC is required to develop a regulatory impact assessment and communicate it back to the central organization a "workflow" is created that maps all the required steps and notifications.
- Information in a submission plan automatically populates tasks for impacted LOCs and generates a tracking function for the central office. Central users, as well as LOC users, are able to see who in an LOC is responsible for a designated activity.
- A central Regulatory user accelerates translation activity associated with regional and local submissions by planning and tracking submission components, releasing each component rather than waiting for the entire submission to be completed.
- A Regulatory lead, who could be located in any location, plans and tracks the individual components that can be included in a submission as part of an integrated plan. Milestone dates for submission planning are integrated with delivery dates of submission components. The submission components detailed in the plans are linked to the documents in the electronic document management system (EDMS) and the plans are dynamically updated via the EDMS metadata. The reverse should also be true. The submission plan should serve as the outline of the submission with content placeholders that inherit the relevant metadata from the plan and populate them through template controls in the content itself.
- When planning is completed, the user seamlessly manages the submission process, starting with tracking the central dispatched submission. Once the LOC receives notification which includes the location of the submission, they continue tracking the progress, including submission and approval dates, statuses and associated commitments. The LOC is able to communicate any change made to the central submission and indicate which module is impacted (e.g. change to Quality Overall Summary impacts are recorded and linked to the changed document or component). A link is made to the submission assembly in the EDMS and to any other pertinent documents that may or may not have been included in the submission (e.g. local internal approvals such as medical or commercial review of product information). The LOC tracks implementation dates for CMC variations as well as for labeling changes.
- Each of the LOCs can manage local regulatory activities within the central capability using an interface that allows personalization based on personal preferences and centrally managed user profiles. An LOC can plan local submissions (including plans for local products) that are not relevant

to the central organization, track the outcomes and report on the status. The LOC can pre- populate plans with local information (e.g. local HA timelines). Other countries can be prevented from seeing that information or a specific local document since it is not relevant to them. An LOC user's home screen identifies and then takes them directly to the activities that are relevant to them.

- A Business Administrator can support the users to easily configure fields to adapt to the changing internal and external environments that include the reconfiguration of business units, country clusters, country assignments, and the movement of personnel – without incremental effort on the part of end users and LOCs. At any time, the system provides the capability to secure the information on a data, content, user, region, product, trial, and submission level (in other words multiple security levels). For example, it is possible to add or remove extra security levels, e.g. blinding trial result data for 3 months.

Tracking

- Multiple regulatory roles interact with the same product in order to track information related to one or more submissions or regulatory activities. Because there is a seamless transition of information from the 1st submission through to post approval management, products in multiple phases of development and registration are always up to date and accessible. A user responsible for managing the regulatory aspects of a clinical trial application (CTA), enters CTA tracking information in the clinical trial management system (CTMS) database. Automatically, selected fields are exchanged between the product and life cycle registration (P&LCR) database and the CTMS. From there, a Regulatory CMC user continues to track CMC information and related submissions. The Regulatory CMC user can select whether the product is a "clinical" product or a "clinical/marketed product. From there the required fields for clinical trial applications are available to the user.
- A Clinical Trial Regulatory lead (CTRL) can create a regulatory planning summary for an entire trial across participating countries. The report includes targeted approval dates, actual approval dates, submission dates, dispatch dates, by country and for all countries (in a Gantt chart type format). The CTRL can see an overall trial submission plan in one report.
- While clinical trials are proceeding, the global team can be preparing for Health Authority meetings. Any meeting correspondence (either to or from the HA) is uploaded to the EDMS and required data populates the P&LCR database (e.g. date of meeting, metadata extracted from content). The P&LCR database contains all information related to HA contacts, regulatory strategy and HA advice and replaces the need for "contact reports". A Regulatory user can enter specific HA requests and related agreements, regardless of whether the information is associated with an application or submission (e.g. for pre-MA meeting there may not be a Marketing Application yet). Meeting outcomes that could impact development plans can be searched across the database. A regulatory user can link specific knowledge gained from the HA meeting into the regulatory intelligence database that collects informal information from HA interactions.
- Once submissions are filed and the Core Dossier is developed, the central publishing team signals the Core Dossier is available to the applicant (e.g. LOCs, CRO, etc.) based on the up to date distribution list and associated plan for the dossier. A regional lead can track the progress of each Core Dossier and the detailed product information from dispatch to the applicant / LOC, preparation of

the local dossier, submission to the HA and ultimately tracking approval. The regional lead and the applicant / LOC can track the relationship of the Core Dossier to the local submission so it is always clear from which Core Dossier a local submission is derived.

- The first point of contact (e.g. a central or regional user or an LOC) enters approval dates into the system. The system notifies a previously identified dependent region or country when the reference country approval is received (e.g. country waiting for a COPP is notified when the reference country has approval). A LOC can enter requests for additional information or HA questions into the system database. A distributor or legal representative who takes the place of an LOC in some countries enters information directly into the system capability.
- In addition to tracking HA questions and the related response submissions, a user responsible for drafting a response can search across the system to see what certain HAs asked about previously filed submissions. The user can search a database of tagged questions and their related responses to find similar types of HA questions and/or responses. The user, for example can search for questions related to a particular dosage form, the eCTD section that corresponds with the inquiry and the key word "impurity". The user can then access previously used language, if applicable, as well as link to the original HA correspondence that contains the questions and answers.
- Once approvals are received, central, regional and local users manage their commitments within the P&LCR system. A local user tracks commitment milestone dates (planned and actual) and overall status (open, closed, etc.) and key milestones. This includes dates (HA or internally defined) by which commitments must be met. A user links directly to related procedural documents (1 click away) from any given task. The commitment owner identifies all responsible groups and/or individuals and the system alerts them as soon as the commitment is entered. The system proactively notifies users of due dates of commitments based on user defined business rules (e.g. CMC needs notifications 6 months prior to due date).

Health Authority Information Management

- When a user is working from the country level configurable page, he/she can be able to see all relevant planning, submission and registration information for one or more products including country filing requirements related to a specific product. This information includes tactical HA information such as requirements for certificates, submission filing and review timelines, and specific submission format details.
- Updates to country filing requirements must be dynamic and the user should always see current information. If a user wants to see more detail on a requirement, he/she can be able to click into the guidance document. Planning and content templates are pre-populated with local country filing requirements. When a regulatory impact assessment is required, the system can automatically "recommend" the type of change anticipated. For example, if a change in shelf life is required for Country A, the system recognizes the key word and associates it with one or more guidance documents for extending shelf life and suggests the type of submission/approval required (e.g. shelf life extension = notification to HA). Any informal HA information is also flagged and linked to relevant information.

Regulatory CMC Activities

- The P&LCR module will have granularity of the information that allows a Regulatory CMC user to search and find detailed product information and the related documents. From the P&LCR database, a Regulatory CMC user can search and find information for any product or active pharmaceutical ingredient (API) such as approved manufacturing sites, registered retest/shelf life, presentations, formulations, company information and manufacturing sites under review by a country. A user can create a compilation of all countries for a given parameter. A user can determine which countries have which documents approved. For example a user can locate approved and submitted specifications and link to the submitted specification in the EDMS. A user similarly can find the information for shelf life/retest date for any country or region.
- A Regulatory CMC user can view the status of selected documents such as: approved to include in a dossier, dispatched to LOCs, partial or full HA approval/acceptance, or implemented.
- A Regulatory CMC user can query the system database to confirm which countries have received a specific version of a document.
- A Regulatory CMC user can search by product, country and dossier to see what information an LOC received. The user can search for "CMC variations only" and the search returns only CMC variations and no other life cycle submissions such as labeling. From the results of the search, the user is able to see the content of the global submission dispatched to the LOC, as well as the local submission sent to the HA. If the approved dossier is different from the submitted dossier, the user is able to view the details of the changes.
- From the home page a CMC user sees their "To Do List" of assigned pending tasks, all of their ongoing projects, any recent activity that has occurred on any of their ongoing projects and any overdue tasks related to their projects including country-specific tasks.
- The Regulatory CMC user can capture the user's assessment of a change control from the Supply chain change control system (e.g. TrackWise). The Supply chain change control system automatically creates a change control record in the system where additional Regulatory details such as the regulatory strategy are tracked. This new change control record in the system is linked to the product and allows the Regulatory CMC user to create a submission plan (or link to an existing planned submission) and to start determining when a submission can be dispatched and which change control(s) would be included in that submission.
- Multiple versions of a submission can be drafted to support global registrations. Since a product can be at different lifecycle stages across the global community it is often necessary to include different change controls in different versions of the dossier. A user can tick off the specific change controls that are to be covered in a version of the dossier. The user is then able to associate the countries needed to receive that version of the dossier. When approval of the dossier is received in a country, all impacted change controls are updated accordingly with regulatory impact/ regulatory rationale and submission/approval information.

vi) Release of Product: Commercial and Clinical Trial

- When a control of change (COC) variation is approved by a local HA, a local regulatory manger enters local (individual) activities related to the regulatory clearance (e.g., local labeling updates completed, custom forms completed, provide additional clarification) into the P&LCR system. A second local regulatory manager gets assigned as a verification reviewer and is notified by the system that a regulatory clearance is required. The verification reviewer checks that all the fields are completed correctly and approves the regulatory clearance. The verification reviewer applies an electronic signature to a summary page that presents regulatory clearance information and all related updated (local) product information.

- The central regulatory CMC user is notified by the P&LCR system when all tasks related to a regulatory clearance are completed by the LOC. The regulatory CMC user is able to see the summary of the information for review prior to release.
- The regulatory CMC user checks a box in P&LCR which signifies that regulatory clearance for that product/ country is completed and the product with the variation could be released for distribution. The system alerts the appropriate supply chain user that regulatory clearance is received.

Labeling

- The following scenarios for managing labeling are representative and are not intended to be an exhaustive list of use cases or requirements.

Primary Labeling Document Authoring

- During the product life cycle authors of Primary Labeling Documents (e.g. Target Label, Company Core Data Sheet (CCDS), European Product Information and United States Prescribing Information) write label content in the English language, using enabling tools such as Adverse Drug Reaction (ADR) dictionaries, patient labeling term bases and competitor labeling information.
- A "lead" author coordinates the contributions of authors to develop the CCDS. During the authoring process, authors from functional areas add content and the lead author manages the tracking and reconciliation of each author's content through the life cycle of the labeling. For example, separate content changes may be managed in parallel (branch versioning) and then combined at a later date.
- The authoring environment must support the evolution of the content and the history and use of the content. Over time, the CCDS content is incorporated into national and regional labeling. Individual Health Authorities often respond to national submissions with required changes which may not be appropriate for other national labeling. Content must be managed and tracked in the CCDS and national labeling as versions are submitted in parallel and sequentially, as needed.
- The lead and contributing authors work on the CCDS through a set of workflows which support the authors by providing alerts / reminders / status for authoring, review and internal approval of each version of the CCDS. Data regarding key milestones are required for audit/inspection requests, such as the author, approver name and status dates for each component or section.
- Authors working on the CCDS may be assigned responsibility for discrete components of the CCDS. At a minimum, each major section of the label is a separate component and each component can be versioned independently of other components. In addition, text variants within each component are allowed to be managed in parallel. Alternatively each component can be further divided into more granular components that may apply to individual markets.

- The CCDS is the "parent" labeling document and the relationship among the components and the "child" labeling document that uses or modifies each component must be tracked and be visible to reporting and text comparison tools. For example, a report should show a link between a Warning in the CCDS and the equivalent Warning in the USPI, EUPI and labeling, even when the text is not exactly the same.

Translation

- The content of the CCDS is used to create regional and national labeling, which is usually non-English. As content is translated and approved by Health Authorities, the text is stored in the labeling system and used as a reference for future labeling to ensure compliance and accuracy in future labeling. Automated and semi-automated translation tools make use of this "translation memory" to assist translators. The "translation memory" also supports terminology management e.g. MedDRA term vs. patient terminology.
- The "translation memory" is updated as needed and the original text and terminology mapping is retained and associated with the labeling that was approved by the Health Authority at the time of approval.
- A user at an LOC is able to share information with one or more external vendors who support the translation process. When translations are outsourced and multiple vendors are used for different languages, an LOC user can coordinate translation tasks with external vendors and share documents.

Initial Submission and Change Implementation

- Labeling managers use the labeling capability, in conjunction with the P&LCR database to manage the development and inclusion of labeling in all national or regional, marketing approval applications, supplements and variations (e.g. supports all the EU procedures including Centralised, Mutual Recognition, Decentralised and National, and US submissions including major supplements and "changes being effected").
- The labeling manager monitors the status and content through the full labeling life cycle including implementation (both electronically via websites, and in printed packaging). This includes all labeling document types e.g. patient labeling (PPI, PL, etc.) and packaging components (carton labeling, etc.) and related derived deliverables such as translations and structured product labeling (SPL).
- Notification of CCDS changes, status and other labeling related information would be available to labeling managers and others through a variety of methods including reports, dashboards, and subscriptions to "push" notification.
- Labeling managers have access to reports and dashboard displays of the implementation status of each labeling content change in each market/LOC through all stages of the life cycle of the change. This includes, but is not limited to, receipt of notification of the change at the LOC, preparation of the national label including translations, submission to the HA, approval by the HA, first run of the packaging components on the production line, and use of the updated labeling in the national or regional market.

- In addition, meaningful differences in the content of national labeling or "deviations" from the CCDS, are identified and tracked. Reporting on national labeling deviations is required by product, by section, by country and ideally by keyword (adverse reaction, indication for example). It is expected that this tracking can be integrated with the P&LCR database and the appropriate artwork capability and can be available to supply chain to ensure release of compliant product / product labeling.

Search

A user can search across the entire system capability for any type of regulatory information, document, submission or document component using enhanced "smart" search capabilities commonly found on commercial e-commerce sites – e.g. the user is able narrow search by relevant metadata; find similar content that might be of interest; autocomplete search terms.

ARCHITECTING AN INTEGRATED REGULATORY INFORMATION MANAGEMENT PLATFORM

An integrated Regulatory Information Management platform can be a key solution to address these requirements. This platform can be used by Global Regulatory professionals across Central and hundreds of Local Operating Companies (LOC). *Representative reference architecture* for such a platform is given in Figure 3.

Figure 3. Reference Architecture of the integrated regulatory information management platform
Note: Highlighted Purple boxes represent functional blocks in scope of RIM Transformation. Green boxes represent suitable products for respective functional blocks.

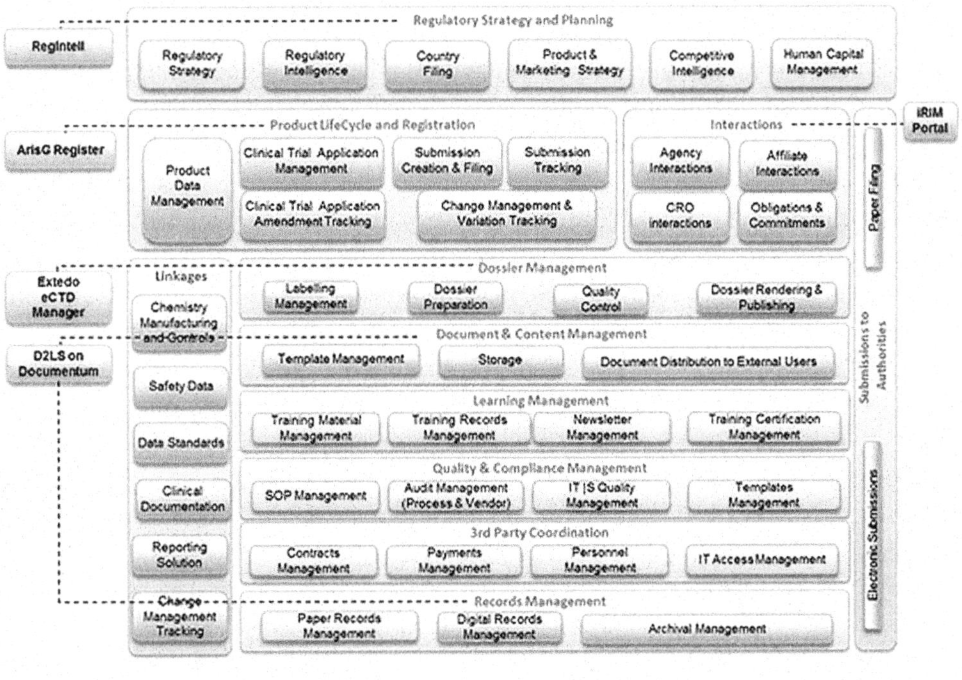

The key tenets/components of the architecture include:

- Regulatory Strategy and Planning layer to understand country filling requirements, be aware of market and competitor trends and create regulatory strategy for pipeline based on latest guidance from Health Agencies
- Product and Lifecycle Registration layer to manage and track registration process of product across its whole lifecycle
- Interaction Layer to manage interaction between central office, affiliates, CROs and Health agencies
- Dossier Management layer to prepare and manage dossiers for submission to Health agencies
- Document and Content Management layer to author and store submission related documents that can be integrated with dossier management layer to publish and submit dossiers to Health agencies
- Learning Management Layer to manage training requirements, materials, records and certifications across different affiliates and central office.
- Quality and Compliance Management Layer to manage SOPs, templates etc. across different stakeholders to establish same business process and avoid gaps
- 3rd party coordination Layer to manage contracts, payment, personnel, access etc. across different parties involved
- Records Management Layer to manage paper and digital records and perform proper archival management depending on record types
- Linkages layer for data standardization, documentation and change management tracking and reporting across different business process involved
- Submission to Authorities Layer to manage submissions, acknowledgements and approvals from health agencies and track it properly with help of Interaction and Product and Lifecycle Registration layer

RIM should encompass one whole complete chain of events where information is flowing through continuously, end to end in a way that makes it easy for the business user to manage and track information. Some companies use RIM to do registration tracking or regulatory document management in isolation, but this application of RIM creates unnecessary and inefficient silos. It creates problems because manual systems are needed to bridge information flow gaps.

The *success measures* of the RIM system will depend on the following aspects:

Given the opportunities to improve regulatory strategy and function via a networked collaboration model, it is recommended to develop RIM as one single networked collaboration platform.

This one platform encompasses the creation of all documents, including content from clinical, quality and manufacturing and it includes authoring and publishing systems to enable complete lifecycle management of the submission. The business reality is that dossiers are developed and maintained either individually or as a group so an intuitive and innovative system needs to accommodate two sets of approvals. When submissions are approved internally, all components should be interconnected to ensure all content gets approved at the same time. When QA is releasing a product, they want the confidence of knowing the product is ready to be released based on accurate product and market information. One integrated system can accomplish this important business goal.

Table 1. Key Success Measures of the integrated regulatory information management platform

Long Term Goals	Short Term Objectives
Demonstrate Compliance: • Enable global source of trusted regulatory information and automation; address critical touch points such as supply release Reduce Total Cost of Ownership: • Reduce or eliminate manual work due to lack of automation and many individual disconnected systems (many systems have been tagged for decommission and resources to be redistributed or reduced) Increase Efficiencies: • Consolidate and transform central, regional, and local systems and processes; improve information exchange (internal, external) • Increase productivity by shifting resources to value-added activities and support of increased workload • Enable transition to offshore routine operations activities • Enable Standardization	• **Transform Product and Life Cycle Registration Management:** One authoritative source and standardized global processes within 2 years • **Improve Regulatory Data Quality:** Finalizing Regulatory Master Data Management scheme along with the proper data and process governance to ensure regulatory data quality. Implementing and operationalizing the Identification of Medicinal Products (IDMP) standard • **Intelligent Regulatory Affairs:** Supporting, gathering, and managing local country filing requirements and regulatory soft intelligence • **Unified View:** Providing an integrated view of all regulatory information from a unified and highly user-friendly information interface (from both the data entry and information consumer perspectives) that will transform how people work and perform their regulatory activities. Establishing information and process connections with regard to key touch points • **Scalability and Flexibility:** Practically and economically scaling up to support all sector divisions. Providing scalable architecture that is either based upon or can easily adopt a cloud solution. Transforming content management and archive layer in the long term • **Improve Decision Making and Focus on 'Core':** Providing an externalization model to supplement publishing resources to add value and reduce cost. Delivering an on-demand capability to aggregate information for real-time executive and operational reporting

In addition, RIM needs to respond to the plethora of correspondence with various agencies in multiple countries. All documents and information should be flowing through the submission gateway and be included in the submission history and tracking. Submission planning and registration needs to be customized to different counties.

DEEP-DIVE INTO THE IMPLEMENTATION OF INTEGRATED REGULATORY INFORMATION MANAGEMENT PLATFORM

This chapter addresses the details of building architecture of integrated Regulatory Information Management platform and key design considerations associated with it.

The chapter is intended for various types of audiences with diverse priorities, objectives and technical perspectives including business analysts, IT architects, IT designers and developers. Given this diversity, this chapter seeks to provide a high level and focused view on the architecture capabilities and definitions.

The key sections of this chapter are:

- Section 1.4 highlights the key architectural guiding principles that Regulatory Information Management platform should adhere to, throughout the design and implementation in order to create a robust, scalable and extensible solution.

- Section 1.5 elaborates an illustration of use of custom off the shelf product in the architecture. It is important to note that use of specific product is for illustration only. Author does not intend to promote any specific product for this purpose
- Section 1.6, 1 .7 and 1.8 elaborates the enterprise view, Logical and technical architecture view of the platform respectively. The architecture views will provide for a comprehensive and layered architecture with all the required functional and system components that will enable a robust, scalable and extensible architecture for the platform. In particular the architecture will capture the key architectural elements for integration, modularity, service reuse and security for the integrated Regulatory Information Management platform. These sections provide in detail the various core components of the Product Architecture – Portal, Security, and Integration Architecture views. For each of the views the respective sections provide elaboration for all the key components of the architecture and its interactions, dependencies. Also the architecture views will demonstrate how the functional scenarios/use cases (and system use cases) are realized through the core components (and its interaction with other components) that take up the architecture.
- Section 1.9 will illustrate an execution strategy for implementation of the platform
- Section 1.10 elaborates consideration of deployment of the integrated Regulatory Information Management platform.
- Section 1.11 explains a recommended support model for the platform
- Appendix A elaborates in detail the key architecture and design decisions, motivation, implications and dependencies. The architecture and design decisions have been rationalized based on the specific functional and non-functional requirements of the integrated Regulatory Information Management platform.

ARCHITECTURE GOALS AND CONSIDERATIONS OF THE RIM PLATFORM

Following are the key architecture requirements and constraints that have a significant bearing on architecture, design and building the integrated Regulatory Information Management platform:

- **Capability Driven Architecture:** The architecture will have the technical and functional capabilities with pure plug and play architecture that can be switched on/off based on the configurations. The architecture will also be providing integration points with various sub systems and have connectors open to adapt to other systems.
- **Modular:** Individual components can be built and integrated in a phased approach, and replaced (Lego-block approach).
- **Reuse and Extensibility:** Extensible architecture enables components to be adopted and customized from existing proven industry solutions, sponsor systems and best-of-breed technology, and architecture patterns. This reduces risk, allows faster build time, and enables utilization of existing knowledge.
- **Service Oriented:** Ability to independently build, assemble, deploy and consume specific business services for varied stakeholder (regulatory users, CMC users, LOC users) needs
- **Multi-stream development:** Allow multiple development tracks to reduce build time, ease integration, and reduced priority conflicts

- **Lose Coupling and Highly Cohesive:** Well-isolated, and tiered layers and component
- **Internationalization and Localization:** The various components within the platform support multiple languages and can easily be configured depending on the need
- **Security:** Architecture will enable a federated security model ensuring message confidentiality and message integrity during message conversation between two or more end points, in compliance with SAML 2.0 authentication scheme enabling web-based authentication and authorization scenarios
- **Compliant with Pharma Regulatory standards:** The architecture is to comply with industry specific regulatory requirements and compliance such as 21 CFR – Part 11 and Safe-harbor certification
- **High Availability:** Application will ensure high availability for stakeholders by adhering to high standards of RPO (recovery point objective) and RTO (recovery Time objective)
- **Scalability:** scalability at the various layers – web, app and persistence layers – adopting hardware, tools and methods that ensure scalability and elasticity such as Load Balancing and Clustering services, virtualization
- **Hosting Options:** Solution is flexible to allow dedicated or cloud based hosting options
- **Build-vs-Buy options:** An optimized mix of standard COTS products seamlessly integrated with bespoke solution components

USE OF OPEN SOURCE AND COMMERCIAL CUSTOMIZED OFF THE SHELF PRODUCT/SOLUTIONS

Industry has witnessed new and innovative products and solutions in the emerging areas of technology like social media, collaboration platform, security solution, analytics and SaaS / cloud solution. The new integrated Regulatory Information Management platform needs to utilize these innovative solutions. As it is evident from the description of functional & architectural goals of the platform, specific point solutions are required in the area of security, portal technology, middleware, database technology, document management and cloud technology etc. This case study mentions specific solutions, available in this area, It does not intend to recommend or promote these solutions. Solutions with similar capabilities can certainly be used to create similar architecture and commission an industrialized solution.

Use of open source technology can assist to bring down the overall cost of ownership. This case study has referenced to certain set of open source technology. Specific steps need to be adopted to ensure that platform remains compliant with regulatory guideline

Product Selection Approach:

A three stage structured approach is adopted for deciding the optimal product stack for Regulatory Information Management solution.

- For RIM Transformation functional areas, products were listed which could potentially meet the requirements for a functional area or across functional areas
- From a potential list of products, a short-listing was done to narrow down to smaller list based on the overall fitment w.r.t. to RIM transformation goals
- There was a detailed evaluation of the products from the final list based on key parameters to come up with the overall product stack.

Following are the parameters which were used for evaluation:

- Requirements Fitment
    - The extent to which product is able to meet the functional requirements with out-of-the-box features and configuration. Requirements which could be met by customization or part of future roadmap were also noted down.
- Product Maturity and Industry Position
    - Maturity of the product in the market w.r.t. commercial releases and industry adoption i.e. how many live and in-progress client installations.
- Fitment with the overall RIM Transformation solution
    - How well the product ready to integrated in larger environment with other new products and existing set-up
- Flexibility
    - Ability to choose a replacement based on changing business and technical priorities and dynamic regulatory environment
- Performance and Scalability
    - Ability to meet the standards of enterprise class regulatory platform for top pharmaceutical organizations and supporting global user base and volume of regulatory submissions. This also included analysis of how ready product is for cloud hosting and integration.
- Price Sensitiveness
    - Inclination to invest in the partnership and overall price competiveness across product license, implementation and support.

Apart from the above, *COTS products* were evaluated for some of the technical components of the integrated Regulatory Information Management solution. For example, security, self-registration form a significant part of the This can be accomplished by using commercial off the shelf solution. The table below outlines different components of such solution.

ARCHITECTURAL PATTERNS APPLIED IN RIM PLATFORM

An architectural style, sometimes called an *architectural pattern*, is a set of principles—a coarse grained pattern that provides an abstract framework for a family of systems. An architectural style improves partitioning and promotes design reuse by providing solutions to frequently recurring problems. Garlan and Shaw define an architectural style as: Even though an architectural pattern conveys an image of a system, it is not the architecture itself. Multiple architecture options may implement the same pattern and share the related characteristics. Patterns are often defined as "strictly described and commonly available". For example, the layered architecture is a call-and-return style because it defines an overall style to interact. When it is strictly described and commonly available, it is a pattern.

Some of the key *architecture patterns* used in best of breed solutions include the following:

Integrated Regulatory Information Management solution requires combination of architectural patterns & design patterns. Both functional as well as the architectural goal need consideration before elaborating in to specific patterns, those are required to commission a platform like this. Detailed descriptions of

Table 2. Regulatory COTS products evaluated and selected

Sr No	Requirements category	Capability	List of Products (Evaluated)	Product Stack	Comments
1	P&LCR	Submission Planning	• Microsoft project • Clarizen • Oracle Primevera • Extedo eCTDmanager – SCPmanager	Extedo eCTDmanager - SCPmanager	Planning as is an integral part of regulatory cycle. Difficult for a generic project management tool to fit-in and integrate.
2	P&LCR	Product and Life Cycle Registration	• Extedo MPDmanager • Aris G Register	Aris G Register	
3	HA Information Management	Spontaneous Health Authority Queries	• Rosetta Pyramid • Irim	Irim	
4	HA Information Management	Health Authority Questions and Answers	• Rosetta Pyramid • iRIM	Irim	
5	General Content Management	Submission Content Management	• Veeva Vault Submissions • D2LS on Documentum • Qumas R&D Solution on Documentum	D2LS on Documentum	
6	Publishing	Publishing	• Extedo eCTDManager • Lorenz Docubridge	Extedo eCTDManager	
7	P&LCR and HA Information Management	Regulatory Information Archive and Correspondence Tracking	• Rosetta Pyramid • Aris G Register	Aris G Register	
8	HA Information Management	Country Filing Requirements	• RegIntell bespoke solution) • • Thomson Reuters Corellis	RegIntell	
9	Label Content Management	Labeling	• PRISYM ID • Lorenz Labelbridge • Virtify • Extedo - XML Content Authoring & Management (e. g. DITA) • CCDS implementation Tracking		No clear winner. Virtify does support SPL only

design consideration are described in Appendix A, along with the specific topic, description, assumptions & constraints, motivation, implications, dependencies, alternatives & recommendation.

The platform architecture is composed of the following key architecture patterns:

- The platform architecture applies the principles of service orientation to model components and services that will directly enable the business processes and functional features of the integrated regulatory information management platform. Given the integrated regulatory information management platform to be used as "the system of engagement" for different stakeholders such as CMP users, LOCs etc. the principles of service orientation enables composing reusable business services for planning, tracking, labelling, translation etc. and thereby ensuring an extensible architecture.

Table 3. Technology COTS products evaluated and selected

Component	COTS Product / Technology	License (commercial / open source)	Description
Access Manager/ Federation	CA SiteMinder	Commercial	CA SiteMinder Secure SSO & Flexible Access Management can provide enterprise-class secure single sign-on (SSO) and flexible identity access management so that integrated clinical colloaboration platform can authenticate investigators, Sponsor users, other site users and control access to applications and services deployed on member companies' trusted network. CA SiteMinder with Federation capability can be used for securing access to Portal and LMS
Policy Store	CA Directory	Commercial	The SiteMinder policy store (policy store) is an entitlement store that resides in an LDAP directory server. The purpose of this component is to store all policy-related objects, including the: • Resources SiteMinder is protecting • Methods used to protect those resources • Users or groups that can or cannot access those resources • Actions that must take place when users are granted or denied access to protected resources
Web agent +Web agent option pack	CA Secure Proxy Server	Commercial	This component is used as web agent and also as proxy engine for Reverse Proxy
IDM	CA Identity Minder	Commercial	IDM is used for provisioning the users and to maintain work flows
Provisioning	CA Provisioning Server	Commercial	Provisioning server can be used by IDM for provisioning the users to end points.
Provisioning Directory	CA Directory	Commercial	Provisioning Directory can be used for storing the users
Reports	CA Reports Server	Commercial	Used by IDM and SiteMinder to generate reports

Table 4. Commonly used Architectural Patterns

Architecture Pattern	Description
Client/Server	Segregates the system into two applications, where the client makes requests to the server. In many cases, the server is a database with application logic represented as stored procedures
Component-Based Architecture	Decomposes application design into reusable functional or logical components that expose well-defined communication interfaces.
Domain Driven *Design*	An object-oriented architectural style focused on modeling a business domain and defining business objects based on entities within the business domain.
Layered Architecture	Partitions the concerns of the application into stacked groups (layers).
Message Bus	An architecture style that prescribes use of a software system that can receive and send messages using one or more communication channels, so that applications can interact without needing to know specific details about each other.
N-Tier / 3-Tier	Segregates functionality into separate segments in much the same way as the layered style, but with each segment being a tier located on a physically separate computer.
Object-Oriented	A design paradigm based on division of responsibilities for an application or system into individual reusable and self-sufficient objects, each containing the data and the behavior relevant to the object.
Service-Oriented Architecture (SOA)	Refers to applications that expose and consume functionality as a service using contracts and messages.

MSDN Library - Application Architecture Guide, 2nd edition (Oct, 2009) – Chapter 3: Architectural Patterns and Styles (Retrieved from: https://msdn.microsoft.com/en-in/library/ee658117.aspx)

Figure 4. Logical Solution Architecture of the RIM Platform

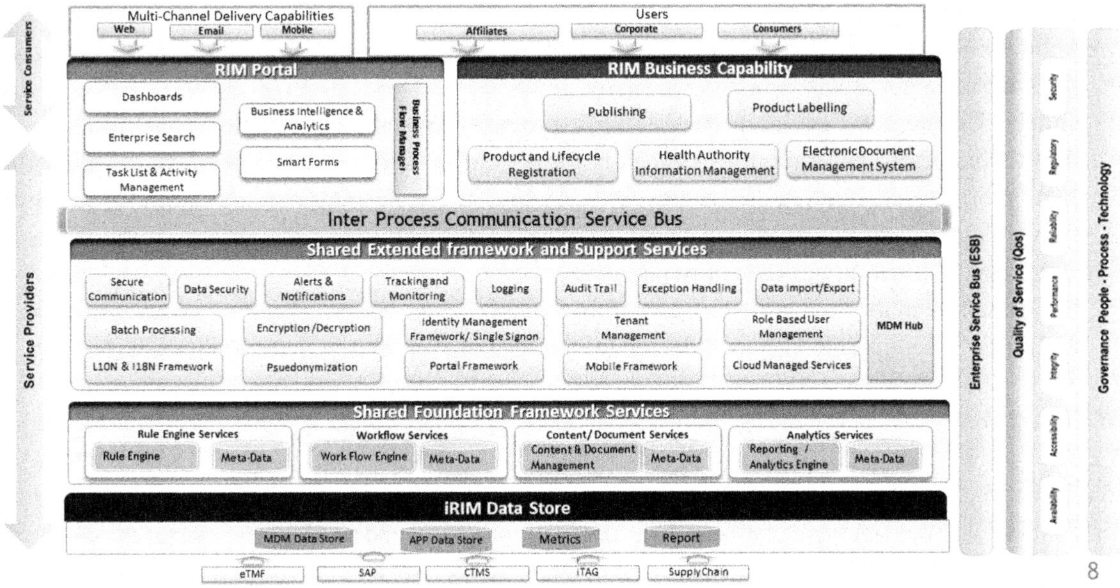

- The platform architecture applies the principles of loosely coupled message orientation to enable service based integration to integrate regulatory information management platform with member systems (such as HA).
- The platform architecture provides for orthogonal services for unified security and governance for the integrated regulatory information management platform

LOGICAL ARCHITECTURE OF RIM PLATFORM

The enterprise portal (‘**iRIM’**) is envisaged to be the entry point to all Clinical, Regulatory, Safety, Sales and Marketing and Manufacturing portals. Based on access levels, the user can be automatically directed to the appropriate portal of the business function to which (s) he belongs. The ‘**iRIM**’ portal can be the hub for all regulatory activities and information sharing at the central, regional as well as country levels.

The key components of the solution architecture include:

- **A Multi-Channel Enabled Enterprise Portal (iRIM):** The portal is a single point entry into the RIM platform. The portal can be based on responsive web design to cater to multiple channels with customized style sheets and themes to enable seamless access to task lists, COTS products and other enterprise apps. The portal is built using a SharePoint based portal platform which is fully equipped with workflows, business process management and offer integration with other products through web services
- **iRIM Business Capability:** The platform is built using accomplished and validated COTS products that provides uniform and standard business processes.
- **Shared Foundation Framework:** The platform is built using common framework components such as a workflow engine, rules engine, Documentum and Reporting services using Informatica.

- **Shared Extended Framework and Support Services:** The platform provides a set of core services that manages the operations on the cloud including Identity Management, Single Sign on, Audit Trail, Data Security, Internationalization, Localization, User Management and Data services for managing master data.
- **Shared Connector Services:** The platform provides extensive integration capabilities via its shared connector platform built specifically for the cloud. The connector platform comes with its own enterprise service bus and data services which facilitate inter process communication and encapsulate the various products. This allows for loose coupling of the various components and improves the flexibility of the entire platform.
- The complete platform can be hosted within *validated regulatory compliant cloud environment.* The cloud is set up to manage validated systems that are highly regulatory in nature.

iRIM Portal

The Regulatory Portal can be implemented using a Customized *Sharepoint based Solution* which is a collaborative work management platform. This platform automates and accelerates sustained conversation and information exchange to attain specific program and project goals. The solution connects customers, employees, and partners on a single cloud, mobile, and on-premise-enabled collaboration tool, offering faster and easier access to experts and expertise, information discovery, and exchange, and ultimately leading to better decision-making.

This workflow management platform leverages a mature, field-tested framework to simplify and encourage collaboration and the relevant exchange of information. This platform solution:

- Enables companies to capture collective knowledge by providing a single, unified framework and interface for simplifying and expediting business- and technology-based conversations and content exchange
- Fosters efficient and meaningful communications by reducing "noise" and minimizing interrupted conversations
- Enables faster and better problem-solving by connecting people with the relevant knowledge and topic experts
- Increases productivity and job satisfaction across distributed teams

Features of the Customized Workflow Management platform solution include:

- Real-time Activity Streams
  - Activity Streams provide real-time updates and aggregate information silos onto a single stream. The stream also enables discoverability of knowledge and people or subject matter experts
- Work Inbox
  - The "Work Inbox" shows what's due immediately and what else is pending – all in a stream-based view. Workers can directly update the status of the work item, attach or download artifacts, and have uninterrupted conversations with team. This is true collaboration in context of work

- Full-featured Knowledge Management
  - The platform's proven, process-centric knowledge management methodology ensures that employees can quickly classify and cross-reference business-related information
- Global Stakeholder Communities
  - The platform's community-based approach provides an integrated communication framework that encourages team engage and interaction. Employees can create work-based or interest-driven communities that are open, moderated, or private, "Follow" others and build their network inside the organization, share status updates with #hashtags and @mentions, post queries, engage in conversations, share documents and photos, discover knowledge and experts – all delivered and managed in a specific work context
- Flexible Search
  - Powerful search mechanisms index both unstructured and structured data enabling teams to get the right information and people at the right time
- Secure web based search
  - All work related artifacts are available on a secure web-based platform which ensures extremely low dependence on resources and enables seamless knowledge transfer among globally distributed members.

Shared Connector Platform

A shared connector platform is implemented by *a customized Cloud hosted Integration (Connector) Solution.*

The key features of the solution include:

- Transfer and synchronization of data between multiple backend systems
- Transfer documents from backend systems to presentation layer (RIM)
- Provide the platform to develop services that are required for RIM
- Provides a layer of service orchestration to build a process workflow that interacts with multiple services
- Route messages to proper consumers
- Filter messages based on the content, etc.
- Mediates between the presentation layer (RIM, etc.) and the backend systems

Foundation Framework

- *Reporting and Analytics*

The Reporting component provides tabular and graphical reports from the source systems. The Analytics component provides features to analyze data including trends and forecast. Key capabilities of the Reporting & Analytics component are,

- Chart based reports
- Map based reports

- Format the search results within the RIM
- Historical data
- Analysis – trends, forecast
- *Document Management using EMC Documentum*

EMC Documentum can be the central Document Management System. RIM system will provide features so that the user can perform document related activities of the RA workflow through the system. These features includes the ability to create a new document, create a new version of an existing document, check-in/check-out, view documents within the RIM, update document metadata, annotation, view document history. RIM will use the services provided by the backend document management system (Documentum) to integrate with the Document Management System.

- *Hosting provided by validated cloud environment*

Cloud Infrastructure Services support planning, building and managing cloud environments with a portfolio of flexible choices. They include our own hosted data center and public or private cloud offerings that host virtual data centers, multitenant enterprise services, and virtual data desktops.

The hosted and managed cloud infrastructure of the current case study runs in world-class data centers and gives enterprises choice and flexibility for managing mission-critical applications in a hybrid cloud environment. Customized Cloud Infra Management solution is used for provisioning, automated monitoring and orchestration of multiple clouds, One can manage and scale applications without adding hardware or hiring specialized skills. The hybrid IT cloud infrastructure can be based on choice of heterogeneous cloud platforms, including:

- Private, public, and hybrid deployment models
- SaaS, PaaS, and BPaaS delivery models

Security Framework

Single sign on to the cloud based system is enabled through use of standard COTS products such as Site Minder. All users can be authenticated against Sponsor's Active Directory before being granted access to the Cloud environment. A copy of the active directory can be maintained within the cloud for managing workflow and communication to various stakeholders.

Other Framework Services

The platform has a set of common reusable services which will manage audit logs, data encryption, internationalization, localization, user role management, cloud infrastructure management, exception handling, alerts and notification services. The data services exposed within the ESB layer will make use of these common services while communicating between various products or with backend systems.

TECHNICAL ARCHITECTURE

The various components of the technical stack are as follows:

1. Multi-Channel Delivery Capabilities: (HTML5, Sencha, Phonegap, RWD) – the platform can be available across multiple channels (desktops, browsers, iPad, mobile) using Responsive Web Design (RWD). The platform envisions future native application support for mobile devices using technologies such as Sencha and Phonegap
2. The iRIM Portal is hosted on Microsoft SharePoint using a Customized Sharepoint based framework solution. This Platform solution connects all other COTS products and frameworks on a single cloud, mobile and on premise collaboration tool providing faster and easier access to information discovery and exchange. Business process monitoring can also be supported using this workflow management platform
3. The iRIM Business Capabilities hosts Commercial Off-The-Shelf (COTS) applications hosted on cloud. These COTS products manage the various business processes of the RIM platform
4. The communication mechanism from the channels to the portal and the business applications can be using HTTP(s) protocol
5. The platform should be equipped with a Customized enterprise Cloud hosted Integration (Connector) solution with an inbuilt ESB. All the portal and business applications can be communicating using this platform, which can provide Data Services and proxy connector services (for pass through communication between COTS products and backend systems). All process communication with happen through these services
6. Support and Extended Services
    a. The communication between presentation services are JSON and SOAP. The plain old java object (POJO) can be used across layers
    b. For the authentication services SSO, LDAP and O-Auth can be used along with Microsoft Active Directory. SSO can be managed through the Siteminder tool which will interface with the Sponsor's SSO system for authentication, access control and managing credentials
    c. ETL Batch should be primarily used for the batch processes
    d. Centrally Managed Security and Data encryption
7. The Shared and Extended Services communicate between Integration framework and the Shared Foundation Services using SOAP.
8. Shared Foundation Services
    a. For the rules and workflow services the workflow management platform solution can be used
    b. EMC Documentum can be the chosen content management system for the RIM platform
    c. Informatica can be used for reporting and analytical services
9. iRIM Data Store
    a. The primary databases used can be Oracle, SQL Server and Informatica.
10. The Sponsor Enterprise Applications (for SAP, Supply Chain) can be able to communicate with the platform with Integration framework using SOAP
11. All the external applications will be able to communicate and integrate with the platform using the Customized Integration Platform ESB.

Table 5 depicts the *key capabilities* considered for the solution architecture along with technology options.

Integration Layer: Enterprise Service Bus (ESB)

The custom developed integration framework is a high-performance, service-oriented integration platform that helps connect any combination of on-premise and cloud-based applications. It also provides a powerful service delivery framework that allows exposing a variety of integration and data services for consumption in the most optimum format

Integration Platform:

- This integration platform will host the integration flows and services to integrate the core RIM systems of Enterprise Content Management (Documentum), Master Data Management and Publishing and Submission Management with the Lines of Business (LOB) Applications and Enterprise Applications
- The integration platform will integrate the core RIM systems hosted in the cloud with the LOB Applications hosted in the cloud
- The integration platform will also integrate the RIM systems hosted in the cloud with the Enterprise Applications hosted on premise.

Table 5. Portal Capability - SW Context Map

Component	Technology Choice	Remarks
Thin-client UI	Responsive Web Desgn using HTML 5 and CSS	1. Configurable navigator flow 2. Plug-n-play architecture
Integration Engine	Customized Integration Frameowrk	1. Multiple protocol handlers available
Persistence Framework	JDBC/ ODBC	1. Part of the Javaz Platform 2. Abstracts database specific interactions
Rules Engine	Customized Sharepoint Frameowk	
Web Services	SOAP and JSON	1. Supports required WS-* Security standards 2. Supports RESTful web services
Database	Microsoft SQL Server Oracle	1. Industrial strength 2. Proven in high volume/transaction 3. Supports Open Standards
Workflow	Customized Sharepoint based workflow management framework	• Prebuilt libraries can be leveraged from the platform
Authentication	Cloud hosted integration platform	
Application Server	IIS JBOSS	
Web Server	IIS APACHE	
Portal	IIS APACHE	

Integration Features:

- The powerful Apache Camel framework serves as the integration backbone for the platform.
- Inbound and outbound connectors for almost all transports
- Very low latency flow execution (Wire formats such as Google Protocol Buffers and Protostuff reduce the payload size by as much as 80% and parsing time in orders of 1000s when compared to traditional formats such as XML)
- Data parsing and manipulation using a Customized DataMorphosis solution
  - Marshal / Un-marshal any non-structured data (CSV, fixed length, delimited, etc.) into structured Java Beans for easy handling
  - Easy to use, graphical data transformation tool that helps map the elements of source and target Java objects using simple drag-and-drop techniques. Support for complex mapping of data elements by using custom code snippets
- Support for real-time as well as batch integration scenarios. Support for optimized handling of large files from various sources such as FTP sites in a streaming mode
- Support for all the enterprise integration patterns, thus allowing for a standards based mechanism for implementing any complex integration scenario
  - Service delivery platform
- Integration Framework will expose business services and data services for a variety of potential consumers through the most optimum message format
  - Workflow and Rules framework will consume the services from the user experience portal application
  - Mobile applications will consume the services in the most optimum data format for mobile devices

Service Delivery Features:

- Build services using simple Java and expose them automatically over multiple data formats
  - Formats supported include SOAP, POX (plain old XML), Fast Infoset, JSON, Smile (Binary JSON) and Protostuff with ability to plug in additional formats as required
  - Various consumers including Web Portals, Mobile Apps, external Web Service clients, internal clients can consume the same service in the formats of their choice
- Rapidly expose Data Services over a variety of data stores
  - Support for data persistence and retrieval from any kind of database (relational, NoSQL, object-based, document-based, graph-based, hybrid, etc.) using the very powerful Spring Data framework as the data abstraction layer
- Build Integration Services for bridging and mediating any internal or external application
  - Integration flows can be exposed through service interfaces for universal consumption
  - Services can be composed into newer services to offer higher level abstractions
- Protect services using sophisticated authentication and authorization standards
  - Standard service interfaces for authentication and authorization with ability to swap providers
  - Out-of-the-box support for authentication using OAuth 2.0, LDAP, Active Directory, enterprise identity stores, etc.

- Out-of-the-box support for authorization using XACML policies, thus offering a robust and flexible attribute based access control model

- Monitoring dashboard for analyzing service usage and performance metrics through various dimensions such as consumer identity, date ranges, response times, et al.

Platform Layer: Master Data Management (MDM)

A master data management (MDM) hub that manages R&D transactional entities and regulatory information on Informatica MDM and provides integrated Data Cleansing and Standardization, Data Modeling and Data De Duplication capabilities. The MDM Hub solution would ensure that Sponsor is able to get a 360 degree view of its critical Regulatory Data Entities. The Hub solution will ensure that despite existence of multiple source systems that lack integration, the business users can be able to access a single version of truth which is accurate and consistent and which requires significantly low cost of maintenance. MDM hub will master the key entities involved in the regulatory process that will form the backbone of the RIM system landscape. The MDM hub can be integrated with data sources, and users can be provided cleansed, consolidated and de-duplicated regulatory entity data –

- Products (Development Products etc.)
- Affiliates (CRO, Authoring Companies etc.)

MDM Hub

Data from the source systems can be loaded into the MDM Hub via the Data Quality Hub using Informatica Power Centre and Informatica Data Quality (IDQ) suites through the following mechanisms:

- **Master Data Identification & Extraction:** The master entity attributes can be identified and extracted from the source systems
- **Generic Delta Detection Framework:** A highly scalable and configurable generic framework can be built to identify & process the delta records for the ongoing incremental batch loads
- **Data Validation and Transformation:** Data validation and transformation rules would be defined and built to implement the relevant MDM specific functional rules
- **Cleansing and Standardization:** Cleansing and standardization rules can be developed for cleansing and standardizing the master attributes for the purpose of data de-duplication.

Informatica MDM tool can be used to develop the hub of MDM solution. The Key features of this component have been explained below:

- **Address Standardization:** Standardization of addresses
- **Data matching and De-Duplication:** De-duplication done on matching attributes using business defined match rules. Consolidation and survivorship rules are configured to define the survivorship mechanism
- **Hierarchy and Relationship Management:** Out of the box rules can be configured to provide the hierarchical view and relationships

- **Cross Reference:** The MDM hub maintains cross reference information and creates linkages between the raw and consolidated data.
- **Auditing and History Management:** The MDM solution would provide the ability to maintain history and audit trail for the changes done to regulatory master data. This feature is highly configurable and easy to maintain.
- **Security:** The MDM solution would have security features like ability to authenticate systems and users. MDM shall have the ability to authorize access to users based on their roles (Data Steward, Business User and Developer)

MDM Data Governance

Business Goals for Data Governance

- Compliance with internal and external regulations for data usage and reduce risk exposure relative to data and its use.
- Business value generated from our data and information assets

Technical Goals for Data Governance

- Establish and enforce standards for data
- Improve data quality; remediate its inconsistencies; share data

The Data Governance Strategy involving Collibra has the following key considerations:

- Analysis on Industry trend and best practices - Evolving Global Standards
- Operational Data Stewardship Structure
- Enterprise-level Data Governance Strategy
- Enterprise-level Metadata and Reference Data Strategy
- Evolving Content Management Technology
- Stakeholder Support

R&D commercial data governance software (Collibra) for regulatory master data fulfils the following objectives:

- **Reference Data Management:** MDM solution provides the ability to store and manage the reference data using Collibra
- **Data Dictionary:** Data Dictionary or business glossary of MDM hub solution can be developed by using Collibra tool, which would have the ability to store technical name of entities attributes, business entities, business concept, rules repository etc.

MDM Data Stewardship

Data Stewardship processes relate to enforcement of policies related to data definition, quality, security and usage of data in accordance with established guidelines. The primary operational support for Data Stewardship includes establishing a stewardship framework at SPONSOR including setting up SOPs and metrics that would enable SPONSOR to conduct data stewardship

Stage 1: Define Baseline

In the first stage, baselines & SOPs for stewardship activities including scope of the Stewardship model are defined in consultation with client. Following high level activities are part of this stage:

- Mapping the Master Data attributes to the source system attributes
- Closely work with the client team in defining and documenting the Stewardship Process
- Define Standard operating procedures to for streamline business process across the organization and tackle any issues

Stage 2: Establish SLA

- Interact with the client team to understand their prioritization and SLA requirements
- Throughput is monitored and measured over the first few weeks and the baseline throughput SLA is defined

Stage 3: Steady State

- All activities are executed within agreed scope
- Match, Merge or Linking /Delinking process continues for incremental data
- Exceptions are handled and documented as per the Business Processes

Following additional activities to be performed as part of Data Stewardship process:

Potential Match

- Search records in the manual merge queue
- Validate the potentially matched records based on the understanding merge the records

Other Stewardship Activities:

- Data Investigation based on user request
- Merged record analysis_

The MDM UI would be built using configurable features of the Informatica Data Director tool to provide the role based Data Stewardship CRUD activities like search, record insertion, profile update, identify duplicates, merge and unmerge operations.

- The data in the MDM hub provides a 360 degree view of the entities through this UI
- It would also have the ability to provide the hierarchical view of alliances amongst the entities
- This will also act as the single point of onboarding interface for in scope master data.
- Approval workflow is another feature which can be leveraged and customized to onboard, merge and unmerge regulatory information.

The de-duplicated data is published as gold copy records in a Publish area (either in database tables or in materialized views) through a batch mechanism. These golden/mastered data is extracted by the data integration layer from the publish layer to feed into the Datamart and other downstream systems

Technology stack of the proposed Data Aggregation and MDM Solution is shown in Table 6.

Presentation Layer: Dashboard, Metrics, Task list, Alert

The Regulatory Portal can be implemented using the Customized Sharepoint based workflow management portal Solution.

Features of the Customized Workflow management portal solution include:

- Real-time Activity Streams
  - Activity Streams provide real-time updates and aggregate information silos onto a single stream. The stream also enables discoverability of knowledge and people or subject matter experts
- Work Inbox
  - The "Work Inbox" shows what's due immediately and what else is pending – all in a stream-based view. Workers can directly update the status of the work item, attach or download artifacts, and have uninterrupted conversations with team. This is true collaboration in context of work

Table 6. MDM and Data Aggregation - SW Context Map

MDM Hub	
Architecture Layer	**Technology Used**
Address Standardization Engine	AddressDoctor
ETL Integration Layer	Informatica PowerCentre
Application Server	Weblogic/Websphere/jBoss Application Server
Database Server	DB2/Oracle Database server
MDM Hub	Informatica MDM
Data Stewardship UI	IDD
MDM Governance	Collibra

- Full-featured Knowledge Management
    - The workflow management platform's proven, process-centric knowledge management methodology ensures that employees can quickly classify and cross-reference business-related information
- Global Stakeholder Communities
    - The platform's community-based approach provides an integrated communication framework that encourages team engage and interaction. Employees can create work-based or interest-driven communities that are open, moderated, or private, "Follow" others and build their network inside the organization, share status updates with #hashtags and @mentions, post queries, engage in conversations, share documents and photos, discover knowledge and experts – all delivered and managed in a specific work context
- Flexible Search
    - Powerful search mechanisms index both unstructured and structured data enabling teams to get the right information and people at the right time
- Secure web based search
    - All work related artifacts are available on a secure web-based platform which ensures extremely low dependence on resources and enables seamless knowledge transfer among globally distributed members.

Deployment of RIM platform

The RIM platform architecture comprises multiple layers. The request from the browser can be redirected to a load balancer. The load balancer will redirect the traffic to one of the web server cluster nodes. The request can be processed by the web server, and then can be directed to another load balancer which will route to one of the application servers based on the current load. Load balancer and clustering of web/application servers would be done to ensure high degree of scalability, performance and availability of the system. The deployment architecture of the integrated RIM platform needs to support all the functional and architectural goals stated earlier including high availability and scalability. The below diagram depicts the deployment architecture for the platform from portal services view.

In order to make specific architectural provisioning for high availability and disaster recovery, uptime, recovery point objective (RPO) and recovery time objective (RTO) need to be set. RPO is the maximum tolerable period in which data might be lost due to an incident. RTO is the duration of time and service level within which the system must be restored after a disaster. Combination of RPO, RTO, availability goal and anticipated load in the system will determine exact number and size of the servers and equipment.

Clustering and load balancing are the features of the deployment architecture. They ensure architectural goal of high availability and disaster recovery.

A cluster is a set of nodes (servers) that communicate with each other and work toward a common goal. In the above depiction of deployment of the RIM platform

- The web server cluster consists of multiple Apache httpd servers
- The app server cluster consists of multiple JBoss Application servers
- The persistence layer has an active-passive Oracle server cluster. In case, the active primary server goes down, the passive secondary server can be promoted as a primary until the primary instance comes alive.

The integrated RIM platform implements a load balancer to moderate the requests hitting the web and app servers. None of the one web or app servers can be fully utilized when the others is are available for utilization. The web browser sends in requests and receives responses directly over the wire using the HTTP protocol. A load balancer is required to process all requests and dispatch them to server nodes in the cluster.

State replication is directly handled by JBoss. When JBoss is run in the all configuration, session state replication is enabled by default and is replicated across all JBoss instances in the cluster. F5 load balancer is recommended to use for the platform.

Following consideration needs to be made for the load balancer:

- Security: A firewall is deployed in front of the Load Balancer
- Availability: Use of at-least two load balancers in a HA configuration to reduce single point of failure.
- Session Persistence: Session persistence should be enabled using SSL
- Algorithm: Least Connections, Round-Robin is generally acceptable
- Web application Firewall: Can be used to apply URL restrictions on the Cloud load balancer access to admin based on source address
- Advanced configurations: Can be used to allow certain types of client traffic to dedicated nodes

Oracle Data Guard ensures high availability, data protection, and disaster recovery for enterprise data. Data Guard provides a comprehensive set of services that create, maintain, manage, and monitor one or more standby databases to enable production Oracle databases to survive disasters and data corruptions. Data Guard maintains these standby databases as transactionally consistent copies of the production database. Then, if the production database becomes unavailable because of a planned or an unplanned outage, Data Guard can switch any standby database to the production role, minimizing the downtime associated with the outage. Data Guard can be used with traditional backup, restoration, and cluster techniques to provide a high level of data protection and data availability. With Data Guard, administrators can optionally improve production database performance by offloading resource-intensive backup and reporting operations to standby systems.

EXECUTION STRATEGY

The primary objective of the RIM platform strategy is to create a global regulatory platform that allows better productivity and collaboration between Sponsor and affiliates. As such, the solution to the RIM platform needs to be approached with a top-down strategy and a bottom-up execution.

A recommended approach for *future state process design* in 5 steps includes:

1. Assess current state through project deconstructions
    - Run 2-3 full-day process deconstructions and ~1-2 full-day cross-functional, cross-team value-stream workshops to deconstruct prioritized current process, identify all critical issues, and understand potential solutions
    - Complete follow-up interviews to further probe processes and identify capabilities

  - Depending on the findings, consider employing additional tools including surveys to probe further
2. Define ideal future state process through clean-sheeting
   - Use a clean sheet approach to design the "ideal" future state (e.g., how would the process look like in a perfect world with no constraints)
   - Consult with key stakeholders to brainstorm and refine the ideal future state process
3. Finalize "reinvented" future state L0 processes
   - Test zero-based, clean-sheeted processes in real-world understanding barriers and constraints
   - Redesign future state processes with global harmonization allowing for country specific differences working with key stakeholders (Exhibits 5)
   - Identify critical changes/ deviations from current state
   - Outline implications for broader changes beyond supporting systems (e.g., SOP changes, training, total cost changes, compliance needs)
4. "Granularize" L1-L4 processes
   - Logically group end-to-end processes by business value and priority and create detailed operational level flows (Exhibit 6)
   - Articulate system implications and initial end user needs and functionality requirements to support future state processes
5. Create L5 system level flows
   - Decompose processes into system level flows and process attributes
   - Create detailed technical requirements to provide as input to architecture and systems engineering teams

A recommended approach for the future state execution strategy as below:

- There is no one product that provides for end to end functionality – each of the Sponsor's processes would prefer for the best of breed solution for their area to make business responses efficient and ensure compliance to the highest level.
- There is a need to stich these together in a logical architecture – providing each his own while giving one unified view.
- This architecture needs to be on cloud and eventually be offered in in "SAAS" model
- Futuristic and in adaptable to change in accordance with the dynamic global regulatory environment.

Based on these inputs we derived the relevant point parameters leading to the end state architecture which is an integrated ecosystem comprising best of the breed products, standard and uniform processes across geographies, leveraging the cloud and move to an completely validated platform.

CONCLUSION

It is evident that the cost of clinical trials continues to escalate. An integrated regulatory Information Management solution can increase operational efficiency, which in turn leads to cost reduction and speed of delivery for the clinical trials. Advent of better digital media, collaboration and cloud technology and acceptance of externalization of IT infrastructure can make the process of collaboration real. Industry wide effort to make the process harmonized shall make adoption of technology much master.

ACKNOWLEDGMENT

My sincere thanks and acknowledgement to Cognizant Technology Solutions for my understanding of Regulatory domain and RIM solution. As an employee of Cognizant, I understood the facts of many of the cases, business problems and the reference architectures used in this chapter. I also thank Mr. Partha Chakraborty (Sr. Director – Cognizant Technology Solutions) for his help, backing and persuasion to get this chapter concluded. I have tapped into his knowledge of clinical and Regulatory domains multiple times in past 10 years years enabling me to acquire the courage to write this chapter.

REFERENCES

Extedo. (n. d.). Retrieved from http://www.extedo.com/products/the-extedosuite-for-regulatory-information-management/

Architectural Patterns and Styles (Ch. 3). (2009, Oct). MSDN Library - Application Architecture Guide (2nd ed.). Retrieved from https://msdn.microsoft.com/en-in/library/ee658117.aspx

KEY TERMS AND DEFINITIONS

API: Active Pharmaceutical Ingredients - refers to a substance or substance combination used in manufacturing a drug product.

CCDS: Company Core Data Sheet: a Primary labelling source document.

CTRL: Clinical Trial Regulatory Lead.

eCTD: Electronic Common Technical Document – is an interface for transfer of regulatory information in a common document format.

P&LCR: Product and Life Cycle Registration – Process/ system for Managing the Registration of Drug Products and their Life-cycle.

RIM: Regulatory Information Management (and Submission) related IT systems used by Pharmaceuticals.

SPL: Structured Product Labeling - is a document markup standard approved by Health Level Seven (HL7) and adopted by FDA as a mechanism for exchanging product and facility information.

APPENDIX

Table 7. Acronyms and Definitions

Acronym	Meaning
HIV	Human Immunodeficiency Virus
HCV	Hepatitis C
BRIC	Brazil, Russia, India, China,
CRO	Contract Research Organization
LoC	Local Operating Company
R&D	Research and Development
HA	Health Authority
EDMS	Electronic document management system
CMC	Chemistry Manufacturing Controls
CTMS	Clinical Trial Management System
CTA	Clinical Trial Application
MA	Marketing Application
COPP	Certificate of Pharmaceutical Products
COC	Control of Change
ADR	Adverse Drug Reaction
USPI	United Surgical Partners International
EU	European Union
SaaS	Software as a Service
RPO	Recovery Point Objective
RTO	Recovery Time Objective
URL	User Resource Locator
SOAP	Simple Object Access Protocol
SAML	Security Assertion Markup Language
SSO	Single Sign On
XML	Extensible Markup Language
CCD	Common Cookie Domain
IDP	Identity Provider
CA	Computer Associates
COTS	Common of the Shelf
SP	Service Provider
IDM	Identity Minder
SOA	Service Oriented Architecture
CMP	Content Management Portal
PaaS	Process as a Service
BPaaS	Business Process as a Service
ESB	Enterprise Service Bus

continued on following page

Table 7. Continued

Acronym	Meaning
LDAP	Lightweight Directory Access Protocol
ETL	Extract Transform Load
SW	Software
CSV	Comma Separated Values
FTP	File Transfer Protocol
MDM	Master Data Management

Product Fitment Analysis of key COTS components of the RIM system architecture (The EXTEDOsuite for Regulatory Information Management (2014) – Retrieved from: http://www.extedo.com/products/the-extedosuite-for-regulatory-information-management/) COTS evaluation: P&LCR (Product and Life Cycle Registration)

Table 8.

Requirements Category	Submission Planning
Capability	Enable Submission Planning
System Description	Submission Planning Solution (SPS) is a globally deployed system for planning submissions. Information from SPS provides input to the central publishing organization. SPS provides insight into the central, regional and local submissions required to register new products and line extensions as well to implement changes (e.g. Company Core Data Sheet (CCDS), Chemistry Manufacturing Controls (CMC) and Safety Reports). The application instance is customized for regulatory purposes.
# of users	Over 1000 users supporting 80+ global products
Chosen COTS	Extedo eCTDmanager – SCPmanager
Alternative List of COTS available	Planisware Microsoft project Clarizen Oracle Primevera SAP for Project Management Liquent Insight
Evaluation Comments	a) Liquent Insight is more a tracking than a planning tool - but may be an option b) Octagon Quantum - Requires high degree of custom development to fit into the customers needs. c) All the others a quite generic project management tools. Very difficult to integrate into something meaningfull d) SAP project Management - "dead on arrival" for the Reg.Affairs community
Chosen COTS Description	Extedo is a key Regulatory Information Management solutions provider for life sciences firms. Extedo eCTDmanager is a full suite of all-in-one submission, publishing and management solution for eCTD and non-eCTD electronic and paper submissions. SCPmanager will enable users to plan submission content on the document level. Users can be able to assign documents to any node of the submission sequence. By adding metadata, such as responsible user and milestones to documents, users can manage submission content to a specific planning level. In addition, SCPmanager provides a graphical overview of planned submission content including timelines, status and resources. Also due to the integration of SCPmanager and eCTDmanager, users can be able to view planned submissions within eCTDmanager

Table 9.

Requirements Category	P&LCR (Product and Life Cycle Registration)
Capability	Enable Product and Life Cycle Registration
System Description	With the introduction of the 2010 pharmacovigilance legislation, marketing authorisation holders (MAHs) are obliged to submit to the European Medicines Agency (EMA) medicinal product data and keep this information up to date. The EMA collects and stores this data in the Extended EudraVigilance Medicinal Product Dictionary (XEVMPD) for the purpose of assisting pharmacovigilance activities and medicines regulation in the Union. The initial deadline for submissions to XEVMPD ended on 02 July 2012. As a global company with numerous medicinal products authorized for use in the European Union, the client was in need of a tool that would enable the company to manage and submit XEVMPD data efficiently and in compliance with current regulatory requirements. With regard to deadlines and time constraints, XEVMPD data entry functionalities and user-friendliness were top priority requirements for the desired solution. Overall, the tool should cover the whole scope of medicinal product data management from maintenance operations (update, editing, and validation of XEVPRMs) to XEVMPD authoring, import and export of data, and data exchange via the gateway. Investments in time and resources for installation, validation and implementation of the solution, as well as for pulling the data together had to be kept to a minimum. Besides these immediate requirements, the company also considered future developments and upcoming regulatory requirements in their choice of an XEVMPD tool. In the intermediate- and long-term, the solution had to ensure a smooth transition to IDMP, provide planning & tracking functionalities and thereby become an integral part of the company's Regulatory Information Management System (RIMS).
# of users	For the specific implementation it was required to find an effective solution that could replace / integrate an existing implementation comprising multiple formal systems containing parts of the product registry. e) MPD/PKB (Medicinal Product Dictionary/Product Knowledge Base) global system containing current product registrations (~18,000 registrations) and serves as the regulatory product dictionary. MPD is the source for XEVMPD information and has the capability to transfer data. f) RCW - Web based custom application used for specific activities such as the maintenance of distribution lists for notification of CMC changes, recording approval status and dates of variations, managing drug substance and drug product information, tracking LOC and Health Authority questions and answers, tracking and fulfilling requests for certifications (e.g. CoPP). g) CMC Affiliate SharePoint - In Regulatory CMC, LOCs are able to request specific documents (certificates, documents required for legalization) through a SharePoint site that helps support the fulfillment of these requests. The CMC Affiliate system links to RCW to support management of HA Q&A. h) PMC Tracker – Master spreadsheet (xls) supporting the management, tracking, and reporting of the status of post-marketing commitments (PMCs) made to Health Authorities. PMC points of contact are responsible for updating the data. Each LOC has their own system (s) to manage certain aspects of the product registry and also local product management (estimated around 250 – Excel, custom solution)
Chosen COTS	Extedo MPDmanager
Alternative List of COTS available	Aris Register TrackWise
Evaluation Comments	a) Trackwise is more of a tracking than a planning tool – and not specifically customized for regulatory purposes b) Aris Register - Requires high degree of custom development to fit into the customers needs.
Chosen COTS Description	EXTEDO's MPDmanager was selectedfor maintenance and submission of its medicinal product data records. MPDmanager can be installed and ready for use within 5 staff days. Due to its intuitive and easy-to-use interface, only minimal training is required. In its functionality as a central data repository, MPDmanager allows the combining of information from different sources: Data under the scope of XEVMPD that was distributed between company sites or even managed by third parties was entered or automatically imported into MPDmanager via Excel and xml-based interfaces. The company then reviewed and analyzed the data, managed additional product attributes according to corporate standards and prepared the records for submission to the EMA in compliance with the XEVMPD standard. MPDmanager provides a holistic approach to regulatory information management that allows the company to enter, import and share all relevant regulatory master information of medicinal product data across the enterprise and easily comply with data submission regulations. With implementation of MPDmanager the clients are able to reduce staff workloads, reduce personnel expenditures, streamline business processes and increase the efficiency and accuracy of data management. This is mainly due to: Automated processes in data collection and maintenance Reuse of content making recurring data entry obsolete Lifecycle management including versioning Quick overview, reference and statistics on product data delivering useful information for regulatory affairs, pharmacovigilance, and marketing departments Also, With the data structure and processes of MPDmanager being geared towards a smooth transition to new standards, moving from XEVMPD to IDMP will be relatively simple.

Table 10.

Requirements Category	General Content Management
Capability	Submission Content Management
System Description	Enable a cross pharma document management system that enables collaborative document and submission management on a set of common standards and technology.
# of users	Over 5000 users from 80+ countries, 4.5 million documents
Chosen COTS	ECM Documentum Front end: QUMAS/D2
Alternative List of COTS available	VeeVa Vault Submissions ECM Documentum Qumas R&D Solution
Evaluation Comments	Recommended best fit in terms of platform stability, scalability and performance
Chosen COTS Description	**Documentum** is an enterprise content management platform, now owned by EMC Corporation. Products selected inlcude: a) Documentum Content Server (core product) Platform that manages content in a repository consisting of three parts: a content server, a relational database, and a place to store files. Items in the repository are stored as an object. The file associated with an object is usually stored in a file system; the object's associated metadata (a file name, storage location, creation date, etc.) is stored as a record in a relational database.[3] b) Documentum Clients Configurable clients such as Documentum D2 and Documentum xCP provide tools to eliminate the need for custom code.

Table 11.

Requirements Category	Publishing
Capability	Publishing
System Description	customer uses a combination of common off-the-shelf products to manage both paper and electronic submissions along with the typical e-submission viewers and validators. > 10,000 submissions/year with an anticipated 20% increase
# of users	10 locations consisting of about 55 publishers.
Chosen COTS	Extedo eCTDManager
Alternative List of COTS available	Lorenz Docubridge Liquent Insight Publisher
Evaluation Comments	Simplify life-cycle management of pharmaceutical submissions Automatically Build and Manage Study Tagging Files (STFs) Easy to use
Chosen COTS Description	EXTEDO's eCTDmanager is an off-the-shelf electronic submission management solution that satisfies all requirements for eCTD and non-eCTD submissions, whether electronic or paper. eCTDmanager enables the user to build, view, validate and publish compliant submissions based on CTD, eCTD, NeeS, eCopy, IMPD, CTA, VNeeS, DMF, ASMF and other submission structures easily. With its powerful hyperlinking and bookmarking engine, it helps users to handle the detection, notification and correction of broken hyperlinks resulting in the creation of high-quality submissions. An integrated validation function ensures compliance of the submissions to ICH and regional specifications. With the eCTDmanager suite we are looking ahead to the future of electronic submission management: Prospective eCTDmanager solutions will be ready for new standards like Regulated Product Submission (RPS).

Table 12.

Requirements Category	Label Content Management
Capability	Labeling
System Description	The customer) uses a variety of tools to manage authoring, review, artwork management, planning, tracking, quality control and reporting. The user community internally is approximately 50 users, however the community of labeling stakeholders is in the thousands, and requires access to labeling information at differing stages of development.
# of users	50 core users + 3000 users who require view access
Chosen COTS	Extedo - XML Content Authoring & Management (e. g. DITA)
Alternative List of COTS available	Lorenz Labelbridge Virtify
Evaluation Comments	With Extedo - Even without the knowledge of XML technology, users can easily build and review submissions, add, edit and delete elements or even set hyperlinks and comments at any time during the submission compilation. Documents can easily be scanned, copied, moved or imported from the file system and from most Document Management Systems The labeling solution of Lorenz is not very mature product and lack a lot of functionality. Virtify does support SPL only – however, there may not be any EU based XML labeling solution. Translation like "TRADOS" is missing
Chosen COTS Description	With EXTEDO users can automatically build and edit Study Tagging Files (STFs). The STF Wizard Window provides a fast and efficient way to create STF sections. The STF Filetags View simplifies tag handling to save you time and avoid errors. Additionally, the STF building feature allows for the automatic assignment of required metadata attributes as submission content files are added to the appropriate non-clinical and/or clinical sections of a submission

Table 13.

Requirements Category	HA Information Management
Capability	Country Filing Requirements
System Description	Regulatory filing requirements for each country are collected. This information includes specifics on formats, required forms and certifications and other local "operational" submission contents related information. A number of informal tools (mainly Excel and Word) are used (e.g.in publishing, regional groups) and stored and shared via SharePoint.
# of users	2000 users globally
Chosen COTS	RegIntel- Custom developed regulatory information management solution
Alternative List of COTS available	RegIntell Thomson Reuters Corellis
Evaluation Comments	No COTS fits all requirements
Chosen COTS Description	RegIntell (Regulatory Intelligence solution for country filing and soft intelligence) is a Documentum/D2 based bespoke solution for country filing, HA correspondence and soft intelligence requirements

Table 14.

Requirements Category	HA Information Management
Capability	Regulatory Information Archive and Correspondence Tracking
System Description	Provide users the ability to access content stored in several repositories including Documentum and a secure file server. • HA submissions: Marketing (NDA, BLA, MAA, NDS) and investigational (CTA, IND) • Correspondence with HAs, including emails and records of contact • Incoming HA correspondence regarding the submissions (e.g. acknowledgements, approvals, minutes) In addition support the tracking of near complete submissions (e.g. Core Dossiers, CTA core packages) that are dispatched to LOCs.
# of users	2000 users globally
Chosen COTS	Extedo MPDmanager
Alternative List of COTS available	TrackWise - Correspondence and Commitment Tracking System Rosetta Pyramid
Evaluation Comments	With implementation of MPDmanager the clients are able to reduce staff workloads, reduce personnel expenditures, streamline business processes and increase the efficiency and accuracy of data management. This is mainly due to: Automated processes in data collection and maintenance Reuse of content making recurring data entry obsolete Lifecycle management including versioning Quick overview, reference and statistics on product data delivering useful information for regulatory affairs, pharmacovigilance, and marketing departments Also, With the data structure and processes of MPDmanager being geared towards a smooth transition to new standards, moving from XEVMPD to IDMP can be relatively simple.
Chosen COTS Description	EXTEDO's MPDmanager provides a holistic approach to regulatory information management that allows the company to enter, import and share all relevant regulatory master information of medicinal product data across the enterprise and easily comply with data submission regulations.

Chapter 9
Personal Diagnostics Using DNA-Sequencing

Udayaraja GK
IGB, Saudi Arabia

ABSTRACT

DNA sequencing is the process to identification of nucleotides order in genome which developed from very broad history, also it is derived from version of the Sanger biochemistry. SOLiD, 454 and Polonator sequencing based on emulsion PCR to amplify clonal sequencing with in-vitro construction of adaptor-flanked shotgun library, PCR amplified in the context of a water-in-oil emulsion. Solexa technology relies on bridge PCR to amplify clonal sequencing features. At the conclusion of the PCR, each clonal cluster contains ~1,000 copies of a single member of the template library. This chapter focused on next-generation sequencing technologies methods, capabilities and clinical applications of DNA sequencing technologies for researchers in molecular biology and physician scientists. This will also provide the power of these novel genomic tools and methods to use personal diagnostic at molecular level.

INTRODUCTION

Next generation sequencing (NGS) technologies have massively penetrated into biological research science in recent years by producing huge amount of data with low cost compare to Sanger sequencing technology. NGS technology enabled depth analysis in microbial research which associated in humans, plants, animals etc. DNA based studies of the human associated with diseases are of high value to address genetic diseases. Genomic analyses of individuals or population studies of whole genome provide insight into the composition and physiological potential of humans disease mechanisms. RNA based studies can extend such studies in order to elucidate the actual metabolic activities and transcriptional mechanisms of the cells under given conditions.

NGS applications can use various analysis based on DNA and RNA; they allow finding answers to questions that could not be addressed before, largely due to technical and financial limitations (Venter et al., 2001; Lander et al., 2001). The establishment of a reference human genome (Hg19) and large-scale human study in the 1000 Genome project are, in conjunction with the use of next generation techniques,

DOI: 10.4018/978-1-4666-8726-4.ch009

triggering advances in many areas of basic and applied science. Apart from personal diagnosis, clinical applications of NGS include the sequencing of cell-free DNA fragments circulating in a patient's bloodstream. Similar to detection of a low-frequency variant, this experiment relied on deep sequencing (very high coverage) of cell-free DNA to detect changes in a small fraction of that DNA population which belonged to the organ donor. This result shows the potential of using NGS as a noninvasive method for detecting solid organ transplant rejection. Similarly, Palomaki and colleagues showed the potential of NGS as a noninvasive method for detecting Down syndrome and other fetal aneuploidies by sequencing the subpopulation of cell-free DNA in a pregnant mother's bloodstream belonging to her fetus. Results from this study showed the promise of an NGS plasma-based DNA test that can detect Down syndrome and other aneuploidies with high sensitivity and specificity. Together, sequencing of cell-free DNA by NGS in both of these examples offers enormous potential to reduce invasive medical procedures (Venter et al., 2001; Durbin et al., 2010).

Genome sequencing data is used in clinical practice of medicine at diagnostic level to identify disease. Continuous drop in sequencing cost, the actual translation of base pair reads to bedside clinical applications has finally begun. Personalized genome-based medicine would be the value of both whole-genome and targeted sequencing approaches in the diagnosis and treatment of diseases. DNA sequencing of cell-free DNA fragments circulating in a patient's bloodstream like DNA from a heart transplant donor's genome can be found in a recipients bloodstream when a transplant recipient is undergoing an acute cellular rejection, as validated by endomyocardial biopsy. Low-frequency variant detection by deep sequencing of cell-free DNA to detect changes in a small fraction of that DNA population which belonged to the organ donor. This result shows the potential of using NGS as a noninvasive method for detecting solid organ transplant rejection (Kapranov et al., 2012; Sucher et al., 2011; Clark et al., 2011).

BACKGROUND

NGS platforms produce a massive amount data (up to terabases) in parallel sequencing method. Often, NGS platforms are classified as second and third generation sequencing technologies. The methods which depends on a PCR step for signal intensification prior to sequencing as next generation sequencing instruments, opposed to single molecule sequencing. Next generation sequencing technology includes the 454 instruments from Roche, the different Illumina platforms and the Life Technologies instruments, i. e. the SOLiD (Sequencing by Oligonucleotide Ligation and Detection) and PacBio RS by Pacific Biosciences. The third generation sequencing instruments like Ion Torrent(PGM=Personal Genome Machine, IonProton) sequencers, MySeq & MsJunior. Next-Generation technology platforms differ in read length, throughput size, data metrics, read depth, etc (Venter et al., 2001; Mardis, 2008).

The sample preparation in NGS is clonal amplification of single strands of target DNA. Isolation of high quality DNA followed by DNA library preparation should be be performed before sequence. All next generation platforms will produce shorter reads comapare to sanger method, this limitation can overcome by development of paired-end/mate-paired sequencing. which can be performed using all three sequencing systems. Paired-end tags (PETs) are shorter sequences originating from the two ends of a target DNA. There are multiple ways of constructing a paired-end library. One is the clone based method, where the target sequence is ligated with adaptors containing MmeI restriction sites immediately next to the target sequence. Following amplification in E. coli, purification and MmeI digestion, the tag containing vector is recircularized, hereby joining the two sequence tags. After subsequent amplification in E. coli, the PET constructs can be purified using restriction digestion. In second method the target DNA fragments are directly circularized with linker oligonucleotides hereby joining the two ends of the target DNA. The linker sequence contains two restriction sites flanking the two ends of the target DNA, enabling restriction digestion to release the tag-linker-tag construct for sequencing. These two methods can create libraries with long inserts (up to 20 kb) between the two sequence tags, which are often referred to as mate pair libraries. Additional to these methods short insert libraries (200-500 bp) can also be paired-end sequenced using Illumina sequencing. Here paired-end libraries are made using adaptors with two different sequencing primers. Paired-end is performed by first sequencing the target DNA utilizing the first sequencing primer. After subsequent product denaturation, bridging, and second strand synthesis, the opposite strand is cleaved providing a template for a round second sequencing utilizing the second sequencing primer. These libraires also supoorts barcodes can for sample multiplex. The DNA sample is fragmented to a target read length size, library type, application type and chemistry used. This fragmentation is usually carried out by mechanical methods of either sonication or nebulisation. Methods of enzymatic sample fragmentation are currently being commercialized, offering the possibility of easily parallelizing the process. Universal DNA adaptors or bases can be ligase-bound at both ends of each DNA fragment, to allow PCR amplification to be performed either with a single pair of primers or to attach library fragments to surfaces by using adaptor-complementary oligos.

Thus, each platform prepares DNA fragments from the initial sample as a fragment library, on which to perform single-strand clonal amplification and sequencing reactions. Barcoded library is the addition of a specific DNA sequence to each fragment that allows sample multiplexing. These can be added during library construction, being attached as adaptors to the DNA fragments by ligation or polymerization during an amplification process with barcoded primers. Barcodes sequence specific to samples to allow to use multiple sample in single run by pooling library together with less reagent consumables. Mate-Pair Library is consists of two DNA fragments separated by an internal adaptor and flanked by universal adaptors. Due to its construction mechanism both fragments are 'mates', being derived from the ends of the same initial fragment, and the distance between them is known. Usage of this different libraries depends on experiments, mate-pair or paired-end libraries for large genomes size to achieve target size, it can use for gap filling etc (Clark et al., 2011; Mardis, 2008; Shendure, 2008).

Next target enrichment step is the selection of genomic regions which needs to sequence and enrichments performed by PCR, circularization and hybridization capture. PCR technique used for processing samples for sequencing multiplex PCR is alternative to the options, in which, the number of reactions is kept as low as possible for a high throughput. Multiplex PCR requires a careful primer design in order to allow several dozen primers per reaction without any primer interactions or non-specific amplifications. Probe hybridization is alternative enrichment technique but target enrichment via PCR does not require such a high quality or quantity of DNA for sequencing regions of up to several hundred kilobases. It provides a very high enrichment ratio with few off-target reads. Multiplexing capacity is being increased by the improvement and automation of primer design, enrichment by PCR of numerous dispersed genomic regions is possible like IonTorrent Ampliseq and Truseq amplicon applications. Enrichment by hybridization is based on knowledge acquired through microarray research and applied to the enrichment of samples for NGS. Thus, a DNA library is hybridized with probes which represent the target region. Non-specific hybrids are removed by washing and the targeted DNA obtained is eluted. Two alternative methods exist: array hybridization and in-solution hybridization. Roche Nimblegen was introduced hybridization enrichment methods, initially array-based, with capture regions of about 4–5 Mb. Later, Agilent Technologies also commercialized a similar approach, with similar performance. Currently, the major commercializers are still Agilent sureSelect and Roche Nimblegen SeqCapEz, but a large number of protocols and vendors are also beginning to offer new capture options, for example Flexgen and MicroArray. Applied Biosystems TargetSeq and Illumina TruSeq are also offering protocols integrated into the sample preparation platform workflow. Capturing DNA by circularization consists in the use of molecular inversion probes or selector probes. It uses DNA probes, which are single trended oligonucleotides consisting of a common linker flanked by target specific sequences. In MIPs the probes are annealed to the initial DNA sample, undergoing circularization in the process (Kilzer, et al., 2010; Mertes, et al., 2010; Schadt, et al., 2010).

Clonal Amplification

NGS technologies require clonally amplified sample prior to its being sequenced after library preparation, due to the need for an amplification of the signal generated in order to enable its detection during the sequencing process. Bridge-PCR, also called two-dimensional PCR, is a variant of traditional PCR, developed during the 1990s and used first by Solexa in its NGS platforms. Bridge amplification allows clonal amplification of a large number of DNA fragments simultaneously, using a solid oligonucleotide coated surface which also call flow cell.

The oligonucleotides are complementary to the linkers added to the DNA sample during library construction. The flow cell is a multi-channel sealed glass device. A different library can be added to each channel or the same one can be used in each, depending on the size of the region to be sequenced. DNA fragments from the library are denatured and then attached to the flow cell surface. Oligonucleotides of the flow cell are linked to the surface at its 5′end, leaving the 3′ end free for the polymerase to act on. The resulting double-stranded-DNA is covalently attached to the surface.

This double-stranded DNA is then denatured and the single strand flips over to hybridize to adjacent primers, thus forming a bridge. This newly closed amplicon is extended by polymerases to form a double-stranded bridge. After denaturing, two copies of covalently bound single-stranded templates are obtained. This cyclical process is repeated several times, producing approximately one million clonal copies of each initial fragment in the form of clonal clusters.No primers are required in the reaction solution and clusters are spatially separated.

Emulsion PCR (emPCR) is elaboration of the classical amplification reaction, in which each DNA fragment is isolated in independent aqueous microreactors surrounded by an oil phase, and templates are amplified on primercoated beads.An emPCR complex therefore contains millions of compartments separated from each other and each acting as microreactors, in which an independent single-molecule PCR reaction is performed. For its application to NGS, emPCR is usually carried out with beads carrying reaction primers on their surfaces, in order to colocalise the clonally amplified fragments of each reaction on the surface of one bead. After the reaction, the emulsion is broken down with organic solvents and the beads are isolated by extraction of the aqueous phase.Platforms includes SOLiD, Ion Torrent Personal Genome Machine (PGM),Proton & Polonator (Mardis, 2008; Smith, et al., 2010; Pettersson, et al., 2009).

MAIN FOCUS OF THE CHAPTER

Sanger Sequencing

Sanger sequencing is high throughput shotgun method, genomic DNA is fragmented, then cloned to a plasmid vector and used to transform E. Coli. A single bacterial colony is picked for each sequencing reaction, and plasmid DNA isolated. Each cycle sequencing reaction takes place within a microliter-scale volume, generating a ladder of ddNTP-terminated, dye-labeled products, which are subjected to high-resolution electrophoretic separation within one of 96 or 384 capillaries in one run of a sequencing instrument. Labeled fragemetns will control by detector and the sequence will generate through four channel emission spectrum. DNA fragments ligated to adapters are subjected to protocols, that results in millions of arrays of spatially immobilized PCR colonies. Each polony contain multiple copies of a single shotgun library fragment. As all polonies are tethered to a planar array, a single microliter-scale reagent volume can be applied to manipulate all array features in parallel.

Sequencing through in vivo or in vitro cloning method contain template DNA molecule and four parallel reactions that contains unmarked deoxynucleotides and one dideoxynucleotide, are marked by its lack of 3' OH to promote chain elongation. The reaction of sequence continues to grow by addition of deoxynucleotides in matching positions. This leads to DNA sequence elongation and is separated with acrylamide gels to get some hundred of bases. Nucleotide specific fluorescent dye allows the chain-termination reaction for the four dideoxynucleotides to be carried out in a single reaction with automatic laser fluorescence detection. This method useful in validation of DNA sequencing that includes single nucleotide polymorphism detection, single-strand conformation polymorphism hetroduplex analysis, and short tandem repeat analysis. But this sequencing method can generate up to 1000 bp and time consuming as well (Lander et al., 2001; Sucher et al., 2011).

454 Pyrosequencing

Pyrosequencing is a sequencing-by-synthesis method. This method developed in 1996 by the Stockolm Royal Institute of Technology and commercialized for NGS in 2005 by 454 Life Sciences. Pyrosequencing is based on detection and quantification of DNA polymerase activity, which is carried out using the enzyme luciferase. Through emulsion PCR beads are attach to the library and deposited in a PicoTiter-Plate which contains millions of wells of 44 μm diameter. Each well fits only one bead, of diameter 28 μm, although not all wells contain a bead.

DNA polymerase and universal primer which complementary to one of the library adaptors will start sequencing reaction and plate contain ATP sulfurilase, luciferase and apyrase. Then, a modified variant of ATP, which acts as a substrate for luciferase) is added to the reaction. This binds to the complementary 3′ positions adjacent to the universal primer previously added to the reaction. The union is catalyzed by DNA polymerase and generates one free pyrophosphate per base added. ATP sulfurilase then uses the re- leased pyrophosphate to generate ATP by combining it with adenosine 5′ phosphosulfate. The ATP is then used by the luciferase to generate visible light with an intensity equivalent to the quantity of bases incorporated. After each ATP-3′ binding and light-generating step, imaging takes place. Collecting degraded non-incorporated nucleotides, permitting another reaction cycle to begin, with incorporation of another dideoxynucleotide. The data output approximately 700-800 Mb of sequence per run, approximately one million reads and a consensus accuracy at 15× coverage of 99.99%. This long read length is useful for performing de novo sequencing on unknown genomes due to its capacity to align complex regions and relatively low computational power that that requires (Mardis, 2008; Shendure, 2008;).

Illumina Sequencing-by-Synthesis

The Illumina sequencing method is similar to Sanger sequencing, but it uses modified dNTPs containing a terminator which blocks further polymerization, only a single base will add by a polymerase enzyme to each growing DNA copy strand. After a bridge-PCR amplification process, sequencing takes place on the solid surface of the flowcell. The terminator also contains a fluorescent label, which can be detected by a camera. Only a single fluorescent color is used, each four bases must be added in a separate cycle of DNA synthesis and imaging. Following the addition of the four dNTPs to the templates, the images are recorded and the terminators are removed. This process consists of the polymerase catalysed addition of reverse-terminator fluorescently labeled bases, then bases are added simultaneously to the reaction and compete to form a union with oligoprimed cluster fragments.

Once a base is added, it prevents addition of subsequent bases, meaning that only one base will be attached in any one cycle. After base incorporation imaging step is carriedout to record cluster-specific fluorescence. Overall workflow is shown in figure 1. Each flowcell lane consist panels or tiles for a given cluster density. Each image represents one tile. Blocking is chemically removed after imaging and the

Figure 1. Illumina Sequencing Workflow

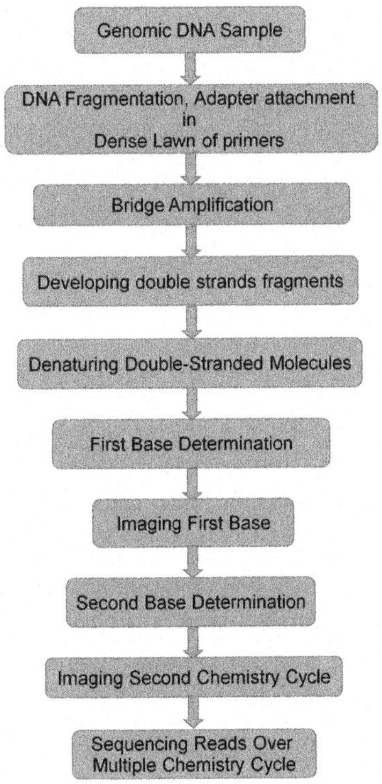

process is restarted. This technology was originally developed by Shankar Balasubramanian and David Klenerman at the University of Cambridge. They founded the company Solexa in 1998 to commercialize their sequencing method. Illumina purchased Solexa in 2007 and has built upon, and rapidly improved the original technology started producing more than 600 Gb data (Mardis, 2008;).

Sequencing-by-Ligation: SOLiD

This technology is sequencing-by-ligation developed based on massively Parallel Signature Sequencing and the aforementioned Polony sequencing by Applied Biosystems, as the SOLiD system. It uses emulsion-PCR is performed to generate clonal amplicons on the surface of beads from DNA library. Beads are enriched and attached to a glass surface in a random pattern and forming a dense array on which sequencing-by-ligation is then performed.

Library adapters get bind to universal primers, followed by fluorescently labeled di-base probes of eight nucleotides in length compete for ligase binding to the DNA fragment adjacent to the universal primer. Di-base probes interrogate the two first bases simultaneously in each ligation cycle and remaining base probes are degenerate.

Three final bases of the probe are cleaved so that the length of the probe is reduced to five nucleotides after ligation and di-base-specific fluorescence is emitted. Imaging step is performed to detect fluorescence after each ligation-cleave. After imaging, a new dibase probe is added and a new ligation cycle begun. By this process, the first two bases of each group of five are interrogated, at three-base interval. After many cycles of ligation, the extended product and the universal primer are cleaved from the DNA fragment, thus resetting the sequencing reaction. A new universal primer, displaced one base towards the adaptor, is added, such that the sequence interrogation performed by the dibase probes is correspondingly displaced by one base. Five different universal primers use to repeat the process five times to obtain every base on each clonally amplified fragment will have been interrogated twice in independent ligations. Achieving maximum sequencing accuracy by two base encoding system.

In an addition, optional sequencing chemistry add-on has been commercialized, introducing a kind of third-base encoding based on the adding new probes in which certain positions are already known. This module increases accuracy up to 99.99%. A SOLiD System sequencing run takes 7–14 days, and currently has a maximum output of 300 Gb per run with approximately 5 billion reads. The read length is 35–75 bp with a paired end library, and multiplex up to 96 barcoded samples (Mardis, 2008; Shendure, 2008; Mardis, 2008). Overall workflow shown in figure2.

Ion-Torrent Semiconductor Sequencing

It is a semiconductor sequencing method for detection of hydrogen ions generated after dNTP (deoxy Nucleoside Triphosphate) addition. Detection is performed by a semiconductor chip. Life Technologies' platform- Personal Genome Machine (PGM) and Ion Proton based on semiconductor sequencing method. Ion Torrent has specific kits for a variety of applications, including DNA fragmentation for small genomes, total RNA-Seq and cancer-specific targeted sequencing. The fragments generated during the library prep are attached to beads and amplified using emulsion PCR. Beads coated with complementary primers are mixed with a dilute aqueous solution containing the fragments to be sequenced along with the necessary PCR reagents. This solution is then mixed with oil to form an emulsion of microdroplets. The concentration of beads and fragments is kept low enough such that each microdroplet contains

Figure 2. SOLiD Sequencing Workflow

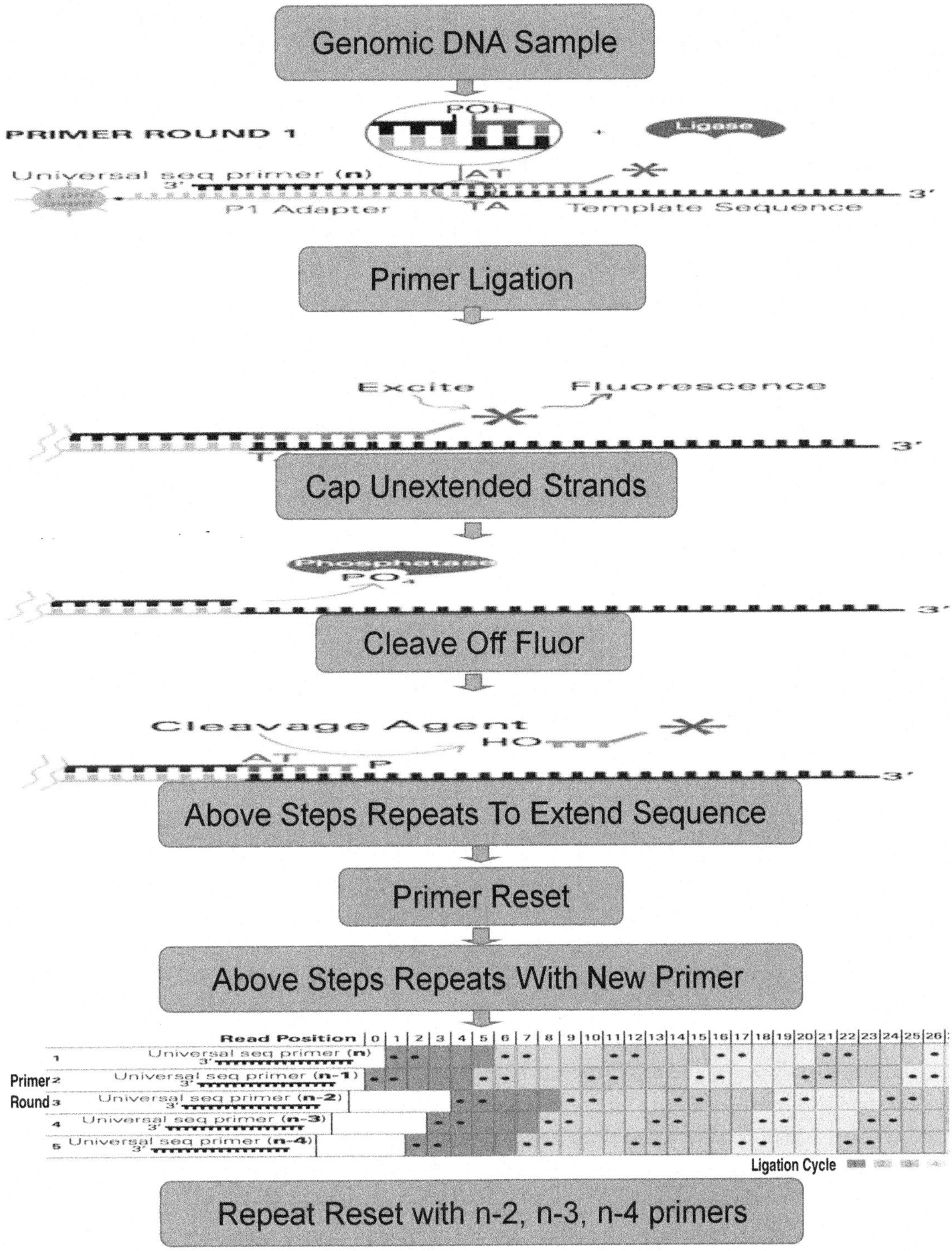

only one of each. Clonal amplification of each fragment is then performed within the microdroplets. Following amplification the emulsion is 'broken' and the amplified beads are enriched in a glycerol gradient. While the emulsion PCR process is effective, it is slow and complicated. Ion Torrent is working to improve emPCR, first through the release of the Ion Xpress kit, which reduced the overall time from 6 hours to 3.5 hours.

They have also launched their "Ion OneTouch" system, which automates much of the process, reducing hands on time to a few minutes per sample.Clonal amplification library is performed through an emulsion PCR with non-paramagnetic beads. The product of the reaction is enriched and then deposited on a microwell surface, with each microwell able to accommodate only one bead and positioned above a sensor plate sensitive to changes in pH. Sequencing initiated by deoxinucleotide (A, T, G, or C) attach to DNA fragments by enzyme polymerase which leads to hydrogen ion release with pH change inside the microwell.Ion semiconductor chips made up of millions of individual reactors with high density arrays in which fluidics will allow reagents to flow over the sensor array. Enzyme polymerase will add nucleotide in DNA fragments leads to chain elongation, which make the proton relase with respective change in pH. This confirm the nucleotide addition in DNA fragments. Voltage change will not occur when no nucleotide is added and double voltage change will occur when two nucleotide is added. This semiconductor technology produce high quality data in short time (Schadt, et al., 2010; Pettersson, et al., 2009).

PacBio RS System

In April 2011, the generation sequencing PacBio RS System introduced from Pacific Biosciences became available. The technology is based on direct observation of a single molecule of DNA polymerase using zero mode waveguide technology. Average read lengths between 700 and 1,000 bp with raw accuracies ranging from 81% to 83% was reported. While certainly an attractive read length for a DNA sequencing system, the raw error rate is inferior to existing sequencing technologies and limits its usefulness. The sequences will be processed using the Pacific Biosciences SMRT Portal software. The new release P6-C4 generate average read length to 10,000 - 15,000 bases, with the longest reads exceeding 40,000 bases. The throughput with the new chemistry is expected to be between 500 million to 1 billion bases per SMRT Cell, depending on the sample being sequenced. By providing an increasing number of longer reads per instrument run, the new chemistry enables users to assemble genomes to a higher quality. The Single Molecule, Real-Time (SMRT) Sequencing able to generate ultra-long reads with unbiased coverage allows researchers to characterise previously undetected structural variants, highly repetitive regions, and distant genetic elements. Sequence prefiltering is used to select the highest quality polymerase reads and the sequences that exceed a given length threshold. Adapter removal clips the synthetic sequence from the reads and yields the collection of subread sequences that will be used in downstream steps. Selection of seed sequences will be performed to ensure that a collection of longer seed sequences can bereliably and effectively used for downstream steps. Sequence assembly is a multistep process that involves polishing the seed sequences with the shorter sequences. The polished seed sequences are assembled and the subread collection is mapped back against the assembled contig to prepare a consensus sequence. Assembly review involves a critical appraisal of what has been assembled and the observed concordance between the individual SMRT cells that were run and the consensus sequence. This analysis produces key figures and summary statistics that include the depth-of-coverage plots. Dot plots are prepared to investigate patterns of repetition within the assembled contigs. A close review of the sequence ends is performed to investigate whether the bacterial genome has been closed and could be circularised (Eid, et al., 2009; Quail, et al., 2012).

NGS Data Analysis and Softwares Tools

Even though each platforms sequencing chemistries are different, all commercialized NGS platforms utilize a similar technical strategy miniaturisation of individual sequencing chemical reactions to overcome the limited scalability of traditional Sanger sequencing, which has been extensively used in somatic and germline genetic studies over the past 30 years and currently remains the GOLD standard for decoding DNA sequences. The miniaturization of individual sequencing reactions, coupled with other technical breakthroughs, including overcoming the bottlenecks of library preparation and template preparation, allows millions of individual sequencing reactions to occur in parallel.

Clonal clusters of an original DNA fragment are sequenced in each miniaturized chemical reaction, and millions of them are spatially arranged so that individual reactions are isolated from one another and can be distinctly detected by digital imaging or other approaches. The results are prodigious volumes of short read sequence data, unprecedented detail and single-nucleotide resolution of sequence complexity, with consequential challenges in storing, MANAGING, analyzing and interpreting such a wealth of data. In a relatively short time span since 2005, NGS technologies have fundamentally changed high throughput genomic research and have opened up many new research areas and novel applications. With the exponential growth of the numbers of NGS related research articles INDEXED on Medline, NGS technologies have demonstrated their enormous potential for researchers working in medicine, biology and life sciences. Along with the development of robust informatics tools for nucleotide variant detection,the ongoing evolution of NGS technologies will continually reduce the cost, simplify the workflow for sample preparation and improve the technical robustness, paving the path for translating NGS technologies into clinical diagnostics and personalized medicine.

A key step in data analysis of a variety of applications utilizing NGS data is often initial alignment or mapping of reads to a reference or assembly of the short reads into larger continuous sequences. Examples of such applications are: genome re-sequencing and subsequent identification of variations, identification of protein binding sites on the DNA combining chromatin immunoprecipitation with NGS, gene expression profiling, identification of the genome-wide methylation pattern.

Each platforms produces reads and raw reads are processed to removing low quality reads and adapter sequence to obtain high quality reads. Based on application, tools can be fit into many general categories including alignment of sequence reads to a reference, base-calling and/or polymorphism detection, de novo assembly from paired or unpaired reads, structural variant detection and genome browsing. A large number of software applications already exist for analyzing NGS data and available computational tools that can be used to face the several steps of workflow to process the data analysis. For variants detection, workflow will align/map high quality reads against reference genome followed by variants call from alignment file and variants annotation will perform by using annotation tools.

Identification of SNP and insertion-deletion (indel) in genomes re-sequencing in order to find variants on respective sequenced genome. The goal of these programs consist in judging the likelihood that a locus is a heterozygous or homozygous variant given the error rates of the platform, the probability of bad mappings, and the amount of coverage. Structural variants detection using high throughput sequencing based on paired-end read mapping, which identifies insertions and deletions by comparing the distance between mapped read pairs to the average insert size of the genomic library. Although this method is able to identify deletions smaller than 1 Kb with high sensitivity, it does not allow the discovery of insertions larger than the average insert size of the library and the exact borders of SVs in complex genomic regions rich in segmental duplication.

There are few commercial tools like CLC genomics workbench, Partek genomics suite, DNAStar, GeneSpring GX from Agilent, Conexio Genomics, ChunLab, DNA Guide, Bioteam, Cypher Genomics etc using for data analysis. Tools are specific to sequencing technologies, such as the short sequence length of Illumina, SOLiD and Helicos reads, the low indel error rate of Illumina reads and the di-base encoding of SOLiD reads. These new tools, named Short read aligners, outperform the performance of traditional aligners in terms of both speed and accuracy. An algorithm for the alignment of short sequence reads produced by HTS technologies must be able to quickly and efficiently align the billions of short reads produced by this technique and permit the alignment of non-unique reads and of reads that do not match exactly the reference genome (Angiuoli, et al., 2011; Caruccio, 2011).

The NGS platforms producing huge volume of data makes huge demands on bioinformatics analysis tools and statistical tools for the analysis. The data throughput of these platforms has crossed terabyte, the sequenced DNA need to translate into biological information for researchers, physicians, pathologist etc. Tools for short-read alignments like Maq, Bowtie, SSAHA, BWA, SOAP2 to use these platforms reads to align against reference genome to detect structural variants. De novo assembly tools like AbySS, ALLPATHS, Edena, SOAPdenovo, Velvet are used to obtain first draft assembly of the genome. RNA sequencing data analysis need some more additional steps like aligning reads against genome along with splice junctions to quantify and splicing isoforms determination.

Tools MAQ, Stampy(seed method), BWA & Bowtie (Burrows-Wheeler transformation) do not allow gap between exons and MapSplice, SpliceMap, TopHat, GSNAP &QPALMA allow gap between exons. Software tools with high-performance computing capacity, in particular for large genome datasets. The research laboratories that are not specialized on NGS data analysis can use cloud computing process over a network and it would be a possible alternative solution for the data analysis [19].

Issues and Recommended Solutions

Bioinformatics algorithms developed for capillary-based sequencing didn't scale up for huge next generation sequencing data. Sequencing reads were shorter and more error-prone. The instruments were expensive, limiting access to the technology and most of the clinical community had no experience with NGS. The cost of DNA sequencing has plummeted much faster than the cost of disk storage and CPU. A run on the Illumina HiSeq2000 provides enough capacity for about 48 human exomes. Even if you don't keep the images, each exome requires about 10 gigabytes of disk space to store the bases, qualities, and alignments in compressed (BAM) format. At three runs a month, each instrument is generating 1.4 terabytes of data files. It adds up quickly. Sequencing platforms sometimes produce reagent issues, upgrade issues, error rates etc. Present NGS platforms produce shorter read lengths (30–400 bp) than Sanger-based method. Shorter read lengths difficult to assemble a genome de novo using such short fragment lengths. Sometimes application of these technologies focus on comparing the density and sequence content of shorter reads to that of an existing reference genome. Those shorter read lengths may not align against reference genome, often leaving repetitive regions of the genome unmappable to these types of experiments. Sequence alignment is also challenging for regions with higher levels of diversity between the reference genome and the sequenced genome, such as structural variants, insertions, deletion, translocations etc (Shendure, 2008; Caruccio, 2011). These issues are combated through the use of longer read lengths or paired-end/mate-pair approaches. Nearly all human genome resequencing conducted today relies on the paired-end or mate-paired approaches of second-generation platforms. Paired-end sequencing is much easier than mate-paired sequencing and requires less DNA, making it the standard means by which human genomes are resequenced. Although more expensive and technically challenging, mate-paired libraries can sample DNA sequence over a larger distance (1.5–20 kb) than paired-end approaches (300–500 bp) and are therefore better suited for mapping very large structural changes. Although these limitations create important algorithmic challenges for the immediate future, we should bear in mind that these technologies will continue to improve with respect to these parameters, much as conventional sequencing progressed gradually over three decades to reach its current level of technical performance (Angiuoli, et al., 2011; Pettersson, et al., 2009).

FUTURE DISCUSSION AND CONCLUSION

DNA Sequencing has been reported in several studies of Personal diagnosis as regular clinical practice. Targeted sequencing panels are already in routine use at many cancer centers; in time, this will likely become exome/genome sequencing. Possibly transcriptome (RNA-Seq) and methylome (Methyl-Seq) as well. Undiagnosed inherited diseases, and rare genetic disorders whose genetic cause is unknown, are two other common-sense applications. There are many hurdles to overcome in order to apply a new technology to patient care. CLIA/CAP certification is a complex, expensive, and time-consuming process. The reporting is more difficult, too. Unlike the research setting in which most NGS results have arisen, a clinical setting requires very high confidence in order to report anything back to the patient or treating physician. This is a good thing, since patient care decisions might be made based on genomic findings. Yet it means that we have a considerable amount of work ahead to ensure that genomic discoveries are followed up, replicated, and otherwise vetted to the point where they can be of clinical use. Reads coverage depth, distribution, and sequence quality determine what information can be retrieved from each sequencing experiment. In theory, an experiment with 100% accuracy and uniform coverage distribution would provide all sequence content information with just coverage depth. For discovery of structural variants, accurate identification of a complete human genome sequence with current platforms, requires approximately sequence coverage to overcome the uneven read distributions and sequencing errors. Numerous research groups have demonstrated the immense discovery power of NGS. The mere fact that dbSNP database of human sequencing variation contain more than 50 million distinct variants tells us something about what pervasive genome sequencing capabilities might uncover. And yet, the variants implicated in sequencing-based studies of human disease are increasingly difficult to "sell" to peer reviewers on genetic information alone. Referees of most high-impact journals want to see some form of functional validation of genomic discoveries. That's a daunting challenge for many of us accustomed to the rapid turnaround, high-throughput nature of NGS. Most functional validation experiments are slow and laborious by comparison. Sequencing technology has been moving with advance methods which can reduce cost and time drastically. Using this NGS technologies will able to address complex biological mechanism even small scale research centers. Advance technology usage will lead to increase research in diseases to understand disease causes in molecular level. Apart from genomic variability, it also help to identify phenotypes associated complex diseases. Because of this huge data generation made demands in mathematical algorithms to support software development to get raw sequence reads into meaningful biological information. NGS technologies open the door for researchers and clinicians to understand the genetic variability underlying in diseases, and provide an unprecedented tool in their investigation. In future, NGS technologies will be a part of regular disease identification method for pathologist, clinicians, physician etc. It will also help to add functional annotation of existing human variability. Understanding these sequencing technologies require novel algorithms and software tools to extract more specific biological information from vast amount of data that are produced from above platforms. Current annotation tools lack in clinical significanc3e which can over come by integrating clinical databases with annotation tools. Developing compressed storage device for sequenced genome data from individuals will be useful for physicians at the time of disease conditions would future need. Organizing markers panels in individuals based on diseases, drug resistances, phenotypic traits may expedite the diagnostic process from screening of huge number of markers.

REFERENCES

Angiuoli, S. V., Matalka, M., Gussman, A., Galens, K., Vangala, M., & Riley, D. R. et al. (2011). CloVR: A virtual machine for automated and portable sequence analysis from the desktop using cloud computing. *BMC Bioinformatics*, 12. PMID:21878105

Caruccio, N. (2011). Preparation of next-generation sequencing libraries using Nextera technology: Simultaneous DNA fragmentation and adaptor tagging by in vitro transposition. *Methods in Molecular Biology (Clifton, N.J.)*, *733*, 241–255. doi:10.1007/978-1-61779-089-8_17 PMID:21431775

Clark, M. J., Chen, R., Lam, H. Y., Karczewski, K. J., Chen, R., & Euskirchen, G. et al. (2011). Performance comparison of exome DNA sequencing technologies. *Nature Biotechnology*, *29*(10), 908–914. doi:10.1038/nbt.1975 PMID:21947028

Durbin, R. M., Altshuler, D. L., Durbin, R. M., Abecasis, G. R., Bentley, D. R., & Chakravarti, A. (2010). A map of human genome variation from population-scale sequencing. *Nature*, *467*, 1061–1073. PMID:20981092

Eid, J., Fehr, A., Gray, J., Luong, K., Lyle, J., & Otto, G. et al. (2009). Real-time DNA sequencing from single polymerase molecules. *Science*, *323*(5910), 133–138. doi:10.1126/science.1162986 PMID:19023044

Kapranov, P., Chen, L., Dederich, D., Dong, B., He, J., & Steinmann, K. E. et al. (2012). Native molecular state of adeno-associated viral vectors revealed by single-molecule sequencing. *Human Gene Therapy*, *23*(1), 46–55. doi:10.1089/hum.2011.160 PMID:21875357

Kilzer, J., Xun, X., Bodeau, J., Breu, H., & Harris, A. (2010). A balanced barcoding system for multiplexed DNA library and SOLiD SAGE Sequencing. *Journal of Biomolecular Techniques*, *21*, 528.

Lander, E. S., Linton, L. M., Birren, B., Nusbaum, C., Zody, M. C., & Baldwin, J. et al. (2001). International human genome, initial sequencing and analysis of the human genome. *Nature*, *409*(6822), 860–921. doi:10.1038/35057062 PMID:11237011

Mardis, E. (2008). Next-generation DNA sequencing methods. *Annual Review of Genomics and Human Genetics*, *9*(1), 387–402. doi:10.1146/annurev.genom.9.081307.164359 PMID:18576944

Mertes, F., Elsharawy, A., Sauer, S., Van Helvoort, J., Van, D. Z., & Franke, A. M. (2011). Targeted enrichment of genomic DNA regions for next-generation sequencing. *Briefings in Functional Genomics, 10*(6), 374–386. doi:10.1093/bfgp/elr033 PMID:22121152

Neveling, K., Collin, R., Gilissen, C., Van Huet, R., Visser, L., & Kwint, M. et al. (2012). Next-generation genetic testing for retinitis pigmentosa. *Human Mutation, 33*(6), 963–972. doi:10.1002/humu.22045 PMID:22334370

Quail, M.A. (2012). Pacific Biosciences and Illumina MiSeq sequencers. *BMC Genomics*, 13–341.

Pettersson, E., Lundeberg, J., & Ahmadian, A. (2009). Generations of sequencing technologies. *Genomics, 93*(2), 105–111. doi:10.1016/j.ygeno.2008.10.003 PMID:18992322

Schadt, E., Turner, S., & Kasarskis, A. (2010). A window into third-generation sequencing. *Human Molecular Genetics, 19*(R2), R227–R240. doi:10.1093/hmg/ddq416 PMID:20858600

Shendure, J., & Ji, H. (2008). Next-generation DNA sequencing. *Nature Biotechnology, 26*(10), 1135–1145. doi:10.1038/nbt1486 PMID:18846087

Smith, A., Heisler, L., St.Onge, R., Farias-Hesson, E., Wallace, I., & Bodeau, J. et al. (2010). Highly-multiplexed barcode sequencing: An efficient method for parallel analysis of pooled samples. *Nucleic Acids Research, 38*, 142. doi:10.1093/nar/gkq368 PMID:20460461

Sucher, N. J., Hennell, J. R., & Carles, M. C. (2011). DNA fingerprinting, DNA barcoding, and next generation sequencing technology in plants. *Methods in Molecular Biology (Clifton, N.J.), 862*, 1322. PMID:22419485

Swerdlow, H., & Gesteland, R. (1990). Capillary gel electrophoresis for rapid, high resolution DNA sequencing. *Nucleic Acids Research, 18*(6), 1415–1419. doi:10.1093/nar/18.6.1415 PMID:2326186

Venter, J. C., Adams, M. D., Myers, E. W., Li, P. W., Mural, R. J., & Sutton, G. G. (2001). The sequence of the human genome. *Science, 291*(5507), 1304–1351. doi:10.1126/science.1058040 PMID:11181995

KEY TERMS AND DEFINITIONS

Amplicons: A piece of DNA or RNA that is the source and/or product of natural or artificial amplification or replication events. It can be formed using various methods including polymerase chain reactions (PCR), ligase chain reactions (LCR), or natural gene duplication.

Aneuploidies: A condition in which the number of chromosomes in the nucleus of a cell is not an exact multiple of the monoploid number of a particular species.

Amplification: A process in a cell by which a particular gene is replicated so that more copies are available to produce a protein for the cell's use.

Colocalise: Refers to observation of the spatial overlap between two (or more) different fluorescent labels, each having a separate emission wavelength, to see if the different "targets" are located in the same area of the cell or very near to one another.

De Novo: Particular biochemical pathways in which metabolites are newly biosynthesized.

Denature: Destroy the characteristic properties of (a protein or other biological macromolecule) by heat, acidity, or other effect which disrupts its molecular conformation.

Deoxynucleotide: The monomer, or single unit, of DNA, or deoxyribonucleic acid. Each deoxyribonucleotide comprises three parts - a nitrogenous base, a deoxyribose sugar, and one phosphate group.

Dideoxynucleotide: Cahin-terminating inhibitors of DNA polymerase, used in the Sanger method for DNA sequencing. They are also known as 2',3' dideoxynucleotides, and abbreviated as ddNTPs (ddGTP, ddATP, ddTTP and ddCTP).

Down Syndrome: A congenital disorder arising from a chromosome defect, causing intellectual impairment and physical abnormalities including short stature and a broad facial profile. It arises from a defect involving chromosome 21, usually an extra copy (trisomy-21).

Endomyocardial Biopsy: Removal of a small sample of heart tissue to check it for signs of damage caused by organ rejection.

Exons: A segment of a DNA or RNA molecule containing information coding for a protein or peptide sequence.

Genome: The complete set of genes or genetic material present in a cell or organism.

Genome Sequencing: Whole genome sequencing (also known as full genome sequencing, complete genome sequencing, or entire genome sequencing) is a laboratory process that determines the complete DNA sequence of an organism's genome at a single time.

Heterozygous: The genetics term heterozygous refers to a pair of genes where one is dominant and one is recessive — they're different. Like all words with the prefix hetero, this has to do with things that are different — specifically genes.

High Throughput Shotgun Method: DNA sequencing is the process of determining the precise order of nucleotides within a DNA molecule. It includes any method or technology that is used to determine the order of the four bases—adenine, guanine, cytosine, and thymine—in a strand of DNA.

Homozygous: Having the two genes at corresponding loci on homologous chromosomes identical for one or more loci.

Ligation: The joining of two DNA strands or other molecules by a phosphate ester linkage.

Multiplex PCR: Modification of polymerase chain reaction in order to rapidly detect deletions or duplications in a large gene.

Nebulisation: A method of administering a drug by spraying it into the respiratory passages of the patient. The medication may be given with or without oxygen to help carry it into the lungs.

NGS: (Next Generation Sequencing) DNA sequencing is the process of determining the precise order of nucleotides within a DNA molecule.

Nucleotide: A compound consisting of a nucleoside linked to a phosphate group. Nucleotides form the basic structural unit of nucleic acids such as DNA.

Oligonucleotide: A polynucleotide whose molecules contain a relatively small number of nucleotides.

PCR: - (Polymerase chain reaction) a laboratory technique used to make multiple copies of a segment of DNA. PCR is very precise and can be used to amplify, or copy, a specific DNA target from a mixture of DNA molecules.

Polyacrylamide gel electrophoresis (PAGE): A technique widely used in biochemistry, forensics, genetics, molecular biology and biotechnology to separate biological macromolecules, usually proteins or nucleic acids, according to their electrophoretic mobility.

Polymerase: An enzyme which brings about the formation of a particular polymer, especially DNA or RNA.

Primer: A strand of nucleic acid that serves as a starting point for DNA synthesis. It is required for DNA replication because the enzymes that catalyze this process, DNA polymerases, can only add new nucleotides to an existing strand of DNA.

RNA Splicing: A modification of the nascent pre-messenger RNA (pre-mRNA) transcript in which introns are removed and exons are joined. For nuclear encoded genes, splicing takes place within the nucleus after or concurrently with transcription.

Sanger Sequencing Technology: A method of DNA sequencing based on the selective incorporation of chain-terminating dideoxynucleotides by DNA polymerase during in vitro DNA replication.

Semiconductor Sequencing: A method of DNA sequencing based on the detection of hydrogen ions that are released during the polymerization of DNA.

Single Nucleotide Polymorphism: (SNP, pronounced snip; plural snips) is a DNA sequence variation occurring commonly within a population (e.g. 1%) in which a single nucleotide — A, T, C or G — in the genome (or other shared sequence) differs between members of a biological species or paired chromosomes.

Sonication: The act of applying sound energy to agitate particles in a sample, for various purposes. Ultrasonic frequencies (>20 kHz) are usually used, leading to the process also being known as ultrasonication or ultra-sonication.

Chapter 10
Pharmacogenomics Genome Wise Association Clinical Studies

Udayaraja GK
IGB, Saudi Arabia

ABSTRACT

Pharmacogenomics deals with drug responses in individual based on genetic variation in genome. Based on genetic variations, drugs may produce more or less therapeutic effect, and same way in side effects also. Physicians can use information about your genetic makeup to choose those drugs and drug doses to get better therapy. Optimizing drug therapy and rational dose adjustment with respect to genetic makeup will maximize drug efficacy and minimal adverse effects. This broken traditional 'trial and error' method of 'one drug fits all', and 'one dose fits all' which contributing to 25–50% of drug toxicity or treatment failures. This will contribute to improve the ways in which existing drugs are used, genomic research will lead to drug development to produce new drugs that are highly effective without serious side effects. This approach to bring personalized medicine more practice and drug combinations are optimized for each individual' genetic makeup.

INTRODUCTION

Human genetic profiling can be done by microarray expression, human SNP arrays and whole genome variants determination by using next generation sequencing technology. Whole genome wide screening approach enables better understanding about genetic profile, and provides possibility to check efficacy design for any animal experimental design. These technologies are made available to simultaneously monitor many cellular transcripts in parallel allows that result in more complete analysis of complex disease states and prediction of therapeutic response. Applications ranging from expression profiling to SNP genotyping, microarray gene expression in oncology are revolutionized with more clinical support to ensure better analysis of role of genetics in drug response.

DOI: 10.4018/978-1-4666-8726-4.ch010

Allelic variation in individual genotype also provides phenotypic relation, like intermediate, ultra-rapid, non-response categories of CYPgenes metabolizer etc. Several genetic events influence a same phenotypic trait, therefore, establishing genotype-to-phenotype relationships can be far from consensual with many enzymatic patterns. Many genes involved in pharmacokinetics process have been described as being highly polymorphic. Genes likes DPD, UGT1A1, TPMT, CDA involved in the pharmacokinetics of 5-FU/capecitabine, irinotecan, 6-mercaptopurine and gemcitabine/cytarabine, respectively. Patients affected by these genetic polymorphisms will experience severe/lethal toxicities upon drug intake, and that pre-therapeutic screening does help to reduce the risk of treatment-related toxicities through adaptive dosing strategies (Evans & Relling, 1999).

Drug adverse effects vary from patient to patient as well as disease to disease. Drug response characteristics are very complex and depends on profound gene-drug interactions. For drugs that have a narrow therapeutic index caused by inactivation of certain polymorphic drug-metabolizing enzyme and expected to have an increased risk of adverse drug reactions. Some drugs require a metabolic activation by polymorphic drug- metabolizing enzyme (for example, codeine) low therapeutic efficacy or treatment failure. For many drugs that have a broad therapeutic window caused by genetic variants which leads to impaired drug metabolism.

Pharmacogenomics researchers have already identified many genes whose variations affect drug responses. They also know where to look for the numerous others they are bound to discover in the future. Pharmacogenomics tests are used to identify the patients are most likely to respond to certain cancer drugs, tests provide tools for physicians to better manage medication selection and side effect amelioration. Pharmacogenomics is also known as companion diagnostics, meaning tests being bundled with drugs. Examples include KRAS test with cetuximab and EGFR test with gefitinib. Beside efficacy, germline pharmacogenetics can help to identify patients likely to undergo severe toxicities when given cytotoxics showing impaired detoxification in relation with genetic polymorphism like canonical 5-FU.

All drugs cannot be personalized, clinical significance in tailored medicine for prodrugs, drugs with a narrow therapeutic index and drugs that target a key molecule or a critical pathway. Drug safety is the first arena in which patients can benefit from pharmacogenetics and pharmacogenomics. Tumor responses to the inhibitors of oncogenic tyrosine kinases are associated with the presence of activating mutations within the genes encoding the target kinases, targeted cancer therapy is thus a promising individualized drug therapy. The U.S. Food and Drug Administration (FDA) recommends genetic testing for certain chemotherapy drugs like mercaptopurine (Purinethol) to patients with acute lymphoblastic leukemia. Some people have a genetic variant that interferes with their ability to process the drug. This processing problem can cause severe side effects and increase risk of infection, unless the standard dose is adjusted according to the patient's genetic makeup.

Using microarray, cellular transcripts expression in disease states and after treatment to identify drug response at molecular level. Microarray is either silicon chip or glass slide contain series of immobilized complementary DNA molecules or oligonucleotide probes. Targeted DNA labelled with a fluorescent, transcripts abundance based on amount of hybridized labelled dye on each microarray. Each gene represented by a probe set with 11-12 pairs of oligos composed of 25 nucleotides. This approach would make it possible to consider the measured intensities as a proxy for actual mRNA concentration. The scanner produce DAT file is the intensity for each pixel on the array and CEL file, which is the summary of probe intensity for every probe on the array. An expression file, which is the Affymetrix Suite compilation of the probes into gene expression values. An intensity of each probes are corrected for background noise normalize combined into probeset expression values. Normalization will remove systematic measurement errors, both within and between arrays. These can be seen in the scatterplot matrices as deviations from linearity between arrays. This should improve our power to detect differential expression. Transcriptional profiling has use to identify class of tumors, response to therapy etc. Many studies used in cases like acute leukemias, diffuse large B-cell lymphomas, and breast tumors (Evans & Johnson, 2001; McLeod & Evans, 2001; Phillips,Veenstra,Oren,Lee & Sadee, 2001).

BACKGROUND

DNA variation in human genome occurs at every 100 to 300 bases leads to amino acid change in protein coding and noncoding part of the genome. Some of these variations may not produce any effect in cell but few can cause pathogenic effect and can also change influence in drug response etc. Most of the population showing monogenic (single-gene) or mendelian inherited pharmacological traits are caused by genetic polymorphisms. Phenotype differences caused by homozygosity of an SNP with 1% has a frequency in the population of one in 10000. Plasma concentration versus time curve (AUC) or the mean steady-state concentration of the drug is a very specific marker of drug elimination. Low concentration leads to rapid elimination and a high plasma concentration will slow down elimination. If the elimination of drug depends on one enzyme, then the drug level is a very specific in vivo marker of the enzyme in question.

Pharmacogenetics of drug metabolizing enzymes and particular the cytochrome P450 (CYP) enzymes has been in focus for almost 30 years. Remainder of this chapter deals with in vivo methodologies and strategies in CYP pharmacogenetics. Drug metabolism takes place in liver which transforms drug into active molecule called as biotransformation, metabolism typically inactivates drugs, some drug metabolites are pharmacologically active sometimes even more so than the parent compound. Metabolism processing by oxidation, reduction, hydrolysis, hydration, conjugation, condensation, or isomerization; whatever the process, the goal is to make the drug easier to excrete. The enzymes involved in metabolism are present in many tissues but generally are more concentrated in the liver. Drug metabolism rates vary among patients. The intensity and duration of pharmacological action of most lipophilic drugs are determined by the rate they are metabolized to inactive products. Cytochrome P450 monooxygenase system is the most important pathway in this regard. This depends on phenotype & genotype of the patients. However, in cases where an enzyme is responsible for metabolizing a pro-drug into a drug, enzyme induction can speed up this conversion and increase drug levels, potentially causing toxicity. Some patients metabolize a drug so rapidly that therapeutically effective blood and tissue concentrations are not reached; in others, metabolism may be so slow that usual doses have toxic effects. Individual drug metabolism rates are influenced by genetic factors, coexisting disorders, and drug interactions. For many drugs, metabolism occurs in 2 phases. Phase I reactions involve formation of a new or modified functional group or cleavage, these reactions are non-synthetic. Phase II reactions involve conjugation with an endogenous substance reactions are synthetic. Metabolites formed in synthetic reactions are more polar and thus more readily excreted by the kidneys and the liver than those formed in non-synthetic reactions. The most important enzyme system of phase I metabolism is cytochrome P-450 (CYP450), a microsomal superfamily of isoenzymes that catalyzes the oxidation of many drugs. The electrons are supplied by NADPH–CYP450 reductase, a flavoprotein that transfers electrons from NADPH to CYP450. CYP450 enzymes can be induced or inhibited by many drugs and substances resulting in drug interactions in which one drug enhances the toxicity or reduces the therapeutic effect of another drug (Evans & Relling, 1999; Collins, 2010, pp. 2–3).

Genetic variation accounts for some of the variability in the effect of drugs with N-acetyltransferases, individual variation creates a group of people who acetylate slowly and those who acetylate quickly, split roughly 50:50 in the population of Canada. This variation may have dramatic consequences, as the slow acetylators are more prone to dose-dependent toxicity. Cytochrome P450 monooxygenase system enzymes can also vary across individuals, with deficiencies occurring in 1 – 30% of people, depending on their ethnic background. Some drugs metabolized by single CYP but not same for all drugs, if the drug has a low therapeutic index and if clinical dose titration is not feasible. For CYP2D6 the possible candidates include tricyclic antidepressants, some antipsychotics, and some antiarrhythmics and for CYP2C9 the possible candidates could be warfarin and phenytoin. However pheno or genotyping for CYP enzymes has never achieved widespread use in clinical practice because, as explained above, the response is not determined alone by a single enzyme or gene.

Individual differences in drug metabolism and therapeutic response based on genetic makeup, cytochrome P450 (CYP) genes, which encode enzymes that influence the metabolism of the most drugs. Various metabolic phenotypes like normal, intermediate, ultra-rapid, and poor is based upon the allelic variation within the individual genotype. However, several genetic events can influence a same phenotypic trait, and establishing genotype-to-phenotype relationships can thus be far from consensual with many enzymatic patterns. Number of evidence suggests that patients affected by these genetic polymorphisms will experience severe/lethal toxicities upon drug intake, and that pre-therapeutic screening does help to reduce the risk of treatment-related toxicities through adaptive dosing strategies. Depending on the type of CYP variation present, the patient's metabolizer phenotype and the type of drug, therapeutic drug response is often suboptimal. Poor metabolizers are unable to metabolize certain drugs efficiently, resulting in a potentially toxic build-up of an active drug or the lack of conversion of a prodrug into an active metabolite. In contrast, in ultra-rapid metabolizers, an active drug is inactivated quickly, leading to a sub therapeutic response, while a prodrug is quickly metabolized, leading to rapid onset of therapeutic effect. Sometimes drugs inhibit metabolizing enzyme like quinidine inhibit CYP2D6 which leads to poor metabolizer (Belle & Singh, 2008; Ermak, 2013, pp. 2).

The CYP2C19 enzyme contributes to the metabolism of a large number of clinically relevant drugs and drug classes such as antidepressants, proton pump inhibitors (PPIs), and the antiplatelet prodrug clopidogrel. Like other CYP450 genes, inherited genetic variation in CYP2C19 and its variable hepatic expression contributes to interindividual phenotypic variability in CYP2C19-substrate metabolism. The CYP2C19 "poor metabolism" phenotype was initially discovered by studies on impaired mephenytoin metabolism and the major molecular defect responsible for the trait is the CYP2C19*2 (c.681G>A; rs4244285) loss-of-function allele. CYP2C19 genotype has since been shown to affect the metabolism of several drugs and clinical CYP2C19 genetic testing is currently available. CYP2C19 is predominantly expressed in the liver and, to a lesser extent, in the small intestine. Constitutive expression of CYP2C19 is largely mediated by hepatic nuclear factors 4 alpha (HNF4alpha, HNF4A) and 3 gamma, and transcriptional activation is mediated by the drug responsive nuclear receptors CAR, PXR, and GRalpha, suggesting regulation by endogenous hormones and by drugs such as rifampicin. In vitro expression studies have recently shown that the GATA-4 transcription factor also upregulates CYP2C19 transcriptional activity by binding to two predicted GATA-specific promoter elements. Reduced CYP2C19 activity among women using steroid oral contraceptives results from transcriptional down-regulation of CYP2C19 expression through binding of ligand-activated estrogen receptor alpha to a specific ERE consensus half-site in the CYP2C19 promoter. Certain selective serotonin reuptake inhibitors have an inhibitory effect on CYP2C19, which may cause drug-drug interactions with co-administered CYP2C19-metabolized drugs. For example, early studies suggested that omeprazole diminished the pharmacodynamic antiplatelet effects of clopidogrel and increased corresponding cardiovascular risks. However, it is currently not clear if identified changes in ex vivo platelet aggregation due to concomitant omeprazole and clopidogrel administration translates into clinically meaningful outcome differences. Although a number of genotyping technologies can be used to interrogate variant CYP2C19 alleles in Clinical Laboratory Improvement Amendments (CLIA)-approved laboratories, two genotyping platforms have been approved by the U.S. Food and Drug Administration (FDA) (Mohd Zaki Salleh, 2013).

CYP2C9 is a phase I drug-metabolizing cytochrome P450 (CYP450) enzyme isoform that plays a major role in the oxidation of both xenobiotic and endogenous compounds. Also identified CYP2C9 as one of several CYP2C genes clustered in four gene group CYP2C8-CYP2C9-CYP2C19-CYP2C18 size of 500kb region on chromosome 10q24. Single nucleotide polymorphism (SNP) identified between the CYP2C8 and CYP2C9 genes, primarily expressed in the liver, and the expression level is reported to be the second highest among CYP isoforms. Only the CYP enzyme CYP3A4 is quantitatively more highly expressed in human liver. It has been estimated that CYP2C9 is responsible for the metabolic clearance of up to 15%-20% of all drugs undergoing Phase I metabolism. CYP2C9 is induced by rifampicin which has been shown consistently to increase the clearance of drugs eliminated by CYP2C9. The clearance of losartan, phenytoin, tolbutamide and S-warfarin is approximately doubled in healthy volunteers or patients treated with rifampicin. CYP2C9 is inhibited by amiodarone, fluconazole, and sulphaphenazole among other drugs . Dangerous drug-drug interaction can arise when an inhibitor such as one of these is added to a therapeutic regime that include drugs with a low therapeutic index, such as S-warfarin, tolbutamine, and phenytoin. For example, there are numerous studies documenting potentiation of the anticoagulant effect of warfarin in patients coadministered with amiodarone. CYP2C9 is the enzyme responsible for the metabolism of the S-isomer of warfarin that is principally responsible for the anticoagulant effect of the drug. The crystal structure of both human CYP2C9 in complex with warfarin and unliganded CYP2C9 showed unanticipated interactions between CYP2C9 and warfarin, revealing a new binding pocket, suggesting that CYP2C9 may simultaneously accommodate multiple ligands during its biologic function. Structural analysis suggested that CYP2C9 may undergo an allosteric change when binding warfarin. The gene coding for the CYP2C9 enzyme is highly polymorphic, including functional variants of major pharmacogenetic importance. Changes in metabolic activity caused by genetic variants in CYP2C9 play a major role in pathogenesis due to ADRs. Patients with low enzyme activity are at risk of ADR, especially for CYP2C9 substrates with a narrow therapeutic window, such as S-warfarin, phenytoin, glipizide, and tolbutamide. A large body of literature investigates two common non-synonymous variants within CYP2C9, leading to poor metabolism phenotypes. Individuals with these variants are at risk of prolonged bleeding time and increased incidence of severe bleeding in warfarin therapy, higher possibility of low blood sugar levels during glipizide and tolbutamide therapy, and more frequent symptoms of overdose in phenytoin therapy (Mohd Zaki Salleh, 2013, pp. 1371; Desta, 2002, pp. 913-58).

The cytochrome P450 2D6 (CYP2D6) is an enzyme of great historical importance for pharmacogenetics and is now thought to be involved in the metabolism of up to 25% of the drugs that are in common use in the clinic. Several years before the gene was cloned, researchers observed that Caucasian subjects responded in a bimodal pattern to certain drugs such as debrisoquine and sparteine, with most patients exhibiting "normal" pharmacokinetics, whereas others seemed to have great difficulty in metabolizing debrisoquine or sparteine. Debrisoquine and sparteine became examples of so-called probe drugs, and were used to phenotype patients. This finding led researchers to conclude that there were common polymorphisms in an as yet unidentified metabolic gene that contributed to the variable pharmacokinetics of these drugs. The protein responsible for the altered metabolism was later purified from human liver microsomes. The gene encoding this protein was initially localized to chromosome 22 and cDNA was cloned from human liver cDNA libraries using an antibody against the rat ortholog. The deduced human protein revealed 73% sequence similarity with the rat protein and by use of human-rodent somatic cell hybrids the gene was localized to human chromosome 22, confirming the earlier study. This gene came to be called CYP2D6, and is part of the cytochrome P450 gene family a group of enzymes that is responsible for Phase I metabolism and elimination of numerous endogenous substrates and a diverse array of drugs. Among the drug-metabolizing CYPs, CYP2D6 is the only non-inducible enzyme, which results in a large contribution of genetic variation to the interindividual variation in enzyme activity. CYP2D6 is highly polymorphic, with over 90 known allelic variants (Tranah, 2005).

CYP2D6 became an object of intense research following its identification as the gene responsible for the altered activity observed with debrisoquine and other drugs. It soon became apparent that there were many different polymorphisms in all parts of the world that impacted CYP2D6 activity. There were alleles that led to a complete loss of CYP2D6 activity, which were common in the initially studied Caucasian populations; however, studies in populations of other ethnic origins revealed reduced function and even hyperfunctional CYP2D6 alleles. A system of assigned patients into four categories based on their ability to metabolize CYP2D6 substrates began to emerge. They are, listed in order of highest functioning to lowest: ultrarapid metabolizers (UM), extensive metabolizers (EM), intermediate metabolizers (IM), and poor metabolizers (PM). An individual's highest functioning CYP2D6 allele predicts his/her phenotypic activity. EMs possess at least one fully functional CYP2D6 allele, and are thought of as phenotypically normal. IMs and PMs are not able to metabolize CYP2D6 substrates as well as their EM counterparts, and may be at increased risk for adverse effects resulting from higher plasma levels of the parent drug, or lack of efficacy resulting from an inability to form an active metabolite. UMs, or ultrarapid metabolizers, possess multiple functional copies of a single CYP2D6 gene. The CYP2D6 copy number has been found to be from 2-13. Each functional copy of CYP2D6 that is present increases the rate of metabolism of CYP2D6 substrates significantly. CYP2D6 allele distributions exhibit significant interethnic differences.

Personalized medicine is to achieve maximum therapeutic efficacy with minimum the risk of drug toxicity for an individuals. Inter-individual genetic profile significant to diseases susceptibility, and response to drugs. After the completion of the Human Genome Project, an explosion of genetic susceptibility to complex diseases and genetic variability will cause differences in drug responses between the individuals. Genomics has become an integral part of modern drug development, and a large number of pharmaceutical companies are using this information to identify novel drug targets, identify patient subpopulations that are likely to benefit from the therapy under development, or for other screening purposes (Collins, 2010, pp. 2–3; Crawley, 2007, pp. 821-822).

Genetic Basis of Variable Drug Response

Drug response involves various phenotypic factors, gene drug interactions is clinically important to dose determining factor. Based on the response, it will be inefficacy, efficacy, resistance, and toxicity. The drug-metabolizing enzymes, three clinically important isoforms cytochrome P450 (CYP)2D6, CYP2C9, and CYP2C19 responsible for major drug metabolism which possess a large number of functionally significant genetic polymorphisms. Drugs those produce a broad therapeutic window possible for carrying genetic variant or variants may exhibit impaired drug metabolism and disposition,but these may be of limited clinical relevance (Takahashi & Echizen,2003).

MAIN FOCUS OF THE CHAPTER

Affymetrix

The GeneChip arrays are chemically attractive between DNA nucleotides adenine (A), guanine (G), thymine (T) and cytosine (C). In this C bond with G and T bond with A. DNA strand matches a complementary strand of RNA, the two strands are complementary and will stick to each other. Also a single base that doesn't match its partner able to make one single strand from sticking to another single strand. Hybridization is use for base pairing attraction to identify each SNP genotype. Human Genome Project made DNA sequence around any SNP can genotype with sequencing methods. The surface of the Affymetrix arrays look like square checkerboard. The 10K array, which is a piece of glass about the size of a thumbnail, has over 400,000 squares. These squares are called features. The features on Affymetrix arrays are incredibly small about 8 microns across. Probe is a unique DNA strand which fit in this square. One square would hold the ATTCATG probe we built, and another square would hold the ATTTATG probe. But, while each square holds one type of probe, there isn't just one probe in each square, but millions of identical copies of the same. These probes one layer at a time, using the same type of manufacturing technology that is used to build computer semiconductors. Multiple probes are synthesized in parallel.

A probe designed to genotype a SNP for extracted the DNA from our sample, DNA can be taken from any biological sample such as blood or saliva. The DNA ready to be genotyped can analyze the SNPs from the DNA you extracted, need to make millions of copies of each piece of DNA containing a SNP. Copying the sequence containing the SNP allows the DNA to be more easily detected on the array and the SNP to be more easily genotyped. The copied of all the SNPs that will be analyzed on the microarray, chemical process to fragment the DNA chains up into millions of short pieces. Then chemical called biotin are attached to each strand and this will act as a molecular glue for fluorescent molecules that will later be washed over the array. Passing the laser on the array will get fluorescent on molecule where the sample RNA has stuck to the DNA probes on the array.

The prepared DNA sample is washed over the array for 14 to 16 hours. The number of molecules involved in this wash are staggering. There are millions of copies of each DNA probe in every square on the chip, and there are also millions upon millions of pieces of tagged DNA from the sample and match will be made. If the sequence of bases in the sample DNA matches that of a DNA probe, then there will be a perfect match and the sample will stick to the probe. The hybridized RNA is tagged with molecular glue (biotin), it's as if each hybridized square on the array has been coated with sticky glue. The glow in the dark molecules (red ball) stick to the biotin glue, the fluorescent molecules are shaken away and the stain only sticks to those places on the array where DNA has bound. After all of this, researchers shine a laser light on the array, causing the stain to fluoresce (Apidianakis, et al., 2005; Boerma, et al., 2002).

Illumina

BeadArray technology is the use of 3μM beads contains a large number of a specific oligonucleotide and a 23 bases that randomly assemble in either a fibre optic bundle substrate or a silica slide substrate. This available on two types, 96-sample array matrix format based on fibre optic bundle and the BeadChip format based on the silica slide. The randomly assembled beads are then decoded through a series of sequential hybridization processes. Illumina SNP genotyping includes the GoldenGate, iSelect, Infinium II and Infinium HD assays. The GoldenGate assay is based on allele specific extension and ligation used to interrogate 384 - 1536 SNPs simultaneously with multiplex also. The Infinium II assay, coupled with BeadChips, allows large scale interrogation of variations in the human genome at unlimited levels of loci multiplexing. Whole genome BeadChips covering SNP content ranging from 240,000 to 1 million and iSelect Infinium for custom Genotyping Panels.

Golden Gate

The activated DNA used for binding parametric particles by robust proces, depending upon the multiplex level, this equates to only 160pg of DNA per SNP genotype call. Assay oligonucleotides, hybridization buffer, and paramagnetic particles are then combined with the activated DNA in the hybridization. Three oligonucleotides are use to sequence for each SNP locus. Two oligos are specific to each allele of the SNP site, called the Allele-Specific Oligos(ASOs). A third oligo that hybridizes several bases downstream from the SNP site is the Locus-Specific Oligo (LSO). All three oligonucleotide sequences contain regions of genomic complementarity and universal PCR primer sites;the LSO also contains a unique address sequence that targets a particular bead type. Up to 1,536 SNPs may be interrogated simultaneously in this manner using GoldenGate technology. During the primer hybridization process, the assay oligonucleotides hybridize to the genomic DNA sample bound to paramagnetic particles, then washing performed to reduce noise by removing excess and mis-hybridised oligonucleotides. These joined, fulllength products provide a template for PCR using universal PCR primers P1, P2, and P3.

Universal PCR primers P1 and P2 are Cy3 and Cy5 labeled. After downstream-processing the single-stranded, dye-labeled DNAs are hybridized to their complement bead type through their unique address sequences. Hybridization of the GoldenGate Assay products onto the Array Matrix or BeadChip allows for the separation of the assay products in solution, onto a solid surface for individual SNP genotype readout. After hybridization, the BeadArray Reader is used to analyze fluorescence signal on the Sentrix Array Matrix or BeadChip, which is in turn analyzed using software for automated genotype clustering and calling (Mohd Zaki Salleh, 2013). Overall workflow shown in figure1.

Figure 1. Illumina GoldenGate Workflow

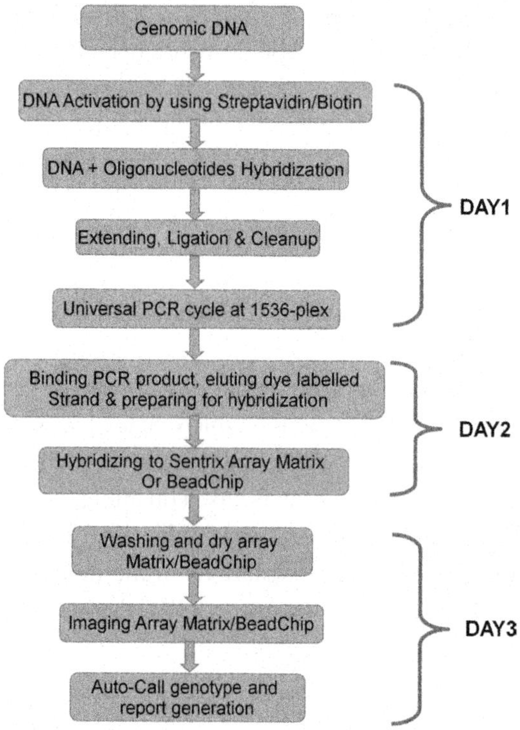

Infinium II Assay Workflow

Genotyping Assay is designed to interrogate a large number of SNPs at unlimited levels of loci multiplexing. Using a single bead type and dual color channel approach, the Infinium II Assay scales genotyping from 10,000 to hundreds of thousands of SNPs per sample. Illumina's optional Laboratory Information Management System (LIMS) and automation ensure positive sample tracking while reducing time required and labor costs.

This amplification has no appreciable allelic partiality. Additionally, a relatively low DNA sample requirement of 750 ng is sufficient to assay over 500,000 SNP loci. The amplified product is then fragmented by a controlled enzymatic process that does not require gel electrophoresis. After alcohol precipitation and resuspension of the DNA, the BeadChip is prepared for hybridization in the capillary flow-through chamber; samples are applied to BeadChips and incubated overnight. The amplified and fragmented DNA samples anneal to locus-specific 50-mers during the hybridization step. One bead type corresponds to each allele per SNP locus. After hybridization, allelic specificity is conferred by enzymatic base extension. Products are subsequently fluorescently stained. The intensities of the beads' fluorescence are detected by the Illumina. BeadArray Reader, and are in turn analyzed using Illumina's software for automated genotype calling. With Illumina's optional Laboratory Information Management System (LIMS) to ensure positive sample tracking, the Infinium II Assay is a robust protocol with a straight forward workflow that can be automated or processed manually (Armstrong, et al., 2002; Mohd Zaki Salleh, 2013). Overall workflow shown in figure2.

Data Processing and Software Tools

Genotyping of specific gene or whole genome by using either Affymetrix or Illumina can be perform to get personal genetic profile of the patient. Individualizing therapy can be provide either through diagnosis of risk factors or markers which present in the patient. Overall process should be genotyping, markers identification and pathogenicity or clinical impacts to predict drug response. Each genotyping platform providing their own genotype call software like genotype console from Affymetrix, Genome Studio from Illumina etc. Apart from those software, some commercial software supporting multiple platforms for data analysis. CLC genomics workbench, Partek genomics suite, DNA STAR, GeneSpring GX from Agilent etc. PharmGKB is pharmacogenetics knowledge base of genomic, phenotype and clinical information curated from pharmacogenetic studies. This data base can use to browse, query, download, submit, edit and process the pharmacogenetic information for registered members and subset is publicly available. Retrieving markers with associated clinical information will provide diagnostic level annotation to use personal medication [23,24].

Issues and Suggested Recommendations

This arrays are high intensive labour requirement for synthesizing, purifying, and storing DNA solutions before microarray fabrication is more expensive. Also during genotyping experiments in the laboratory, sequence homologies between clones representing different closely related members of the same gene family may result in a failure to specifically detect individual genes and instead may hybridize to spot designed gene to happen cross hybridization. In situ oligonucleotide array formats are expensive specialized equipments to carry out the hybridization, staining of label, washing, and quantitation process. Ready made in situ oligonucleotide array format are still expensive, although there has been reductions in cost as the market of microarrays has expanded. Short-sequences used on the array confer high specificity, they may have decreased sensitivity/binding compared with glass cDNA microarrays. Such low sensitivity however is compensated for by using multiple probes. Array designs are not flexible, there are occasions when the production of the array, hybridization and detection equipment are restricted to centralized manufacturer facilities, thus limiting the researcher's flexibility. Similarly, the cost and time needed to manufacture the in situ oligonucleotide array format makes it uneconomical for an average laboratory to synthesize its own chips. Multiplexing can reduce per individual cost, integrating data analysis up to secondary level may reduce complexity of the analysis, drug specific markers arrays or disease specific markers may useful to focus more [20,21].

For example anticoagulant drug Warfarin is one of the most commonly using via oral treatment for a range of thrombotic conditions, including deep vein thrombosis, pulmonary embolus and atrial fibrillation. The aim of treatment with Warfarin is to prevent arterial and venous thromboembolism by thinning the blood to a safe and effective level. Ideally patients recieving Warfarin treatment should have stable INR's within a specified and disease specific range: for example, in order to prevent recurrent myocardial infarction patients should have INR's of between 2.5 to 3.5. Despite the longstanding and widespread use of Warfarin, both loading and maintainance dose, for any particular patient, displays wide variability. While the majority of such variation is thought to be due to lifestyle and environmental factors, Warfarin action and metabolism also has a pharmacogenetic component. CYP P450 genes are important in drug response. However, in this case, it is variation in the CYP2C9 gene that is important

Figure 2. Illumina Infinium Workflow

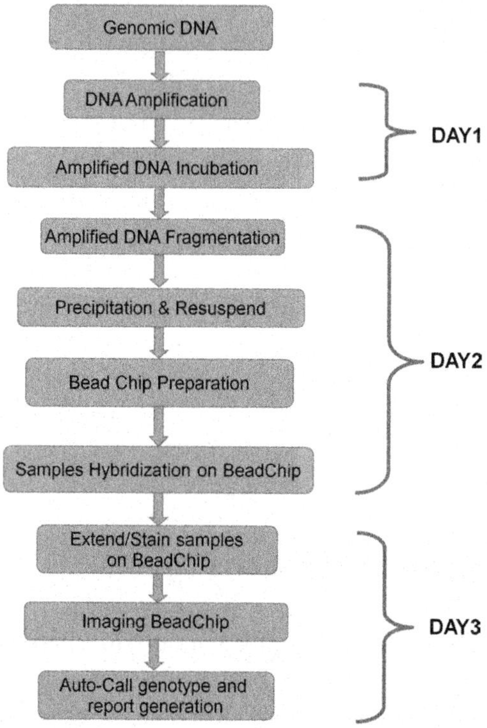

as well Single Nucleotide Polymorphism variation in the VKORC1 gene. As in the case of thiopurine metabolism, mutations of the CYP2C9 gene can create phenotypes that are extremely sensitive to Warfarin and, hence, are at an increased risk of ADR. Thus, in a similar way to thiopurine treatment, Warfarin pharmacogenetics can be used to improve safety in drug prescription. However, in this case, another gene is also implicated in drug metabolism at the other end of the spectrum: here SNP variation in the VKORC1 gene appears to mediate the efficacy of Warfarin treatment. As noted above Warfarin has a narrow theraputic index. That CYP2C9 variants mediate the safety of Warfarin treatment, and VKORC1 independently mediate its efficacy, are not the only complications for clinicians in making both, inital, and long-term, prescribing decisions. The effect of Warfarin is heavily infulenced by a number of lifestyle and environmental factors, phenotypic factors like age, weight, alcohol consumption, drug reigeme compliance, and diet. These account for around 80% of variation in Warfarin patient INR and, given the serious consequences of both over and under anticoagulation, require that this drug therapy is accompanied by intensive and costly pharmacovigilance regimes [33, 34, 35].

FUTURE DISCUSSION AND CONCLUSION

Based on the Pharmacogenomics research on CYP enzymes states the role of genetic factors for variation in drug response can predict. Genetic profile will help to determine some extent of drug response but a single gene or by a group of genes not enough to get predict fully. With full human genome it is possible for the treatment to the individual patient on the basis of the patient genetic profile. Genotyping as an aid to select the right dose of the right drug in the individual patient would be of theoretical use if the response is mainly determined by a single gene or a limited group of genes, and if all of the environmental and constitutional influences have a more limited influence, and besides are known in detail and can be measured in the individual patient. Coming years in the CYP field studying pharmacogenetics or genomics will lead to new important insights and discoveries that will ultimately lead to the development of new and better drugs and to the rational use of drugs that are already on the market. At present, after not reached in patients & physician may be lack of pharmacogenomics knowledge and gap between sequencing & data analysis. In future, integrating knowledge data base with software tools will provide on-spot clinical significance of markers after genotype call.

REFERENCES

Affymetrix Technical Note. (2005). Guide to Probe Logarithmic Intensity Error (PLIER) Estimation, Retrieved from http://www.affymetrix.com

Apidianakis, Y., Mindrinos, M. N., Xiao, W., Lau, G. W., Baldini, R. L., Davis, R. W., & Rahme, L. G. (2005). Profiling early infection responses: Pseudomonas aeruginosa eludes host defenses by suppressing antimicrobial peptide gene expression. *Proceedings of the National Academy of Sciences of the United States of America*, *102*(7), 2573–2578. doi:10.1073/pnas.0409588102 PMID:15695583

Armstrong, S. A., Staunton, J. E., Silverman, L. B., Pieters, R., den Boer, M. L., & Minden, M. D. et al. (2002). MLL translocations specify a distinct gene expression profile that distinguishes a unique leukemia. *Nature Genetics*, *30*(1), 41–47. doi:10.1038/ng765 PMID:11731795

Belle, D. J., & Singh, H. (2008). Genetic Factors in Drug Metabolism. *American Family Physician*, *77*(11), 1553–1560. PMID:18581835

Bloche, G. M. (2004). Race-based therapeutics. *The New England Journal of Medicine*, *351*(20), 2035–2037. doi:10.1056/NEJMp048271 PMID:15533852

Boerma, M., van der Wees, C. G., Vrieling, H., Svensson, J. P., Wondergem, J., & van der Laarse, A. et al. (2005). Microarray analysis of gene expression profiles of cardiac myocytes and fibroblasts after mechanical stress, ionising or ultraviolet radiation. *BMC Genomics*, *6*(1), 6. doi:10.1186/1471-2164-6-6 PMID:15656902

Cawley, S., Bekiranov, S., Ng, H. H., Kapranov, P., Sekinger, E. A., & Kampa, D. et al. (2004). Unbiased mapping of transcription factor binding sites along human chromosomes 21 and 22 points to widespread regulation of noncoding RNAs. *Cell*, *116*(4), 499–509. doi:10.1016/S0092-8674(04)00127-8 PMID:14980218

Collins, F. S. (2010). The Future of Personalized Medicine. *NIH Medline Plus*, *5*(1), 2–3.

Comer, J. E., Galindo, C. L., Chopra, A. K., & Peterson, J. W. (2005). GeneChip analyses of global transcriptional responses of murine macrophages to the lethal toxin of Bacillus anthracis. *Infection and Immunity*, *73*(3), 1879–1885. doi:10.1128/IAI.73.3.1879-1885.2005 PMID:15731093

Crawley, & LaVera (2007). The Paradox of Race in the Bidil Debate. *Journal of the National Medical Association*, 99, 821-822.

Dalen, P., Dahl, M. L., Ruiz, M. L., Nordin, J., & Bertilsson, L. (1998). 10-Hydroxylation of nortriptyline in white persons with 0, 1, 2, 3, and 13 functional CYP2D6 genes. *Clinical Pharmacology and Therapeutics*, *63*(4), 444–452. doi:10.1016/S0009-9236(98)90040-6 PMID:9585799

Derek, V. B. (2010). Cytochrome P450 2C9-CYP2C9. Pharmacogenetics and genomics. *Pharmacogenetics and Genomics*, *20*(4), 277–281. PMID:20150829

Desta, Z., Zhao, X., Shin, J.-G., & Flockhart, D. A. (2002). Clinical significance of the cytochrome P450 2C19 genetic polymorphism. Clinical pharmacokinetics. *Clinical Pharmacokinetics*, *41*(12), 913–958. doi:10.2165/00003088-200241120-00002 PMID:12222994

Ermak, G. (2013).. . *Modern Science & Future Medicine*, *2*, 164.

Evans, W. E., & Johnson, J. A. (2001). Pharmacogenomics: The inherited basis for interindividual differences in drug response. *Annual Review of Genomics and Human Genetics*, *2*(1), 9–39. doi:10.1146/annurev.genom.2.1.9 PMID:11701642

Evans, W. E., & McLeod, H. L. (2003). Pharmacogenomics – drug disposition, drug targets, and side effects. *The New England Journal of Medicine*, *348*(6), 538–549. doi:10.1056/NEJMra020526 PMID:12571262

Evans, W. E., & Relling, M. V. (1999). Pharmacogenomics: Translating functional genomics into rational therapeutics. *Science*, *286*(5439), 487–491. doi:10.1126/science.286.5439.487 PMID:10521338

Hall, A. M., & Wilkins, M. R. (2005). Warfarin: A case history in pharmacogenetics, Heart. *BMJ (Clinical Research Ed.)*, *91*, 563–564.

P450 Drug interaction Table. (2010). Indiana University Division of Clinical Pharmacology. Retrieved from http://medicine.iupui.edu/clinpharm/ddis/table.asp

Ingelman-Sundberg, M. (2004). Pharmacogenetics of cytochrome P450 and its applications in drug therapy: The past, present and future. *Trends in Pharmacological Sciences*, *25*(4), 193–200. doi:10.1016/j.tips.2004.02.007 PMID:15063083

Levy, R. H., Thummel, K. E., Trager, W. F., Hansten, P. D., & Eichelbuam, M. (Eds.). (2000). *Metabolic Drug Interactions* (pp. 29–30). Philadelphia: Lippincott Williams & Wilkins.

Martin Lewis, P., Smart, G., & Webster, A. (2006). False Positive, The commercial and clinical development of pharmacogenetics. Retrieved from http://www.york.ac.uk/media/satsu/res-pgx/FalsePositive2006.pdf

McLeod, H. L., & Evans, W. E. (2001). Pharmacogenomics: Unlocking the human genome for better drug therapy. *Annual Review of Pharmacology and Toxicology*, *41*(1), 101–121. doi:10.1146/annurev.pharmtox.41.1.101 PMID:11264452

Meyer, U. A. (2004). Pharmacogenetics – five decades of therapeutic lessons from genetic diversity. *Nature Reviews. Genetics*, *5*(9), 669–676. doi:10.1038/nrg1428 PMID:15372089

Oscarsson, M., Ingelman-Sundberg, M., Daly, A. K., & Nebert, D. W. (Eds.). (2001). Human cytochrome (p. 450). CYP. Retrieved from http://www.imm.ki.se/CYPalleles

Owen Ryan, P. (2009). Cytochrome P450 2D6, Pharmacogenetics and genomics. *Pharmacogenetics and Genomics*, *19*(7), 559–562. doi:10.1097/FPC.0b013e32832e0e97 PMID:19512959

Phillips, K., Veenstra, D., Oren, E., Lee, J., & Sadee, W. (2001). Potential role of pharmacogenomics in reducing adverse drug reactions. *Journal of the American Medical Association*, *286*(18), 2270–2279. doi:10.1001/jama.286.18.2270 PMID:11710893

Salleh, M. Z. (2013). Systematic Pharmacogenomics Analysis of a Malay Whole Genome: Proof of Concept for Personalized Medicine. *PLoS ONE*, *10*, 1371. PMID:24009664

Sanderson, S., Emery, J., & Higgins, J. (2005). CYP2C9 gene variants, drug dose, and bleeding risk in warfarin-treated patients: A HuGEnet systematic review and meta-analysis. *Genetics in Medicine*, *7*(2), 97–104. doi:10.1097/01.GIM.0000153664.65759.CF PMID:15714076

Scordo, M. G., Pengo, V., Spina, E., Dahl, M. L., Gusella, M., & Padrini, R. (2002). Influence of CYP2C9 and CYP2C19 genetic polymorphisms on warfarin maintenance dose and metabolic clearance. *Clinical Pharmacology and Therapeutics*, *72*(6), 702–710. doi:10.1067/mcp.2002.129321 PMID:12496751

Scott Stuart, A. (2011). PharmGKB summary: very important pharmacogene information for cytochrome P450, family 2, subfamily C, polypeptide 19. Pharmacogenetics and genomics.

Takahashi, H., & Echizen, H. (2003). Pharmacogenetics of CYP2C9 and interindividual variability in anticoagulant response to warfarin. *The Pharmacogenomics Journal*, *3*(4), 202–214. doi:10.1038/sj.tpj.6500182 PMID:12931134

Tranah, G. J., Chan, A. T., Giovannucci, E., Ma, J., Fuchs, C., & Hunter, D. J. (2005). Epoxide hydrolase and CYP2C9 polymorphisms, cigarette smoking, and risk of colorectal carcinoma in the Nurses' Health Study and the Physicians' Health Study. *Moleclar Carcinogenics*, *44*(1), 21–30. doi:10.1002/mc.20112 PMID:15924351

Weinshilboum, R. (2003). Inheritance and drug response. *The New England Journal of Medicine*, *348*(6), 529–537. doi:10.1056/NEJMra020021 PMID:12571261

Wikoff, W. R., Frye, R. F., Zju, H., Gong, Y., Boyle, S., Churchill, E., & Cooper-Dehoff, R. M. (2013). Pharmacometabolomics reveals racial differences in response to atenolol treatment. *PLoS ONE*, *8*(3), 1–8. doi:10.1371/journal.pone.0057639 PMID:23536766

KEY TERMS AND DEFINITIONS

Acetylators: An organism capable for metabolic acetylation.

Acute Lymphoblastic Leukemia: Also called acute lymphoblastic leukemia, is a cancer that starts from white blood cells called lymphocytes in the bone marrow (the soft inner part of the bones, where new blood cells are made).

Allele: (short for allelomorph) A variant of a gene were the DNA sequence differs between two or more variants.

Allelic Variation: The presence or number of different allele forms at a particular locus (locus or loci = place) on a chromosome (allelic variation is sometimes used more loosely to describe the overall diversity present).

Antiarrhythmics: Drugs that are used to treat abnormal heart rhythms resulting from irregular electrical activity of the heart.

Antidepressant: A substance that is used in the treatment of mood disorders, as characterized by various manic or depressive affects.

Antipsychotics: A class of medicines used to treat psychosis and other mental and emotional conditions.

Cellular Transcription: The process in a cell by which genetic material is copied from a strand of DNA to a complementary strand of RNA (called messenger RNA).

Chemotherapy: A category of cancer treatment that uses chemical substances, especially one or more anti-cancer drugs

Complementary DNA: (c-DNA) DNA synthesized from a messenger RNA (mRNA) template in a reaction catalysed by the enzymes reverse transcriptase.

Conjugation: A process by which one bacterium transfers genetic material to another through direct contact.

Cytotoxics: The quality of being toxic to cells

Detoxification: The physiological or medicinal removal of toxic substances from a living organism, including, but not limited to, the human body and additionally can refer to the period of withdrawal during which an organism returns to homeostasis after long-term use of an addictive substance.

Electrophoresis: Separations technique that is based on the the mobility of ions in an electric field.

Endogenous: Those that originate from within an organism, tissue, or cell

Genetic Makeup: The genes that determine what you look like and what physical characteristics you have.

Genetic Polymorphism: The occurrence in the same population of two or more alleles at one locus, each with appreciable frequency, where the minimum frequency is typically taken as 1%

Genetic Profiling: Science of identifying individual genetic characteristics by way of DNA analysis

Genetic Variation: Variation in alleles of genes, occurs both within and among populations.

Genome: The complete set of genes or genetic material present in a cell or organism.

Genome Variants: Variations of genomes between members of species, or between groups of species thriving in different parts of the world as a result of genetic mutation.

Genotype: Genetic makeup of a cell, an organism, or an individual usually with reference to a specific characteristic under consideration.

Genotyping: Process of determining the genetic constitution – the genotype – of an individual by examining their DNA sequence.

Germline: The cellular lineage of a sexually reproducing organism from which eggs and sperm are derived; also - the genetic material contained in this cellular lineage which can be passed to the next generation.

Homozygosity: A cell is said to be homozygous for a particular gene when identical alleles of the gene are present on both homologous chromosomes. The cell or organism in question is called a homozygote.

In Situ: In the normal location. An in situ tumor is one that is confined to its site of origin and has not invaded neighboring tissue or gone elsewhere in the body.

International Normalized Ratio (INR): A state of achieving anticoagulation for prothromboin time.

Lipophilic Drugs: Drugs that are able to dissolve much more easily in lipid (a class of oily organic compounds) than in water.

Metabolites: The products of enzyme-catalyzed reactions that occur naturally within cells.

Microarray Expression: Provides a comprehensive portrait of the transcriptional world enabling us to view the organism as a 'system' that is more than the sum of its parts

Monogenic: Involving or controlled by a single gene

Next Generation Sequencing: Method for sequencing genomes at high speed, at low cost and with great accuracy.

Oligonucleotide: Short, single-stranded DNA or RNA molecules that have a wide range of applications in genetic testing, research, and forensics.

Oncogenic: Causing development of a tumour or tumours

PCR: Polymerase chain reaction, or PCR, is a laboratory technique used to make multiple copies of a segment of DNA. PCR is very precise and can be used to amplify, or copy, a specific DNA target from a mixture of DNA molecules.

Pharmacodynamics: The study of what a drug does to the body, whereas pharmacokinetics is the study of what the body does to a drug.

Pharmacogenetics: The branch of pharmacology concerned with the effect of genetic factors on reactions to drugs.

Pharmacokinetics: The branch of pharmacology concerned with the movement of drugs within the body.

Phenotype: The set of observable characteristics of an individual resulting from the interaction of its genotype with the environment.

Phenotypic Trait: An obvious and observable trait; it is the expression of genes in an observable way. An example of a phenotypic trait is hair color, there are underlying genes that control the hair color, which make up the genotype, but the actual hair color, the part we see, is the phenotype.

Polymorphisms: The presence of genetic variation within a population, upon which natural selection can operate.

Probe: A fragment of DNA or RNA of variable length (usually 100-1000 bases long) which is used in DNA or RNA samples to detect the presence of nucleotide sequences (the DNA target) that are complementary to the sequence in the probe.

Scatterplot Matrices: A great way to roughly determine if you have a linear correlation between multiple variables. This is particularly helpful in pinpointing specific variables that might have similar correlations to your genomic or proteomic data.

SNP Arrays: A type of DNA microarray which is used to detect polymorphisms within a population.

SNP: A Single Nucleotide Polymorphism (SNP, pronounced snip; plural snips) is a DNA sequence variation occurring commonly within a population (e.g. 1%) in which a single nucleotide — A, T, C or G — in the genome (or other shared sequence) differs between members of a biological species or paired chromosomes

Therapeutic Index: A comparison of the amount of a therapeutic agent that causes the therapeutic effect to the amount that causes death (in animal studies) or toxicity (in human studies).

Thrombosis: Local coagulation or clotting of the blood in a part of the circulatory system.

Thromboembolism: Obstruction of blood vessel by a blood clot that has become dislodged from another site in the circulation.

Tumors: A swelling of a part of the body, generally without inflammation, caused by an abnormal growth of tissue, whether benign or malignant.

Chapter 11
Role of Epigenetics in Cancer Genomics

Amit Nagal
GVK Biosciences Private Limited, India

ABSTRACT

Epigenetics is the study of changes in organisms caused by modification of gene expression rather than alteration of the genetic code itself. ChIP-seq, is a method used to analyze protein interactions with DNA. It is a type of epigenetic analysis technique. Chromatin immunoprecipitation coupled with massive parallel sequencing (ChIP-seq) is gaining popularity day by day because of its clinical significance. It is a very effective tool in diagnosis of disease such as cancer. ChIP-seq is found to be very effective tool in understanding basic regulatory mechanism, cell differentiation study and studying disease processes with the decreasing cost of sequencing, ChIP-seq has become an indispensable tool for studying gene regulation and epigenetic mechanisms. The Present review explores epigenetic methods, pipeline and its role in cancer.

INTRODUCTION

Chromatin is the combination of DNA and proteins in eukaryotic cells. Genome-wide mapping of protein-DNA interactions and epigenetic marks and their modifications is essential for a full understanding of transcriptional regulation and cell differentiation. Chromatin states can influence transcription directly by altering the packing of the DNA

ChIP-seq is a technique to interpret protein interactions with DNA. Antibodies are used to select specific proteins or nucleosomes which enriches for DNA-fragments that are bound to these proteins or nucleosomes. These selected fragments can be either hybridized to a microarray (ChIP-ChIP) or sequenced on modern NGS platform (ChIP-seq).ChIP-seq combines chromatin immunoprecipitation (ChIP) with massively parallel DNA sequencing to identify the binding sites of DNA-associated protein. It was first introduced by David and Barbara (2007). However, NGS provides relatively high resolution, low noise, and high genomic coverage with compared to other technology like ChIP-ChIP assays (ChIP followed by microarray hybridization).

DOI: 10.4018/978-1-4666-8726-4.ch011

BACKGROUND

In chromatin immunoprecipitation, cells are lysed and protein-DNA interactions are crosslinked to form covalent bonds by formaldehyde or other chemical reagents which explained by Solomon and Varshavsky (1998). Then the crosslinked DNA is sheared by sonication or DNA-cutting enzymes (e.g., micrococcal nuclease, often called MNase) into 150–500 bp-long fragments. Those DNA fragments crosslinked with the DNA-binding factor of interest are immunoprecipitated using an antibody specific to the factor. ChIP can be applied to a wide range of DNA binding factors, including TFs, transcription co-activators, co-repressors, chromatin regulators, and modified histones. After reverse cross-linking the protein-DNA complexes, the pulled-down DNA fragments are PCR amplified and then subjected to massively parallel sequencing it has been practiced that (Metzker, 2010, pp. 31-46) .Finally, when the resulting ChIP-seq reads are mapped back to the genome, the locations of the factor-DNA interactions can be identified.

MAIN FOCUS OF THE CHAPTER

The Role of Epigenetic in Clinical and Cancer Genomics

The importance and role of ChIP-seq was well described by various scientists. It has been found that STAT1 has DNA association. In this study ChIP-seq was used to map the signal transducers and activators of transcription 1 (STAT1) targets in interferon γ (IFNγ)-stimulated and unstimulated human cervical cancer HeLa S3 cells. By using ChIP-seq, scientists identified 41,582 and 11,004 putative STAT1-binding regions in stimulated and unstimulated cells, respectively. Out of that 34 loci known to contain STAT1 interferon-responsive binding sites, ChIP-seq found 24 (71%). ChIP-seq targets were enriched in sequences similar to known STAT1 binding motifs. This report demonstrated the high coverage and accuracy of the ChIP-seq approach. The performance of ChIP-seq was then compared to the alternative protein–DNA interaction methods of ChIP-PCR and ChIP-ChIP. The other study suggested that Yeast genes seem to have a minimal nucleosome-free promoter region of 150bp in which RNA polymerase can initiate transcription. However, ChIP-seq was used to compare conservation of TFs in the forebrain and heart tissue in embryonic mice. In this study scientist identified and validated the heart functionality of transcription enhancers, and determined that transcription enhancers for the heart are less conserved than those for the forebrain during the same developmental stage.

Several scientist recently discovered global binding maps of androgen receptor (AR) and commonly over-expressed transcriptional corepressors including histone deacetylase 1 (HDAC1), HDAC2, HDAC3, etc., in prostate cancer cells. Their results surprisingly revealed that HDACs are directly involved in androgen-regulated transcription and wired into an AR-centric transcriptional network via a spectrum of distal enhancers and/or proximal promoters. Moreover, they show that these corepressors function to mediate repression of AR-induced gene transcription that promotes epithelial differentiation and inhibits metastasis.

ChIP-seq analysis of SOX2 revealed a consensus sequence of wwTGywTT. An integrated expression profiling and ChIP-seq analysis show that SOX2 is involved in the BMP signaling pathway, steroid metabolic process, histone modifications, and many receptor-mediated signaling pathways such as IGF1R and ITPR2 (Inositol 1,4,5-triphosphate receptor, type 2).

ChIP-seq DATA ANALYSIS

Data analysis is a very important step. A number of computational and statistical tools have been developed for ChIP-seq data analysis.

Reads Mapping and Quality Control

From NGS platform raw data has been generated in various formats so one of the very popular formats is fastq format, containing short DNA sequence and quality scores. In general, the first step of ChIP-seq analysis starts with mapping these raw reads to the reference genome. There are many algorithms have been developed to quickly map millions to hundreds of millions of short sequencing reads.

These tools are well described by the authors Langmead and Salzberg (2012), Krawitz and Robinson (2010). Burrows-Wheeler algorithm was very vital in such analysis tools such as Bowtie, BWA, and SOAP2 works on it, which was originally developed by Li and Durblin (2009) as a data compression technique.

However, choosing appropriate mapping tools depends on sequencing platform, speed, and hardware compatibility. Quality control (QC) can be conducted for reads mapping. In order to simplify analysis, usually only reads mapped to one unique location in the genome (called uniquely mapped reads) with minimum allowed mismatches (e.g., up to two mismatches) are kept for downstream analysis.

In General ChIP-seq read is mapped to multiple locations on the genome, a general solution is to randomly assign one of the locations to it. Often, the ratio of the number of uniquely mapped reads over the total number of ChIP-seq reads can be an assessment of library quality. Another useful QC measure is the number of redundant reads that are mapped to the same genomic coordinates, because high redundancy rate suggests PCR amplification bias from limited ChIP material. Studies suggest that the ratio of redundant reads over all mapped reads should ideally be below 50%.

Peak Modeling and Identification

The sequence reads mapped to the genome are subject to peak calling to detect regions with significant enrichment of ChIP signals with compare to control. ChIP-seq experiments with single-end sequencing, DNA fragments are sequenced from the 5' ends; as a result, bimodal distributions, surrounding the true binding site, are formed from reads mapped on the + and – strands respectively Therefore, to precisely detect the correct binding site, some peak callers empirically model the distance between the + and – strand modes and than extend the tags towards their 3' direction by the estimated distance and its summit represents the most probable binding location than next peak callers calculate the statistical significance. Various peak callers tools such as MACS developed by Zhang and Li (2008), Sissr developed by Jothi and Zhao (2008) found to be very effective for data analysis.

The detected peaks also need to be checked in terms of quality, where false discovery rate (FDR) or fold change against the background (e.g., control tag counts in the same region) is often used. FDR is defined as the expected proportion of false positive peaks in a list of detected peaks explained by Benjamini and Hochberg (1995). Several peak callers provide empirical or model-estimated FDR or q-value (minimum FDR at a given p value cut-off) for each peak, and 5% FDR is the most commonly accepted value for peaks of good quality. The empirical FDR can be calculated as the number of control peaks passing certain cut-off divided by the number of ChIP-seq peaks passing the same cut-off. Model-estimated FDR can be computed through permutation or random sampling. Fold change, the ratio of tag counts between IP and control in the peak region, is also an intuitive measure of peak quality. In studies A fold change of 5 is generally recommended as a reasonable cut-off, and an enough number (e.g., >50%) of peaks with over 20-fold is an indicator of good ChIP enrichment.

Visualization Tools

Visualization tools are very useful for analysis and interpretation of large genomic data set. ChIP-seq data views either as signal profiles or as called peaks on a genome browser. The most widely used visualization environment is the University of California Santa Cruz (UCSC) genome browser which was developed by Kent and Haussler (2002). In addition it also provides other important genomic information, including tracks for gene annotation (e.g., refseq or UCSC known genes), evolutionary conservation, annotated SNPs and data from NIH funded genomics consortia such as ENCODE [14]. As a web server, UCSC genome browser has limitations in response speed, which are mostly related to the process capability of the server and Internet connection. Others visulaizer such as IGV (http://www.broadinstitute.org/software/igv/home) and IGB (http://www.bioviz.org/igb/) are very effective visualization tools.

Issues

There are few issues and limitation in ChIP-seq experiment:

1. Reads are not uniformly distributed.
2. Sometime Artifact can be found instead of peak.
3. There may be mapping ambiguity due to repetitive regions
4. The value of any ChIP data, including ChIP–seq data, depends crucially on the quality of the antibody used because the quality of different antibodies is highly variable and can also vary.

Solutions and Recommendations

Reads are not uniformly distributed. Normally Control (i.e., Input DNA) can be used for addressing this limitation. To overcome mapping ambiguity Normalization is mandatory process. For Antibody quality issues Rigorous validation should be done .For example in histone modifications, the reactivity of the antibody with unmodified histones or nonhistone proteins should be checked by western blotting cross-reactivity with similar histone modifications (for example, dimethylation compared with trimethylationat the same residue) should be checked by using two independent antibodies in combination with RNA interference against enzymes that are predicted to add the modifying group

FUTURE DISCUSSION AND CONCLUSION

The NGS market is currently dominated by three different platforms: the FLX pyrosequencing system from 454 Life Sciences (a Roche company), the Illumina Genome Analyser (developed initially by Solexa), and the AB SOLiD system (now Life Technologies). On all three platforms, DNA fragments are sequenced in parallel, producing large numbers of relatively short sequence ''reads'' or ''tags''.The throughput varies from hundreds of thousands of reads. It is important to note that these technologies are evolving at a tremendous pace, with ever increasing numbers and lengths of sequence reads. The three major systems differ significantly in the approaches used to produce massive amounts of sequences. Next-generation sequencing technologies are revolutionizing genomics research and beyond by enabling the much more rapid and cost-effective generation of massive amounts of sequences compared to traditional Sanger sequencing. This technological breakthrough provides an opportunity for regular research institutes and departments to engage in ambitious projects which so far have only been conceivable for large genome centers.

There is a great impact of epigenetics in cancer. The importance of epigenetics is increasing as development of NGS methods for surveying DNA methylation, mapping of transcription factor occupancy, modified histones and epigenetic regulators. The most well studied epigenetic mark, DNA methylation, can be interrogated at the whole genome level by bisulphite sequencing.

Epigenetics based MN therapies also useful in identifying novel biomarker in recent studies. ChIP-seq technique was used to analyze the variations in a methylated histone (H3K9me3) in peripheral blood mononuclear cells from 10 MN myeloid neoplasms patients and 10 healthy subjects. There were 108 genes with significantly different expression in the MN patients compared with the normal controls. In MN patients, significantly increased activity was seen in 75 H3K9me3 genes, and decreased activity was seen in 33, compared with healthy subjects. Five positive genes, DiGeorge syndrome critical region gene 6 (DGCR6), sorting nexin 16 (SNX16), contactin 4 (CNTN4), baculoviral IAP repeat containing 3 (BIRC3), and baculoviral IAP repeat containing 2 (BIRC2), were selected and quantified. There were alterations of H3K9me3 found in MN patients. These may be candidates to help explain pathogenesis in MN(myeloid neoplasms) patients. Such novel findings show that H3K9me3 may be a potential biomarker or promising target for epigenetic-based MN therapies.

The various studies suggest that ChIP-seq has great potential in cancer and clinical genomics and It can be a very vital application in personal healthcare.

The challenges of ChIP-seq require novel experimental, statistical, and computational solutions. Ongoing advances will allow ChIP-seq to analyze samples containing far fewer cells, greatly expanding its applicability in areas such as embryology and development where large samples are prohibitively expensive or difficult to obtain. Nano-ChIP-seq can analyze a sample as small as 10,000 cells. The number of false positive peaks can be reduced both experimentally and computationally. Another way to eliminate massive amounts of false positive peaks is to limit the regulatory binding sites to nucleosome-depleted regions, which are accessible for regulator binding. These regions are mapped by DNase I hypersensitivity sequencing (DNase-seq) and similar techniques: scientist. found that 94% of the human transcription factor binding sites fell into DNase hypersensitivity regions with only a few exceptions like the transcription factors ZNF274, KAP1, and SETDB1, which also bind to closed chromatin .False positive peaks are also due to unrealistic p-values (and hence FDRs) coming from unrealistic statistical models used in most methods.

ChIP-seq will also detect indirect DNA binding by the protein (via another protein or complex), so predicted sites not containing the motif may also be functional. Finally, binding does not necessarily imply function, so it will remain necessary to use additional information (such as expression or chromatin conformation data) to reliably infer the function of individual binding events.

In the analysis of ChIP-seq data several steps are involved. Effective analysis of ChIP-seq data requires sufficient coverage by sequence reads (sequencing depth). The required depth depends mainly on the size of the genome and the number and size of the binding sites of the protein. For mammalian transcription factors (TFs) and chromatin modifications such as enhancer-associated histone marks, which are typically localized at specific, narrow sites and have on the order of thousands of binding sites, 20 million reads may be adequate (4 million reads for worm and fly TFs). Proteins with more binding sites (e.g., RNA Pol II) or broader factors, including most histone marks, will require more reads, up to 60 million for mammalian ChIP-seq. Most Importantly, control samples should be sequenced significantly deeper than the ChIP ones in a TF experiment and in experiments involving diffused broad-domain chromatin data. This is to ensure sufficient coverage of a substantial portion of the genome and non-repetitive autosomal DNA regions. To ensure that the chosen sequencing depth was adequate, a saturation analysis is recommended. The peaks called should be consistent when the next two steps (read mapping and peak calling) are performed on increasing numbers of reads chosen at random from the actual reads. Saturation analysis is built into some peak callers (e.g., SPP [9]). If this shows that the number of reads is not adequate, reads from technical replicate experiments can be combined. To avoid over-sequencing and estimate an optimal sequencing depth, it is important to take into account library complexity. Several tools are available for this purpose. For example, the preseq package allows users to predict the number of redundant reads from a given sequencing depth. and how many will be expected from additional sequencing. Similarly, the ENCODE software tools offer a quality metric called the PCR bottleneck coefficient (PBC), defined as the fraction of genomic locations with exactly one unique read versus those covered by at least one unique read.

Before mapping the reads to the reference genome, they should be filtered by applying a quality cutoff. The remaining reads should then be mapped using one of the available mappers such as Bowtie, BWA and SOAP. Recent versions support gapped alignment (e.g., Bowtie2.It is important to consider the percentage of uniquely mapped reads reported by the mapper. The percentage varies between organisms, and for human, mouse, or Arabidopsis ChIP-seq data, above 70% uniquely mapped reads is normal, whereas less than 50% may be cause for concern. A low percentage of uniquely mapped reads often is due either to excessive amplification in the PCR step, inadequate read length, or problems with the sequencing platform, but with some ChIPed proteins it may be unavoidable (e.g., if the protein binds frequently in repetitive DNA). The read mappers are designed to allow a (user-settable) number of mismatches in the reads, and it is important to choose this parameter to be appropriate with the NGS platform being used (consult the manufacturer). A final potential cause of high numbers of "multi-mapping" reads is that the protein binds frequently in regions of repeated DNA.

A pivotal analysis for ChIP-seq is to predict the regions of the genome where the ChIPed protein is bound by finding regions with significant numbers of mapped reads (peaks). It is strongly recommended that mapped reads from a control sample be used (e.g., from input DNA), although some peak callers can use GC content or mappability as information necessary to assess the level of non-specific or background binding. Duplicate reads (same 5′ end) can be removed before peak calling to improve specificity. Although some peak callers support both single and paired-end reads (e.g., MACS), others are specifically designed to improve sensitivity and specificity in paired-end sequencing (e.g., SIPeS]). Existing peak callers have many user-settable parameters that can greatly affect the number and quality of the peaks called. For instance, the enrichment metric for most peak callers, such as p-value or FDR, could be hugely affected by the statistical model used, the sequencing depth, or the actual number of binding sites in the genome. Thus, using the same p-value or FDR threshold does not ensure that the numbers of peaks called are comparable across libraries and different peak callers.

Comparative ChIP-seq analysis of an increasing number of protein-bound regions across conditions or tissues is expected with the steady raise of NGS projects. For example, temporal or developmental designs of ChIP-seq experiments can provide different snapshots of a binding signal for the same TF, uncovering stage-specific patterns of gene regulation.

There are two alternatives have been proposed. The first one is qualitative implements hypothesis testing on multiple overlapping sets of peaks. The second one is quantitative proposes the analysis of differential binding between conditions based on the total counts of reads in peak regions or on the read densities, i.e., counts of reads overlapping at individual genomic positions. The direct calculation of differentially bound regions between treatment samples without controls (i.e., using one of them as a control) is not recommended because highly enriched regions could be identified due to artefacts or different chromatin structure and not due to true binding events.

In order to increase sensitivity for detecting differentially bound regions more relaxed thresholds can be used to find peaks at each condition. Then, depending on the biological question, the sets of peaks called in any of the conditions can be considered separately, or collapsed into one or more meaningful lists of consensus peak regions. One can use the qualitative approach to get an initial overview of differential binding. However, peaks identified in all conditions will never be declared as differentially bound sites by this approach based just on the positions of the peaks. The quantitative approach works with read counts (e.g., DBChIP) or read densities (e.g., MAnorm) computed over peak regions, and has higher computational cost, but is recommended as it provides precise statistical assessment of differential binding across conditions (e.g., p-values or q-values linked to read-enrichment fold changes). It is strongly advised to verify that the data fulfill the requirements of the software chosen for the analysis. For instance, DIME assumes that a significant proportion of peaks are common to the conditions under comparison, MAnorm assumes that peaks that are common in both conditions do not change significantly, while other methodologies may expect a constant number of peaks across conditions. Importantly, with some tools only two conditions can be submitted simultaneously for comparison (e.g., MAnorm), and some may perform better depending on the protein ChIPed (e.g., ChIPDiff for histone marks and POLYPHEMUS for RNA Pol II).

To analyze peak annotation several approaches has been performed. The aim of the annotation is to associate the ChIP-seq peaks with functionally relevant genomic regions, such as gene promoters, transcription start sites, intergenic regions, etc. In the first step, one uploads the peaks and reads (in an appropriate format, e.g., BED or GFF for peaks, WIG or bedGraph for normalized read coverage) to a genome browser, where regions can be manually examined in search for associations with annotated genomic features. If comparable data (e.g., ChIP-qPCR) is available, it can be compared with the ChIP-seq peaks and reads manually in the browser as well. A systematic analysis can also be performed using tools in packages such as BEDTools to compute the distance from each peak to the nearest landmark (e.g., TSS), or to identify the genes within a given distance of a peak. The output of such "location analyses," obtained for instance using CEAS or the Bioconductor package ChIPpeakAnno, can be further correlated with expression data (e.g., to determine if proximity of a gene to a peak is correlated with its expression) or subjected to a gene ontology analysis (e.g., to determine if the ChIPed protein is involved in particular biological processes). Gene ontology analysis can be done using DAVID, GREAT, or GSEA. Sometimes, the reads densities relative to a specific annotated feature are plotted and compared across different samples, thus revealing protein-binding pattern differences between them.

Motif analysis is useful for much more than just identifying the causal DNA-binding motif in TF ChIP-seq peaks. When the motif of the ChIPed protein is already known, motif analysis provides validation of the success of the experiment. Even when the motif is not known beforehand, identifying a centrally located motif in a large fraction of the peaks by motif analysis is indicative of a successful experiment.

Motif analysis can also identify the DNA-binding motifs of other proteins that bind in complex or in conjunction with the ChIPed protein, illuminating the mechanisms of transcriptional regulation. Motif analysis is also useful with histone modification ChIP-seq because it can discover unanticipated sequence signals associated with such marks. Table S4 and [48], [49] list a small sample of the publicly available tools for motif analysis.

The final step in ChIP-seq data analysis is Motif identification and analysis. Motif analysis is applied to the genomic regions identified by peak-calling algorithms. Hence, the first step in motif analysis is to assemble a set of genomic sequences in FASTA format corresponding to all the significant ChIP-seq peaks. The second step in motif analysis is motif discovery and it is advisable to input the peak sequences to two or more of the many algorithms able to discover sequence motifs in unaligned DNA sequences, as the algorithms have complementary strengths and weaknesses. Some motif discovery algorithms form part of pipelines that perform several motif analysis steps (e.g., MEME-ChIP and peak-motifs), including word-based motif discovery algorithms and motif enrichment algorithms that can identify motifs present in only a small fraction of the peaks. Following motif discovery, comparing the discovered motifs with known DNA motifs using motif comparison software is useful to confirm the presence of the ChIPed TF motif if its (or its TF-family) binding motif is known. The results will also provide hints about other TFs that bind near the ChIPed TF. Next, central motif enrichment analysis will determine if other known DNA motifs are enriched near the centers (or summits) of the ChIP-seq peaks. It can also be useful to perform local motif enrichment analysis on regions centered on genomic landmarks such as transcription start sites overlapped by ChIP-seq peaks. Additionally, a motif spacing analysis detects preferred distances and arrangements of pairs of motifs that can be indicative of physical interactions between TFs. Finally, motif prediction maps and visualizes the genomic locations of the motifs in each of the ChIP-seq regions. In this step, the discovered or enriched motifs are used to scan the ChIP-seq peak regions, and the coordinates of the matches are uploaded to a genome browser for visualization.

It has taken several years for the field to develop objective ways to quantify key aspects of success in Immunoprecipitation enrichment, library building, and final sequencing. Poor datasets that have high false-negative rates in peak calling are a predictable pitfall that has significant downstream consequences for some kinds of biological and computational analyses. In estimating data quality, the traditional approach of visual inspection at a limited number of sites (often previously well-characterized using low-throughput approaches) is inefficient, subjective, and ultimately can be deceptive. It is also possible (and commonly observed in practice) that sites, the biological importance of which has been defined by independent functional assays, can decrease to below the sensitivity threshold of a poor or mediocre ChIP-seq experiment.

There are other technology in Epigenetics like ChIP-on-chip (also known as ChIP-chip) is a technology that combines chromatin immunoprecipitation with DNA microarray . It is used to investigate interactions between proteins and DNA in vivo. Specifically, it allows the identification of the cistrome, sum of binding sites, for DNA-binding proteins on a genome-wide basis. Whole-genome analysis can be performed to determine the locations of binding sites for almost any protein of interest.The most prominent representatives of this class are transcription factors, replication-related proteins, like ORC, histones, their variants, and histone modifications. The main goal of ChIP-on-chip is to locate protein binding sites that may help identify functional elements in the genome. For example, in the case of a transcription factor as a protein of interest, one can determine its transcription factor binding sites throughout the genome. If histones are subject of interest, it is believed that the distribution of modifications and their localizations may offer new insights into the mechanisms of regulation. One of the long-term goals ChIP-on-chip was designed for is to establish a catalogue of (selected) organisms that lists all protein-DNA interactions under various physiological conditions. This knowledge would ultimately help in the understanding of the machinery behind gene regulation, cell proliferation, and disease progression. Hence, ChIP-on-chip offers not only huge potential to complement our knowledge about the orchestration of the genome on the nucleotide level, but also on higher levels of information and regulation as it is propagated by research on epigenetics. However this technology has its own Strengths and Weaknesses. Using tiled arrays, ChIP-on-chip allows for high resolution of genome-wide maps. These maps can determine the binding sites of many DNA-binding proteins like transcription factors and also chromatin modifications.

Although ChIP-on-chip can be a powerful technique in the area of genomics, it is very expensive. Most published studies using ChIP-on-chip repeat their experiments at least three times to ensure biologically meaningful maps. The cost of the DNA microarrays is often a limiting factor to whether a laboratory should proceed with a ChIP-on-chip experiment. Another limitation is the size of DNA fragments that can be achieved. Most ChIP-on-chip protocols utilize sonication as a method of breaking up DNA into small pieces. However, sonication is limited to a minimal fragment size of 200 bp. For higher resolution maps, this limitation should be overcome to achieve smaller fragments, preferably to single nucleosome resolution. The statistical analysis of the huge amount of data generated from arrays is a challenge and normalization procedures should aim to minimize artifacts and determine what is really biologically significant. So far, application to mammalian genomes has been a major limitation, for example, due to the significant percentage of the genome that is occupied by repeats. However, as ChIP-on-chip technology advances, high resolution whole mammalian genome maps should become achievable. Antibodies used for ChIP-on-chip can be an important limiting factor. ChIP-on-chip requires highly specific antibodies that must recognize its epitope in free solution and also under fixed conditions. To overcome the problem of specificity, the protein of interest can be fused to a tag like FLAG or HA that are recognized by antibodies.

The Workflow of a ChIP-on-chip experiment divided into three parts: The first is to set up and design the experiment by selecting the appropriate array and probe type. Second, the actual experiment is performed in the wet-lab. Last, during the dry-lab portion of the cycle, gathered data are analyzed to either answer the initial question or lead to new questions so that the cycle can start again.

In the first step, the protein of interest (POI) is cross-linked with the DNA site it binds to in an in vitro environment. Usually this is done by a gentle formaldehyde fixation that is reversible with heat. Further, the cells are lysed and the DNA is sheared by sonication or using micrococcal nuclease. This results in double-stranded chunks of DNA fragments, normally 1 kb or less in length. Those that were cross-linked to the POI form a POI-DNA complex.

In the next step, only these complexes are filtered out of the set of DNA fragments, using an antibody specific to the POI. The antibodies may be attached to a solid surface, may have a magnetic bead, or some other physical property that allows separation of cross-linked complexes and unbound fragments. This procedure is essentially an immunoprecipitation (IP) of the protein. This can be done either by using a tagged protein with an antibody against the tag (ex. FLAG, HA, c-myc) or with an antibody to the native protein.The cross-linking of POI-DNA complexes is reversed (usually by heating) and the DNA strands are purified. For the rest of the workflow, the POI is no longer necessary.After an amplification and denaturation step, the single-stranded DNA fragments are labeled with a fluorescent tag such as Cy5 or Alexa 647. Finally, the fragments are poured over the surface of the DNA microarray, which is spotted with short, single-stranded sequences that cover the genomic portion of interest. Whenever a labeled fragment finds a complementary fragment on the array, they will hybridize and form again a double-stranded DNA fragment. After a sufficiently large time frame to allow hybridization, the array is illuminated with fluorescent light. Those probes on the array that are hybridized to one of the labeled fragments emit a light signal that is captured by a camera. This image contains all raw data for the remaining part of the workflow.

This raw data, encoded as false-color image, needs to be converted to numerical values before the actual analysis can be done. The analysis and information extraction of the raw data often remains the most challenging part for ChIP-on-chip experiments. Problems arise throughout this portion of the workflow, ranging from the initial chip read-out, to suitable methods to subtract background noise, and finally to appropriate algorithms that normalize the data and make it available for subsequent statistical analysis, which then hopefully lead to a better understanding of the biological question that the experiment seeks to address. Furthermore, due to the different array platforms and lack of standardization between them, data storage and exchange is a huge problem. Data analysis divided into three major steps: During the first step, the captured fluorescence signals from the array are normalized, using control signals derived from the same or a second chip. Such control signals tell which probes on the array were hybridized correctly and which bound non specifically.

In the second step, numerical and statistical tests are applied to control data and IP fraction data to identify POI-enriched regions along the genome. The following three methods are used widely: Median percentile rank, Single-array error, and Sliding-window. These methods generally differ in how low-intensity signals are handled, how much background noise is accepted, and which trait for the data is emphasized during the computation. In the third step, these regions are analyzed further. If, for example, the POI was a transcription factor, such regions would represent its binding sites. Subsequent analysis then may want to infer nucleotide motifs and other patterns to allow functional annotation of the genome.

Currently the ChIP-Seq protocol is well developed and relatively easy to perform: ChIP has been performed for more than a decade and construction of ChIP-Seq library is no rocket science either. Post-sequencing, one run of Illumina Genome Analyzer produces hundreds of gigabytes of data per run and basic pipeline analysis on our server takes more than a day. Nonetheless, even though performing ChIP-Seq does require significant computational and data storage resources, basic analysis is relatively standard. The difficult part of ChIP-Seq experiment is more advanced data analysis. It has now been made easier like peak calling tools, but use of these tools requires more computer expertise than molecular biologists typically possess. Thus development of user-friendly tools will likely make ChIP-Seq results more usable. Creative processing of ChIP-Seq data requires close collaboration between biologists and bioinformaticians.

Discovery of all binding sites of transcription factor in a given cell type is a step to understanding the transcriptional network that regulates gene expression and function of this cell. Further, since number of tags found at a specific target site reflects strength of interaction, these data can be used to determine which target sites have higher affinity and will be occupied first upon the start of transcription factor expression. Target data for a number of transcription factors combined will enable the construction of predictive models of transcription control, gene expression and interaction of transcription factors and co-factor proteins. Many researchers will likely use ChIP-Seq results to examine just several individual binding sites. When comparing binding between different samples it is relatively easy to compare localization of binding. Much more care should be taken when comparing binding levels at the same site between different samples: enrichment in a given ChIP experiment depends on many intractable parameters, likely including a phase of a moon. Thus, change in tag number at certain sites might result not only from real binding changes but also from experimental irregularities. On a positive side ChIP-Seq offers an internal control in the form of average tag density profiles for the control sites.

One of the drawbacks of the ChIP approach is the inability to distinguish between a binding event happening in the whole cell population and an event happening in only a few cells at a time. To this end, it is important to learn to reduce the number of cells required for ChIP-Seq. The ability to perform ChIP-Seq on a single cell level will provide answers to many interesting questions. Now almost any method used for analysis of protein–DNA interactions at a single locus can be combined with sequencing for genome-wide coverage. In fact the new Sequencing techniques can be outright easier and retire older, often radioactivity based techniques. A prime example is DNase footprinting .This method is used to identify regions of open chromatin, which are considered candidate areas for enhancers, promoters and other genomic elements. In the past, DNase treatment was followed by indirect end-labeling or LM-PCR and gel separation or Southern blotting. Now for genome-wide coverage one can simply sequence the ends of DNase-cut fragments. Another application is mapping of nucleosome positions using footprinting with micrococcal nuclease. This approach allowed us to discover specific nucleosomal organization of open promoters: unlike closed promoters, the first several nucleosomes in open promoters are positioned at highly specific distances from the transcription start sites. The analysis of nucleosome positioning will likely be much easier with the use of paired-end sequencing.

Currently number of tags produced by a single lane of Illumina GAII is sufficient for mapping most of narrowly distributed chromatin modifications, such as H3K4me1/2/3, but is not sufficient for mapping H3K27me3, which occupies larger share of genome or nucleosomes. Improvements in current sequencing technologies and development of new ones by Helicos, Pacific Biosciences will make mapping of such modifications easier. Longer reads, which are being promised by Illumina and ABI, will make a larger share of the genome mappable. Further, single molecule sequencing techniques promoted by several companies will eliminate the need for pre-sequencing PCR thus solving problems of biased coverage. There are many data analysis tools available for ChIP-seq technology. In this technique selecting a peak detection algorithm is central to ChIP-seq experimental studies. Though the algorithmic details may seem arcane to many biologists, computational analysis is the key to leveraging meaningful information about biology from sequence-based data. We demonstrate that eleven ChIP-seq analysis programs of varying algorithmic complexity identify protein binding sites from common empirical datasets with remarkably similar performance with regards to sensitivity and specificity. A more complete analysis of the origin of noise and improved metrics for determining the noisiness of datasets would certainly benefit future in ChIP-seq experiments.

The programs differed most significantly in the spatial resolution of their estimates for the precise binding region. These tools would be an excellent choice especially for applications such as de novo motif discovery in regions with multiple motifs, where it is important to accurately minimize sequence search space. ChIP-seq experiments may provide the most reliable manner of filtering false positives from true binding sites, a practice already encouraged by several groups such as the ENCODE consortium .It can be concluded that rather than focus solely on algorithmic development, equal or better gains could be made through careful consideration of experimental design and further development of sample preparations to reduce noise in the datasets.

There are other tools implemented in R like ChIPpeakAnno which enables batch annotation of binding sites identified from ChIP-seq, ChIP-chip, CAGE or any technology that results in a large number of enriched genomic regions for any species with existing annotation data within the statistical programming environment R. It allows users to pass their own annotation data such as different ChIP preparation and a dataset from literature, or existing annotation packages, such as GenomicFeatures and BSgenome, provides flexibility while the tight integration to the biomaRt package enables up-to-date annotation retrieval from the BioMart database. The main advantage of ChIPpeakAnno is the flexibility to plug in with other annotation packages, ChIP-chip analysis packages, other fast moving deep-sequencing analysis capabilities and infrastructure and statistical analysis tools in Bioconductor. Another advantage of ChIPpeakAnno is that it enables comparison between a set of peaks with any annotation feature objects, between two sets of peaks from replicate experiments or transcription factors within a complex and determination of the significance of the overlap. Chromatin immunoprecipitation (ChIP) followed by high-throughput DNA sequencing (ChIP-seq) has become a valuable and widely used approach for mapping the genomic location of transcription-factor binding and histone modifications in living cells. Despite its widespread use, there are considerable differences in how these experiments are conducted, how the results are scored and evaluated for quality, and how the data and metadata are archived for public use. These practices affect the quality and utility of any global ChIP experiment. ENCODE and modENCODE consortia have developed a set of working standards and guidelines for ChIP experiments. This guidelines address antibody validation, experimental replication, sequencing depth, data and metadata reporting, and data quality assessment. All data sets used in the analysis have been deposited for public viewing and downloading at the ENCODE and modENCODE portals.

Guidelines for reporting ChIP-seq data has been well described. To facilitate data sharing among laboratories, both within and outside the Consortium, and to ensure that results can be reproduced, ENCODE has established guidelines for data sharing in public repositories. Raw data can be submitted to the Short Read Archive (SRA) and ChIP results are submitted to GEO.

REFERENCES

Benjamini, Y. & Hochberg, Y. (1995). Controlling the false discovery rate: a practical and powerful approach to multiple testing. *The Journal of statistical society, 57,* 289-300.

David, J. (2007). Genome-wide mapping of in vivo protein–DNA interactions. *Science, 316*(5830), 1497–1502. doi:10.1126/science.1141319 PMID:17540862

Jothi, R., Cuddapah, S., Barski, A., Cui, K., & Zhao, K. (2008). Genome-wide identification of in vivo protein-DNA binding sites from ChIP-Seq data. *Nucleic Acids Research, 36*(16), 5221–5231. doi:10.1093/nar/gkn488 PMID:18684996

Kent, W. J., Sugnet, C. W., Furey, T. S., Roskin, K. M., Pringle, T. H., Zahler, A. M., & Haussler, D. (2002). The human genome browser at UCSC (online advance copy). *Genome Research, 12*(6), 996–1006. doi:10.1101/gr.229102. Article published online before print in May 2002 PMID:12045153

Krawitz, P., Rödelsperger, C., Jager, M., Jostins, L., Bauer, S., & Robinson, P. N. (2010). Microindel detection in short-read sequence data. *Bioinformatics (Oxford, England), 26*(6), 722–729. doi:10.1093/bioinformatics/btq027 PMID:20144947

Langmead, B., & Salzberg, S. L. (2012). Fast gapped-read alignment with Bowtie 2. *Nature Methods*, *9*(4), 357–359. doi:10.1038/nmeth.1923 PMID:22388286

Langmead, B., Trapnell, C., Pop, M., & Salzberg, S. L. (2009). Ultrafast and memory-efficient alignment of short DNA sequences to the human genome. *Genome Biology*, *10*(3), 25–30. doi:10.1186/gb-2009-10-3-r25 PMID:19261174

Li, H., & Durbin, R. (2008). Mapping short DNA sequencing reads and calling variants using mapping quality scores. *Genome Research*, *18*(11), 1851–1858. doi:10.1101/gr.078212.108 PMID:18714091

Li, H., & Durbin, R. (2009). Fast and accurate short read alignment with Burrows-Wheeler transform. *Bioinformatics (Oxford, England)*, *25*(14), 1754–1760. doi:10.1093/bioinformatics/btp324 PMID:19451168

Li, H., & Durbin, R. (2009). Fast and accurate short read alignment with Burrows-Wheeler transform. *Bioinformatics (Oxford, England)*, *25*(14), 1754–1760. doi:10.1093/bioinformatics/btp324 PMID:19451168

Metzker, M. L. (2010). Sequencing technologies the next generation. *Nature Reviews. Genetics*, *11*(1), 31–46. doi:10.1038/nrg2626 PMID:19997069

Raney, B. J., Cline, M. S., Rosenbloom, K. R., Dreszer, T. R., Learned, K., & Barber, G. P. et al. (2011). ENCODE whole-genome data in the UCSC genome browser. *Nucleic Acids Research*, *39*(Database), 871–875. doi:10.1093/nar/gkq1017 PMID:21037257

Solomon, M. J., Larsen, P. L., & Varshavsky, A. (1988). Mapping protein-DNA interactions in vivo with formaldehyde: Evidence that histone H4 is retained on a highly transcribed gene. *Cell*, *53*(6), 937–947. doi:10.1016/S0092-8674(88)90469-2 PMID:2454748

Zhang, Y., Liu, T., Meyer, C. A., Eeckhoute, J., Johnson, D. S., & Bernstein, B. E. et al. (2008). Model-based analysis of ChIP-Seq (MACS). *Genome Biology*, *9*(9), 137–145. doi:10.1186/gb-2008-9-9-r137 PMID:18798982

KEY TERMS AND DEFINITIONS

Epigenetic: It alter the physical structure of DNA.

Immunoprecipitation: It is the technique of precipitating a protein antigen out of solution using an antibody that specifically binds to that particular protein.

PCR: Polymerase chain reaction.

SNP: Single Nucleotide Polymorphism.

Sonication: It is the act of applying sound energy to agitate particles in a sample, for various purposes.

Chapter 12
Impact of Human Exome Sequencing on Clinical Research

Anu Acharya
Mapmygenome India, India

Shibichakravarthy Kannan
Ocimum Biosolutions, India

Brajendra Kumar
Ocimum Biosolutions, India

Jasmine Khurana
Mapmygenome India, India

Sushma Patil
Mapmygenome India, India

Geethanjali Tanikella
Mapmygenome India, India

ABSTRACT

Recent advances in human exome sequencing and the associated advantages have made it a technology of choice in various domains. The savings in time, cost and data storage compared with whole genome sequencing make this technology a potential game changer in clinical research settings. Recent advances in NGS have made it feasible to use exome sequencing in clinical research for identifying novel and rare variants that can lead to change in protein structure and function which may finally culminate into a totally different phenotype. If whole exome is not desired the same technology can be used for studying target exonic regions to investigate causative genes for a specific phenotype associated with disease. Exome sequencing has emerged as an effective and efficient tool for the translational and clinical research. There is a demand for systematically storing variant information in large databanks. Meaningful information from the exome-seq data can be combined with other data. This can be correlated with clinical findings within a clinical trial setting for a better study outcome.

INTRODUCTION

A Brief History of NGS

The major breakthrough in the field of genetics was due to the Human Genome Project. Completion of this project was possible because of the revolutionary Sanger's sequencing method. Sanger's chain termination method and Maxim-Gilbert's sequencing method have laid the foundation for advanced se-

DOI: 10.4018/978-1-4666-8726-4.ch012

quencing techniques such as next generation sequencing, leading to progress in genomics and proteomics. Sanger's sequencing method utilized the basic principle of DNA replication, which dictates that for template elongation, 3' -OH of the previous nucleotide should be free. Therefore, in the reaction mixture scientists introduced dideoxynucleotide, which lacked free 3' -OH group. Thus, the chain is terminated whenever dideoxynucleotide is incorporated, giving rise to templates of various basepair lengths. At that time, there was no procedure available to differentiate between all four nucleotides. Therefore, four different mixtures had to be prepared containing one of ddATP, ddGTP, ddCTP, ddTTP and other three remaining nucleotides, polymerase enzyme and radio-labeled primer. After PCR reaction, all the four reactions were loaded in different wells of agarose gel and depending on the size, strands were separated (ABI, 2007). Agarose gel had its limitation with the formation of ssDNA loops. Therefore, denaturing polyacrylamide gel electrophoresis was used instead. Use of radioactively labeled primer and four different reactions made it laborious. Thus, non-radioactive based first generation sequencing came into picture. Leroy Hood's laboratory made some modification of Sanger's method in mid 1980s by introducing fluorescent dyes instead of radioactively labeled primers (ABI, 2007). This made it possible for all the reactions to take place in one reaction mix. Further, the method was simplified by introduction of capillary electrophoresis, thus, making the system more flexible for increasing the sample size and significantly sped up the analysis (Zhou et al., 2010a). This resulted in higher instrument throughput with 96 samples on ABI 3730 platform and 384 samples on Amersham MegaBACE in one run (Zhou et al., 2010a).These advances have helped in completing the Human Genome Project in record time, way ahead of schedule. These first -generation sequencers can achieve sequencing length up to 1000 bp, with raw accuracy as high as 99.999%, at a cost as little as $0.50/Kilobase and throughput close to 600000 bp/day (Zhou et al., 2010a). Even with this advancement, it became necessary to invent new methods to reduce cost and increase speed of analysis. Next Generation Sequencing technology was developed to address these issues (Zhou et al., 2010b).

Exome Capture Kits

Selection of appropriate target capture kit is important for qualitative and quantitative identification of variants in the target regions in human exome (protein coding regions). The quality analysis and association of correct variants to the human diseases is a crucial objective of exome sequencing in clinical research.

Depending on sequencing platform and specific application, it is important to opt for appropriate exome capture kit based on the coverage of different exome database, target coverage efficiency, GC bias, sensitivity in single nucleotide variant detection, sensitivity in small indel detection, and technical reproducibility (Chilamakuri et al., 2014; Parla et al., 2011).

Some popular commercially available exome capture kits in use by researchers and clinically certified labs are:

- *Agilent's SureSelect Target Enrichment Kit* is compatible with all major NGS platforms and is capable of detecting the most SNPs and small Indels compare to other platforms (Agilent Technologies).
- *NimbleGen SeqCap EZ Exome v3* presents the highest number of probes, being the only technology with an overlapping probe design, thus giving it the highest probe density technology comparatively (Chilamakuri et al., 2014).

- *TruSeq Exome Enrichment Kit* by Illumina offers isolation of exonic regions of interest in the human genome using hybrid selection. It is the only platform which is designed to enrich UTRs as well (Illumina a).
- *The Nextera Rapid Capture Exome and Expanded Exome* has been designed to allow optimal alignment with Illumina sequencing systems. It delivers high enrichment rates, coverage uniformity and reproducibility (Illumina b).
- *Ion Ampliseq Exome RDY Kit* can be used with oligo pools for ultrahigh multiplex PCR exome enrichment on a dried-down plate. This kit is compatible with the Ion Library Equalizer Kit for ultimate ease in library normalization along with the Ion Proton sequencer (Life Technologies).
- *Ion TargetSeq Exome Kit* is compatible with both Ion Proton and PGM sequencing instruments. It provides enrichment of exons and other important regions within the human genome, using solution phase DNA-probe capture technology for targeted resequencing applications (Life Technologies).

Current Platforms Available In Market

High throughput sequencing technologies commercially available today include the sequencing by synthesis method offered by Illumina, Pyrosequencing technique from 454/Roche, semiconductor based detection system from Ion Torrent (Life Technologies), Sequencing by ligation method used by SOLiD (Life Technologies), DNA nanoball sequencing from Complete Genomics, single molecule sequencing (Heliscope sequencing) from Helicos Biosciences and Single molecule real time sequencing (SMRT) technique being offered by Pacific Biosciences (Quail et al., Liu et al., 2012).

The main advantage of next generation sequencing is that there is no necessity of in-vitro cloning of DNA for sequencing purpose and it is accomplished by cyclic array based sequencing by synthesis. Although the principle behind each platform differs, all of them depend on clonal amplification of DNA on a large scale and thus making massively parallel sequencing a possibility. This reduces the time taken and the error rate in sequencing.

Sequencing Platforms

Illumina

Bridge amplification method is used for template amplification. Single stranded DNA with adaptor sequence is attached to flow cell consisting of single stranded oligonucleotide anchors complementary to the adaptors. Polymerase based extension takes place. Reversible terminators are utilized for sequencing the strands. Addition of fluorescent nucleotide is detected by high resolution image of the entire flow cell. Thus, the assembly of all the images represents the complete sequence (Morozova & Marra, 2008)

Illumina offers a wide range of high throughput sequencers that differs in cost and the yield of reads, quality of reads, and turnaround time for sequencing. Hence, researchers can make a choice among the MiSeq, HiSeq or NextSeq sequence platforms based on the application and experiment design. Illumina platform is also being used in CLIA and CAP accredited labs today for translational genomics and clinical research. Illumina platforms are reported to be used in various other applications as well, such as (Illumina c):

- DNA sequencing,
- SNP discovery and structural variation analysis
- Gene Regulation Analysis
- Quantitative and qualitative transcriptome analysis
- Cytogenetic analysis
- ChIP-seq and Methylation Analysis
- Small RNA Analysis
- De novo metagenomics and meta transcriptomics

Advantages of Illumina Platform

- A short-read multiplex sequencing method for reliable, cost-effective and high-throughput genotyping in large-scale studies
- 70 -90% of base pairs have quality score of Q30 with the choice of platform.
- Illumina reversible terminator chemistry handles homopolymer sequencing errors better than the pyrosequencingtechnique (Hurd & Nelson, 2009).
- Illumina allows sequencing of reverse strand of each molecule (Kircher & Kelso, 2010).

Limitations of Illumina Platform

- High substitution errors are reported in Illumina platforms (Kircher & Kelso, 2010).
- Requires high concentration of DNA

454 GS20 Pyrosequencing-Based Instrument

This technique involves a step of emulsion PCR using millions of micro beads each containing varying numbers of DNA templates, thus helping in massively parallel sequencing. This technique uses light emitting enzymes for detection of nucleotide addition. Once a nucleotide is added, pyrophosphate molecule is released, which is used to produce ATP. The ATP is used up in chemical reactions that emit light. The intensity of light can be used to decipher how many nucleotides have been added per reaction. All the 4 nucleotides are added in successive order, so that the correct nucleotide sequence can be detected (Roche).

Advantages of 454 Technology

- Generating reference assemblies after de novo sequencing
- High accuracy in annotating genes in whole genome assemblies
- Easily detect SNPs, mutations and structural rearrangements by comparing sequence with existing database
- Used for comparative population studies to identify horizontal gene transfers and orthologs
- Discovery of novel mutations associated with the disease
- Detect drug-resistant viral mutations
- Carry out GWAS studies
- Use of ribosomal RNA for viewing microbial diversity and identification of new strains of pathogenic organisms from clinical samples (Roche; Morozova & Marra, 2008).

Limitations of 454 Technology

Primary issues faced with this technique are the high error rates seen when dealing with homopolymers and high costs. Pyrosequencing initially had a disadvantage of generating short run reads; but now, the technology has improved making long reads possible (Roche; Morozova & Marra, 2008).

ABI/ SOLiD- Massively Parallel Sequencing by Ligation

Emulsion PCR is used for DNA amplification. Amplified products are then transferred onto a glass surface where sequencing occurs by sequential rounds of hybridization and ligation with 16 dinucleotide combinations labeled by four different fluorescent dyes. Nucleotide is identified by analyzing the color from two successive ligation reactions. Main advantage of this technique is that it can differentiate between sequence error and polymorphism (Morozova & Marra, 2008).

SOLiD technique can be used for following applications:

- At the genome level, de novo sequencing, targeted resequencing, and whole genome resequencing is possible.
- Epigenetics plays a major role in functioning of gene regulation. In this aspect, SOLiD can be used for identification of protein binding sites using Chromatin Immunoprecipitation Sequencing (ChIP-Seq).
- Identification of transcription binding sites as well as histone modification on a genome-wide scale.
- It can be used for Gene Expression Profiling, Small RNA Analysis, and Whole Transcriptome Analysis (Morozova & Marra, 2008).

Advantages of SOLiD Technique

- It offers 99.94% accuracy as each base is interrogated twice
- It is not hindered by homopolymers (Hurd & Nelson, 2009)

Limitations of SOLiD Technique

One of the major issues noted is palindromic sequences (Huang, Chen, Chiang, Chen, & Chiu, 2012).

Ion Torrent

The introduction of Ion-torrent (by Life Technologies, Inc) in the market, delivered in the form of Personal Genomic Machine (PGM), raised the NGS platform for biomedical and clinical applications.

It follows "sequencing by synthesis" method where the DNA polymerase incorporates dNTPs complementary to the template nucleotide, would release hydrogen ions. The hydrogen ions would trigger ISFET (Ion-Sensitive Field Effect Transistor) ion sensor and produces electrical signal. Basically, hydrogen ion release would change the pH of the solution which is further detected by ISFET and a potential change (ΔV) is recorded. The same base stretch would result in higher electronic signal. The series of electrical pulses so produced are translated into DNA sequence by the computer software (AllSeq; Liu et al., 2012).

Advantages of Ion Torrent

- Low-cost instrument for high-throughput applications.
- No nucleotide labeling required.
- No fluorescence or camera scanning required
- It gives fast runs(<3 hours) and has smaller instrument size

Limitations of Ion Torrent

- Technically difficult to enumerate long repeats or homopolymers
- Throughput is less in comparison to other high-throughput sequencing technologies.
- Biased coverage observed when AT-rich genome gets sequenced (Quail et al., 2012b).

Third Generation Sequencing

Recently, there have been staggering achievements in the sequencing domain because of the necessity to address the limitations of current technologies and to expand the scope for novel approaches. Today, we witness the dawn of a new era in sequencing with so-called third generation technologies. One of the biggest advantages is sequencing without PCR amplification step, thereby eliminating the necessity to amplify DNA and avoid amplification-induced errors in sequence. These technologies have literally made Metagenomics, sequencing the genomes of all microorganisms (even rare strains) in any biological sample a possibility.

Helicos Method

The Protocol used for this method is as follows: DNA is broken down to 100-200bp each. Poly-A tail is attached to 3' end of each fragment along with fluorescently labeled nucleotides. These fragments are then hybridized onto surface of flow cell containing immobilized oligo-T nucleotides, which is complementary to the Poly-A primer. Laser is used to illuminate the surface of the flowcell and capture the fluorescent signal emitted. It records addition of every nucleotide on a single molecule as opposed to Illumina's cluster based system. Thus, billions of unique fragments are independently sequenced at the same time. Thus, this technique has 100% accuracy with single molecule (base) level resolution (Zhou et al., 2010a; Morozova & Marra, 2008).

SMRT Method

SMRT sequencers use zero mode waveguides, which are small pores surrounded by film and silicon dioxide that enable detection of single molecule while DNA polymerase replicates the chain inside the well. It uses fluorescently labeled nucleotides (fluorescence attached to terminal phosphate instead of the nucleotide) for detection of nucleotide addition. DNA polymerase cleaves of the fluorescent label during base addition, emitting light, which is captured by the nano-photonic chamber. The main advantage of this technique is its simplicity, faster genome assembly compared with other platforms and the ability to generate longer read lengths. Attributes such as long reads, modified base detection and high accuracy make SMRT a useful technology and an ideal approach to the complete sequencing of small genomes (Zhou et al., 2010a; Roberts, Carneiro, & Schatz, 2013).

A PARADIGM SHIFT IN CLINICAL RESEARCH

Sanger's sequencing method is expensive and laborious and therefore, could not be easily used for research purpose. NGS has opened the doors for clinical research on a large scale. NGS could easily identify disease causing mutations in Mendelian disorders. Thus, this information could further be used for management of the disease. NGS has also helped in finding the cause in case of undiagnosed diseases which is exemplified by Undiagnosed Disease Program (Gahl et al., 2011) has also made population based studies easier and thus can identify single nucleotide polymorphisms (SNPs) associated with increased risk of acquiring a particular disease during one's lifetime and also protective SNPs which may decrease the chance of acquiring such diseases. Epigenetic studies are being conducted on a large scale to identify methylation status of genes to study the intricate mechanisms of gene regulation and its relationship to gene function. These studies try to identify the role of various transcription factors and their binding sites. Thus, NGS has started to generate a clearer picture regarding which mutations are responsible for what disease, how the genes are regulated and what controls their function (Kato, 2009).

Whole Exome Sequencing vs. Whole Genome Sequencing

Human genome consists of ~ 3×10^9 bases having coding and noncoding sequences. Approximately 1% of the genome codes for a protein; this region is known as the exonic region of the gene. There has been a debate on whether to go for exome sequencing or whole genome sequencing. Most of the researchers agree that exome sequencing wins over whole genome sequencing mainly because most of the Mendelian diseases are linked to mutations in the exonic region. It has been suggested that 85% of the disease causing mutations are located in the exonic region (Choi et al., 2009) thus emphasizing the importance of exome sequencing. More research has been carried out on mutations in the exonic region causing diseases compared with whole genome sequences. Therefore, it becomes easier to interpret the exome sequencing results than the whole genome sequencing.

More research needs to be done in case of the function of non-protein coding region of the gene to improve the interpretation of result of whole genome sequencing. Whole genome sequencing costs almost 6 times more than exome sequencing, thus another advantage over whole genome sequencing (Biesecker, Shianna, & Mullikin, 2011). Fourth factor is quicker turn over time. As in case of whole genome sequencing, more number of nucleotides need to be sequenced and also more data is generated for analysis, the time required is much more than exome sequencing. Thus, exome sequencing is advantageous over whole genome sequencing. However, in few cases like when any mutation in the exome could not be found responsible for causing a particular disease, whole genome sequencing will be required.

Exome Sequencing Benefits and Limitations

Of the ~$3x10^9$ bases in the human genome, about $3x10^7$ base pairs (1%) (30Mb) are coding sequences. It is the well known from the studies that 85% of known disease-causing variants occur within 1% of the genome that accounts for exome only. For this reason, sequencing of the complete coding regions (exome) has the potential to uncover the causes of large number of rare, mostly monogenic, genetic disorders as well as predisposing variants in common diseases and cancers (Teer & Mullikin, 2010).

Studies that focus on exome and genome sequencing have revealed rare variants that are causative for well studied classical Mendelian diseases as well as variants that play major role in complex diseases including Ehler-Danlos Syndromes, Cystic Fibrosis and more. Whole-exome sequencing is the cost effective alternative to whole-genome sequencing. The aim of whole exome sequencing is to provide efficient and effective genetic data for best treatment and cost-effective diagnosis (Lyon et al., 2011).

Better Targeting of Study Subjects

In clinical research, whole exome sequencing is being extensively used to identify the causative variants in several monogenic disorders, many genetically heterogeneous conditions and complex disorders such as hearing loss, cardiovascular diseases, hypertension, obesity, diabetes, and cancer.

For an effective clinical trial, it is mandatory to have stringent inclusion criteria that meet the study objectives. Many large-scale randomized clinical trials have failed to show significant primary end points, mainly due to lack of homogeneous study population and unintentional biased selection. On the other hand, some of the successful Phase III studies have huge economic impact and cost several millions of dollars to the study sponsor. Pre-screening the study subjects based on genetic information will enable researchers to exclude study subjects who may not respond well to the proposed treatment. ExomeSeq is the method of choice to screen for known variants that confer such resistance or altered drug response phenotypes. Effectively, the study design can be modified to investigate a smaller sample size with significant primary and secondary end points. In case of cancer related drugs, it is also ethical to stop giving unnecessary medication to study subjects if we already know that it is not going to benefit a subset of the population who carry a specific mutation. Such new methods to assess variants across the entire genome or exome, may facilitate rational patient stratification for clinical trials and permit more individualized prognostic information and treatment decisions in clinical care (Su, Broach, Connor, Gerhard, & Simmons, 2014).

Targeted Resequencing Advantages

If whole exome is not desired, the same technology can be used for studying target exonic regions. Targeted re-sequencing of the exonic region of interest enables researchers to investigate causative gene for a specific phenotype associated with disease. Cancer genomics laboratories across the world have started to sequence the whole exome of cancer cell lines in an attempt to identify the SNPs, INDELs, structural variants, breakpoints for chromosomal rearrangements, classifying synonymous and non-synonymous mutations (Rabbani, Tekin, & Mahdieh, 2014).

The first clinical use of Exome sequencing was for treatment of a young boy with inflammatory bowel disease. Exome sequencing identify the underlying mutation, i.e., a single point mutation in the X-linked inhibitor of apoptosis (XIAP) gene, guided the treatment which involved a bone marrow transplantation (Worthey et al., 2011).

Whole exome sequencing has improved health care by influencing disease management and drug discovery. It has shown great achievement in personalized medicine for customizing the health of each individual. Whole exome sequencing data is applied for various purposes in different disorders: diagnosis, screening procedures and research.

Identify Previously Unexplored and Novel Mutations

For a newly identified variant, its absence in the population is verified, the presence of the same and other variants in the same gene in other patients or families with the same disease are usually used to confirm the novel and earlier unexplored mutation. Identification of function of the newly identified variant in biological pathways is difficult.

Identification of the variants causing the disease brings research into clinical practice. Disease-causing variants with large pathogenic effect (high penetrance), mostly seen in single gene disorders, are the first group of classified variants. These variants are mainly rare. Although some variants are in handful of individuals who are associated with rare or uncommon diseases, they are categorized as likely disease-causing variants with less certainty due to incomplete penetrance. NGS approach would verify these variants, which is helpful in management of individuals carrying such variants. Other group is the variants with higher frequency and lower penetrance in cases than controls based on genome-wide association studies. These variants could be detected with DNA chip genotyping and NGS approaches. Whole exome sequencing could identify these variants and is used in clinical management of individuals. For example, in Familial Hypercholesterolemia, dietary management is lifesaving in individuals known to carry such causal variants. High penetrance variants detected by whole exome sequencing are important in diagnosis of the patients and healthy carriers.

NGS has improved our understanding of the genetic pathology of the diseases. Different genes and causal variants have been discovered by whole genome sequencing and whole exome sequencing. The aim is to provide efficient and effective genetic data for best treatment; however, accurate, fast and cost-effective diagnoses of the patients are a major concern. Similarly, the preclinical individuals at risk of having the disease could be identified.

Clinical Limitations

Whole exome sequencing attempts to examine the important coding regions of approximately 20,000 genes in the genome. However, the technical ability to capture and sequence the exome is limited, and currently 85%-92% of the entire exome can be evaluated.

Pathogenic variants may be present in a portion of the genes not covered by the test and therefore would not be identified. Thus, the absence of reportable findings for any gene does not mean there are no pathogenic variants in that gene.

Certain types of mutations are not detected. Only single base pair changes or small insertions or deletions of DNA are detected. Large deletions, duplications, or rearrangements, mitochondrial genome mutations, tri-nucleotide repeat expansions, genes with pseudogenes, mutations involved in tri-allelic inheritance, and many epigenetic defects may not be detected by this approach.

The clinical utility of whole exome sequencing depends on the accuracy of clinical information provided by the referring physician and the predicted inheritance pattern. DNA sequencing from family members often improves interpretation of test results (Jiao et al., 2011; Majewski, Schwartzentruber, Lalonde, Montpetit, & Jabado, 2011; ACMG Board of Directors, 2012).

EXOME SEQUENCING AS A TOOL FOR CLINICAL RESEARCH

High throughput sequencing technology has empowered clinical researchers to dig deep in the origins of complex and rare diseases as well as Mendelian diseases. Recent advances in NGS chemistry, instrumentation, methodology and analytic tools have made it possible to scan through entire chunks of coding and non-coding exons (~1% of genome, i.e., 30 million bases) on the human chromosomes (whole exome) at once with the significant level of accuracy. Whole exome sequencing can efficiently cover >95% of these exons, which eventually contain 85% of the variants responsible for disease manifestation -- especially single gene disorders (Mendelian diseases) and SNPs responsible for predisposition of diseases (Rabbani, Tekin, & Mahdieh, 2014).

With drastic reduction in the cost of sequencing and analysis over time, exome sequencing has even defied Moore's law and has become inevitable as well as technology of choice today for clinical researchers globally. Today, it is possible to sequence the whole genome sequence of an individual within $5000-$10,000 and further decline in cost is expected (Taber, Dickinson, & Wilson, 2014), same time there are corporate and academic facilities offering sequencing of whole coding regions in humans within the range of $1000 – $1500 (price taken randomly from science exchange portal) depends upon the expected depth of coverage. Targeted exome sequencing is offered (competitively for limited time) by some vendors as low as $450 for minimum 30X coverage and $650 for 100X coverage using Illumina Nextera Rapid Capture kit and sequencing on Illumina HiSeq 2500 platform showing the increase in business competency. Targeted re-sequencing of coding regions of interest has evolved as regular practice in clinical research for screening lethal alleles and novel variants in the targeted candidate genes/exons of interest specifically. Whole exome sequencing in clinical research could help in improving tumor classification, diagnosis and management for patients with malignant tumors (neoplasms) (Taber, Dickinson, & Wilson, 2014).

Exome sequencing has overcome the limitation of GWAS and linkage based study in terms of experiment cost, feature prediction and accuracy of findings and had given a new direction in fast developing bio-marker panels and genetic testing due to the ability of testing multiple genes in one test (Zhang, Li, & Zhang, 2011).

It is known that a human genome contains 4 million sequence variants whereas the exome itself has 13,000 sequence variants approximately. Understanding the clinical importance of these variations is the biggest challenge in order to use this technology in clinical practices. Researchers are targeting efforts to collect common structural variants in order to differentiate rare variants responsible for known complex diseases and to trace the molecular pathway involved in the pathogenesis of the phenotype (Marian, 2012).

Notably, variant calling and gene annotation of whole exome sequencing data of each individual exome contains about 10000 non-synonymous variants depending on ethnicity and calling methods. A normal individual has been estimated to have 50–100 mutations in the heterozygous state that can cause a recessive Mendelian disorder when being homozygous (Rabbani, Tekin, & Mahdieh, 2014).

Applications of Exome Sequencing in Clinical Research

Exome sequencing has overcome the challenges with linkage disequilibrium and positional cloning methods. The technology is being used extensively in clinical cancer research for screening pathogenic and structural variants and finding *de novo* variants, polygenic diseases and rare Mendelian disorders including dominant as well as sporadic clinical cases. Exome sequencing is in practice in clinical research for discovery studies involving individual, family and population level samples. There is speculation of a new era in personalized treatment where exome sequencing will be the technology of choice to handle the challenge of predicting the relevant pathogenic variants from pool of known tens of thousands of variants for gene mapping. Few instances where this technology has been assessed in characterization of diseases are listed below. (Dolled-Filhart et al., 2012)

Exome Sequencing as Diagnostic Tool for Autosomal Dominant Disease

In 2012, researchers from University College London and University of Oxford made use of targeted exome-capture method as diagnosis tool to detect de novo mutations in the genes LDLR, APOB, responsible for causing Familial Hypercholesterolaemia (FH). Increase in levels of LDL-cholesterol in this condition leads to early coronary heart disease. Using Agilent Human All Exon assay, exomes of 48 definite FH patients were captured and sequenced on Illumina HiSeq 2000 platform. Mean coverage for the gene of interest were observed as 23x, 36x,56x, and 93x for the genes PCSK9, LDLRAP1, LDLR, and APOB respectively. Using the popular GATK pipeline and SAMtools, variants were called and filtered.

Analysis of the exome-seq reads revealed 17 mutations in LFLR gene including 3 copy number variants, 2 APOB mutations were predicted and one heterozygous mutation in LDLRAP1 were found associated with FH disease. Two LDLR novel variants and 5 APOB variants with no known effect were also detected. This study evaluate the utility of this technology and high throughput sequencing in disease diagnosis with the realization of challenges which need to be overcome (Futema, Plagnol, Whittall, Neil, & Humphries, 2012).

Identification of Recurrent Somatic Mutation in Prostate Cancer on Large Cohorts

In a study conducted by scientists at Weill Cornell Medical College (USA) in 2012, exome sequencing was performed to analyze the impact of somatic base-pairs substitutions in prostate tumorigenesis.

Exome captured from large cohorts of 112 prostrate tumor/normal pair were sequenced to generate pair end reads. High throughput sequencing of the genomic DNA from the samples yielded mean depth of coverage of 118x per samples with 89.2% of the target coverage was >=20x. Binary alignment (.bam) format file containing the mapping information of reads with reference to the hg19 genome build were further processed through the 'Picard' tool for quality control. The analysis was performed using the 'Genepattern' tool through the 'Firehose Pipeline'. Broad Institute GATK tool was used for mutation calling and the MutSig algorithm was applied to predict the significantly enriched genes or gene sets for the mutations reported in the cohorts. Overall, 5,764 somatic mutations were identified in tumor DNA, which were not found in the healthy prostate. After applying the statistical parameters while validation, 218 somatic mutations were confirmed. Somatic point mutation and indels were annotated further using the 'Oncotator' application.

Several genes involved in mutations were identified including FOXA1, MED12, THSD7B, SCN11A, and ZNF595. These genes were not previously known to undergo somatic alteration in prostate cancer. This study of tumors and metastases across multiple cohorts revealed novel recurrent mutation in many genes including MED12 and FOXA1 genes. Heterozygous SPOP substitutions were found to occur most frequently (6-15% frequency across localized and advanced prostate cancer) and hence may define a new molecular subtype of prostate cancer (Barbieri et al., 2012).

Novel EMD Variants Identified in Chinese Family with Dilated Cardiomyopathy

In a recent study, researchers from Central South University and Jilin University, China performed whole exome sequencing on two male individuals out of 8 males and 4 females with reported symptoms of familial dilated cardiomyopathy (DCM) from a pedigree consisting of 73 family members of five generations in Jilin province in China. The disease was considered to be X-linked with a frequency of 0.0001 and 95% penetrance after the linkage analysis.

Whole exome sequencing performed on the genomic DNA samples generated 4.5 billion bases of sequences on an average with 65x mean depth of coverage. Variants were called in from 97% of the targeted bases, which showed sufficient threshold coverage. Filtering parameters were applied on the list of predicted SNPs, belonging to the 73 genes, which were screened for hereditary cardiomyopathies. Total of 9 variants satisfy the filtering condition to pass but only one was co-segregated with the disease phenotype. The functional analysis did not show any change between the wild-type and variant GPR50 protein indicating that the variant is unlikely to be sufficient to cause DCM in the pedigree. Scientists further re-examined twenty exons belonging to 10 genes on the linked loci, which initially showed sequence failure due to high GC-content and confirmed a 14-bp deletion in the *EMD* exon 1, causing a frameshift and a premature stop codon at position 81, and generating a truncated 26-amino acid polypeptide. No record for this novel deletion mutation in *EMD* exon 1 is found in the Leiden Open Variation Database (LOVD), which results in almost a complete loss of emerin protein in the pedigree. The study suggests that EMD gene should be considered for screening in the patients with DCM of unknown etiology (Zhang et al., 2014).

Technical Limitations of Exome Sequencing

Recent advances in next-generation technology have drastically reduced the error rate in sequencing reads but even low experimental errors can affect the quality of data analysis.

The capability of whole exome sequencing to generate enormous amount of data is simultaneously challenging to handle and misleading, if not processed, filtered, interpreted and reported efficiently. Hence, this potential and promising technology for clinical needs requires disciplined guidelines in order to find a needle in the haystack. If used properly, exome sequencing can turnout as a tool for personalized healthcare, considering varying clinical setting for different patients with similar ailment. Analytical sensitivity is the biggest concern to overcome with respect to the implementation of exome sequencing for clinical research (Zhang et al., 2014; Machini, Douglas, Braxton, Tsipis, & Kramer, 2014).

Cumulative mistakes at every step of a clinical study, right from sample preparation to annotation and interpretation of result increases the risk of misleading information.

Challenges in Human Exome Sequencing for Clinical Research

Sampling the Genetic Material

It is very crucial to maintain sufficient coverage and depth of sequenced region /template according to application requirements. Insufficient amount or low quality of genetic material can incorporate amplification-based errors, which in turn affect depth of coverage required for data analysis. For example, maintaining required library yield and coverage from FFPE and low input samples is a big challenge due to limited or partially degraded genomic material. Improving the methodology to perform quality whole exome sequencing from DNA derived from FFPE tumor tissues would be a major achievement to use successfully the technology's potential in clinical research (Burke, Trinidad, & Clayton, 2013).

Non-Reliable and Error Free Clinically Certified Platform for Bulk Sequencing

Errors can occur at various stages of sequencing due to limitation of sequencing platforms available. Depending on the objective of the experiment and type of application (whole exome sequencing or targeted re-sequencing) the correct platform has to be chosen; otherwise, this can induce homo polymers, extended repeats, indel errors, etc. These errors can lead to incorrect interpretation of results.

For example, few sequencing platforms are prone to errors in counting bases in homopolymer runs leading to multiple repeat units, resulting in substitution and indel type miscalls, calling base after homopolymer run and GC bias, etc.

Although whole exome sequencing is cost effective compare to whole genome sequencing, the reliability of output data depends on the efficient capture of all exons of interest; otherwise, there are chances of missing clinically significant mutations. In some cases, exome testing or analysis may be targeted to particular genes of clinical interest for a given application.

Apparently, in marker gene experiments, sequencing error can mislead in identification of actual biological feature due to variations in samples introduced by amplification and sequencing errors.

Ion torrent PGM and Illumina Miseq are the two platform currently certified for use in clinical studies although with a minimum error rate compared with other existing platforms.

Lack of Standard Parameter

Whole exome sequencing and targeted re sequencing with particular clinical settings does not follow any standard SOP that could decide on depth of sequencing, reads information (SE, PE, size, average quality, etc.), number of controls, populations based data for annotation, etc. An average considerable depth used for whole exome sequencing is 60X currently; however, it is not a statutory parameter value for best performance claiming best sensitivity and accuracy of data.

Regulatory Guidelines and Ethical Issues

Despite having common consent by the clinical researchers on using whole exome sequencing for clinical studies, it is important to address the major ethical aspects foremost. Although there is some policy statement defined for clinical implementation of NGS technology by ACMG (American College of Medical Genetics and Genomics), currently the three major ethical concerns that draw global attention are informed consent, data handling and reporting the results in case of incidental finding and un-annotated novel variant findings with unknown significance (ACMG Board of Directors, 2012; Green et al., 2013; Pinxten & Howard, 2014).

DATABASES AND UTILITIES

Creating a Database of Known Mutations

Next generation sequencing technologies are expected to advance at an unprecedented pace in the following years. The amount of data that is generated from everyday NGS experiments is overwhelming the current IT infrastructure. NGS is big data in the true sense and necessitates high performance computing as well as terabytes of data storage that keeps growing every day. Several universities and research organizations have invested enormous amount of time and energy to provide the NGS research community with the tools and IT infrastructure needed to store and analyze such big data.

Next Generation Sequencing Catalog (NGS Catalog, http://bioinfo.mc.vanderbilt.edu/NGS/index.html), is one such comprehensive NGS resource, which is a continually updated database that collects, curates and manages available human NGS data obtained from published literature. NGS Catalog deposits publication information of NGS studies and their mutation characteristics (SNVs, small insertions/deletions, copy number variations, and structural variants), as well as mutated genes and gene fusions detected by NGS. It also features user data upload, NGS general analysis pipelines, and NGS software (Xia et al., 2012).

Leiden Open Variation Database (LOVD) is ideal for scientists who want an open source database for storing and displaying variants on the web. LOVD provides a flexible, open source tool for gene-centered collection and display of DNA variations (Fokkema et al., 2011). Several research groups have published their own variations database using this tool and this information is freely available to the general public. LOVD features integration with the Mutalyzer sequence variant nomenclature checker, allowing for direct nomenclature checking of sequence variants during the submission process (Wildeman, Van Ophuizen, Den Dunnen, & Taschner, 2008). Currently there are more than 80 such LOVD public databases on various diseases such as Parkinson's, colon cancer, prostate cancer, Globin gene server, Tuberous Sclerosis and much more (Fokkema et al., 2011).

Locus-specific databases (LSDBs) are curated compilations of sequence variants in genes associated with disease and contain extensive information provided by the literature and benefit from manual curation by experts. Although a cancer genome contains several thousand somatic mutations it has always been a challenge to identify the handful of these mutations that are truly oncogenic. The TP53 LSDB is one such database that enables scientists to draw a model of gene mutation analysis, which may ultimately prove useful for clinical practice. Unlike centralized databases that could change the accessibility of data, with interfaces optimized for different types of users and adapted to the specificity of analysis, LSDBs can offer a more focused approach to address one primary area of research (Soussi, 2014).

Some of the main limitations of LSDBs are that they are highly heterogeneous in terms of quality, content, and format. Although several database management systems (such as the Universal Mutation Database, the Leiden Open Variation Database, and the MUTbase) have been developed to standardize the current data via a framework for LSDB curation, these systems are not inter-compatible (Auerbach et al., 2011). The human Genome Variation Society has not had much success in implementing its published guidelines on database structure and content or on variant nomenclature. The international nomenclature for publishing DNA variants is used in less than 20% of publications, leading to data that are often useless for inclusion in LSDBs (Den Dunnen & Antonarakis, 2000).

The Human Variation Database provides an open-source (PostgreSQL) database for the storage and analysis of thousands of next-generation sequencing variations, a Java API to perform common functions, such as generation of standard experimental reports and graphical summaries of modifications to genes, and libraries to allow adopters of the database to develop their own queries quickly (Fejes, Khodabakhshi, Birol, & Jones, 2011).

The Center for Inherited Disease Research (CIDR) from Johns Hopkins University has developed CIDRVar, a Next-Generation Sequencing Database linking samples, variants, and annotations. CIDR's NGS production data analysis toolbox, CIDRSeqSuite, generates sample variant annotation reports using ANNOVAR and more than 50 flat-file databases (Newcomer et al., 2013).

Gene Panels

Clinical genetics has undergone a significant evolution from traditional Sanger Sequencing or RT-PCR methods that look at single variants or single genes or gene panels to a whole new approach of Clinical Exome Sequencing. Exome sequencing is ideal for studying complex diseases such as diabetes, epilepsy, or cancer as it enables simultaneous evaluation of millions of sequences concurrently. A recent study showed that Whole Exome Sequencing is often the best balance of cost and effectiveness, offering a success rate as high as a 25 percent in solving hereditary disease mysteries (Yang et al., 2013).

The Radboud University Medical Center, Netherlands was amongst the first to implement two-step exome sequencing in clinical genetic diagnostics and to evaluate patient experiences with gene panels based on exome sequencing, using quantified psychological variables: acceptance, psychological distress, expectations of heredity and unsolicited findings. They concluded that most adults accepted the results and were satisfied with gene panels based on diagnostic exome sequencing, few reporting distress (Sie et al., 2015).

The Emory Genetics Laboratory is currently offering a comprehensive panel of testing known as Medical EmExome. It is the next level in clinical whole exome sequencing with enhanced coverage of ~4600 medically relevant and known disease-associated genes. This is the highest coverage offered by any clinical exome sequencing performed in a CLIA-/CAP-certified laboratory. There is also an option for the clinician to choose an EGL gene panel relevant to the patient's phenotype to ensure coverage of all exons, at no additional cost. They also support clinical research for extended exome and genome testing for the discovery of novel disease genes (Emory Genetics Laboratory, 2014).

Mutation Panels

Current off-the-shelf exome kits used for clinical exome sequencing cover 92% of the exome. Traditionally, gene discovery has been done in research laboratories; however, now with the ability to sequence nearly the entire coding region of the human genome, it is possible for clinical laboratories to use this information to identify a previously unrecognized cause of disease. Several commercial entities have now started developing mutation panels to address the Pharmacogenomics aspect of therapeutics. A popular example is the ALK mutation panel for lung cancer patients who are being considered for Crizotinib therapy (Pfizer Inc) (Zhang et al., 2014). Similarly, a comprehensive tyrosine kinases mutation panel is being developed to address the needs of tyrosine kinase inhibitors (TKIs) in the treatment of hematological diseases and gastrointestinal stromal tumors (Novartis AG) (Gleeson et al., 2015).

Exome Sequencing for Target Discovery

Exome Sequencing has generated considerable interest in target discovery and diagnosis of unidentified genetic conditions. A recent study analyzed whole-exome sequencing data from 12 patients with unexplained and apparent genetic conditions, along with their unaffected parents. The study identified in 6 out of 12 cases causal mutations in four genes known to cause Mendelian disease (TCF4, EFTUD2, SCN2A and SMAD4) and one gene related to known Mendelian disease genes (NGLY1). Notably, EFTUD2 was not yet known as a Mendelian disease gene at the time of publication but was nominated as a likely cause based on the observation of de novo mutations in two unrelated probands. Additionally the authors were able to identify homozygous mutations in EFEMP1 as a likely cause for macular degeneration in another patient (Need et al., 2012).

Exome Sequencing for Biomarker Discovery

Biomarker discovery is traditionally performed by in-silico approaches followed by experimental validation of key findings. Large scale high throughput screening of several biologically relevant proteins (targets) is usually accomplished by siRNA approach or animal models. Now with the advent of NGS all such experiments have been given a face lift by adding two more layers of OMICs information (genomics and epigenomics). Methyl Seq and CHIP Seq have predominant role in this area to determine the gene regulation of putative biomarker. ExomeSeq will be necessitated to finally validate the biomarkers and perform functional genomics experiments. Popular algorithms such as SIFT or POLYPHEN enable bioinformaticians to predict the effect of non-synonymous SNP on the amino acid sequence of the protein and other changes like silencing, truncation or improper protein folding and other functional consequences (Panda & Suresh, 2014).

CONCLUSION

NGS technology is here to stay, what with the rapidly reducing costs and other innumerable benefits. Exome sequencing can deliver on the promise of quick turnaround time with targeted results. An example of this is Mapmygenome's exome sequencing for cardiovascular diseases and other panels. Such technologies are heralding a paradigm shift in medicine – in the field of disease prediction and diagnostics, notably in cardiovascular and cancer genomics. However, to increase the impact of these technologies in clinical practice, massive investments are currently required.

REFERENCES

ABI. (2007). A History of Innovation in Genetic Analysis. Retrieved from http://tools.lifetechnologies.com/content/sfs/posters/ABI6247_SOLiD_Timeline_v4_ONLINE.pdf

ACMG Board of Directors. (2012). Points to consider in the clinical application of genomic sequencing. *Genetics in Medicine*, *14*(8), 759–761. doi:10.1038/gim.2012.74 PMID:22863877

Agilent Technologies. (n. d.). Agilent SureSelect Target Enrichment. Retrieved from http://www.genomics.agilent.com/article.jsp?pageId=2094

AllSeq. (n. d.). Life Technologies – Ion Torrent. Retrieved from http://allseq.com/knowledgebank/sequencing-platforms/life-technologies-ion-torrent

Auerbach, A. D., Burn, J., Cassiman, J. J., Claustres, M., Cotton, R. G., & Cutting, G. et al. (2011). Mutation (variation) databases and registries: A rationale for coordination of efforts. *Nature Reviews. Genetics*, *12*(12), 881–881. PMID:22025002

Barbieri, C. E., Baca, S. C., Lawrence, M. S., Demichelis, F., Blattner, M., & Theurillat, J. P. et al. (2012). Exome sequencing identifies recurrent SPOP, FOXA1 and MED12 mutations in prostate cancer. *Nature Genetics*, *44*(6), 685–689. doi:10.1038/ng.2279 PMID:22610119

Biesecker, L. G., Shianna, K. V., & Mullikin, J. C. (2011). Exome sequencing: The expert view. *Genome Biology*, *12*(9), 128. doi:10.1186/gb-2011-12-9-128 PMID:21920051

Burke, W., Trinidad, S. B., & Clayton, E. W. (2013). Seeking genomic knowledge: The case for clinical restraint. *The Hastings Law Journal*, *64*(6), 1650. PMID:24688162

Chilamakuri, C. S. R., Lorenz, S., Madoui, M. A., Vodák, D., Sun, J., & Hovig, E. et al. (2014). Performance comparison of four exome capture systems for deep sequencing. *BMC Genomics*, *15*(1), 449. doi:10.1186/1471-2164-15-449 PMID:24912484

Choi, M., Scholl, U. I., Ji, W., Liu, T., Tikhonova, I. R., & Zumbo, P. et al. (2009). Genetic diagnosis by whole exome capture and massively parallel DNA sequencing. *Proceedings of the National Academy of Sciences of the United States of America*, *106*(45), 19096–19101. doi:10.1073/pnas.0910672106 PMID:19861545

Den Dunnen, J. T., & Antonarakis, S. E. (2000). Mutation nomenclature extensions and suggestions to describe complex mutations: A discussion. *Human Mutation*, *15*(1), 7–12. doi:10.1002/(SICI)1098-1004(200001)15:1<7::AID-HUMU4>3.0.CO;2-N PMID:10612815

Dolled-Filhart, M.P., Lordemann, A., Dahl, W., & Haraksingh, R.R., Ou-yang, & Lin, J. C. H. (. (2012). Opportunities & Challenges With Personalized Exome Sequencing. *Personalized Medicine.*, *9*(8), 805–819. doi:10.2217/pme.12.97

Emory Genetics Laboratory. (2014). Six Medical EmExome Service Levels. Retrieved from http://geneticslab.web.emory.edu/about/news-and-events/news/2014/01/medical-emexome/

Fejes, A. P., Khodabakhshi, A. H., Birol, I., & Jones, S. J. (2011). Human variation database: An open-source database template for genomic discovery. *Bioinformatics (Oxford, England)*, *27*(8), 1155–1156. doi:10.1093/bioinformatics/btr100 PMID:21367872

Fokkema, I. F., Taschner, P. E., Schaafsma, G. C., Celli, J., Laros, J. F., & den Dunnen, J. T. (2011). LOVD v.2.0: The next generation in gene variant databases. *Human Mutation*, *32*(5), 557–563. http://www.lovd.nl/2.0/ doi:10.1002/humu.21438 PMID:21520333

Futema, M., Plagnol, V., Whittall, R. A., Neil, H. A. W., & Humphries, S. E. (2012). Use of targeted exome sequencing as a diagnostic tool for Familial Hypercholesterolaemia. *Journal of Medical Genetics*, *49*(10), 644–649. doi:10.1136/jmedgenet-2012-101189 PMID:23054246

Gahl, W. A., Markello, T. C., Toro, C., Fajardo, K. F., Sincan, M., & Gill, F. et al. (2011). The national institutes of health undiagnosed diseases program: Insights into rare diseases. *Genetics in Medicine*, *14*(1), 51–59. doi:10.1038/gim.0b013e318232a005 PMID:22237431

Gleeson, F. C., Kipp, B. R., Kerr, S. E., Voss, J. S., Graham, R. P., & Campion, M. B. et al. (2015). Kinase genotype analysis of gastric gastrointestinal stromal tumor cytology samples using targeted next-generation sequencing. *Clinical Gastroenterology and Hepatology*, *13*(1), 202–206. doi:10.1016/j.cgh.2014.06.024 PMID:24997326

Green, R. C., Berg, J. S., Grody, W. W., Kalia, S. S., Korf, B. R., & Martin, C. L. et al. (2013). ACMG recommendations for reporting of incidental findings in clinical exome and genome sequencing. *Genetics in Medicine*, *15*(7), 565–574. doi:10.1038/gim.2013.73 PMID:23788249

Huang, Y. F., Chen, S. C., Chiang, Y. S., Chen, T. H., & Chiu, K. P. (2012). Palindromic sequence impedes sequencing-by-ligation mechanism. *BMC Systems Biology*, *6*(Suppl 2), S10. doi:10.1186/1752-0509-6-S2-S10 PMID:23281822

Hurd, P. J., & Nelson, C. J. (2009). Advantages of next-generation sequencing versus the microarray in epigenetic research. *Briefings in Functional Genomics*, elp013. PMID:19535508

Illumina a. Nextera Rapid Capture Exome and Expanded Exome Kits. (n. d.). Retrieved from http://www.illumina.com/products/nextera-rapid-capture-exome-kits.html

Illumina b. Systems. (n. d.). Retrieved from http://www.illumina.com/systems.html

Illumina c. Truseq Exome Enrichment Kit. (n. d.). Retrieved from http://support.illumina.com/sequencing/sequencing_kits/truseq_exome_enrichment_kit.html

Jiao, Y., Shi, C., Edil, B. H., de Wilde, R. F., Klimstra, D. S., & Maitra, A. et al. (2011). DAXX/ATRX, MEN1, and mTOR pathway genes are frequently altered in pancreatic neuroendocrine tumors. *Science*, *331*(6021), 1199–1203. doi:10.1126/science.1200609 PMID:21252315

Kato, K. (2009). Impact of the next generation DNA sequencers. *International journal of clinical and experimental medicine, 2*(2), 193.

Kircher, M., & Kelso, J. (2010). High-throughput DNA sequencing–concepts and limitations. *BioEssays*, *32*(6), 524–536. doi:10.1002/bies.200900181 PMID:20486139

Life Technologies. (n. d.). Ion AmpliSeq™ Exome RDY - OT2 Kit 1x8. Retrieved from http://www.lifetechnologies.com/order/catalog/product/4489837

Liu, L., Li, Y., Li, S., Hu, N., He, Y., Pong, R., ... & Law, M. (2012). Comparison of next-generation sequencing systems. *BioMed Research International, 2012*.

Lyon, G. J., Jiang, T., Van Wijk, R., Wang, W., Bodily, P. M., & Xing, J. et al. (2011). Exome sequencing and unrelated findings in the context of complex disease research: Ethical and clinical implications. *Discovery Medicine*, *12*(62), 41. PMID:21794208

Machini, K., Douglas, J., Braxton, A., Tsipis, J., & Kramer, K. (2014). Genetic counselors' views and experiences with the clinical integration of genome sequencing. *Journal of Genetic Counseling*, *23*(4), 496–505. doi:10.1007/s10897-014-9709-4 PMID:24671342

Majewski, J., Schwartzentruber, J., Lalonde, E., Montpetit, A., & Jabado, N. (2011). What can exome sequencing do for you? *Journal of Medical Genetics*, *48*(9), 580–589. doi:10.1136/jmedgenet-2011-100223 PMID:21730106

Marian, A. J. (2012). Challenges in medical applications of whole exome/genome sequencing discoveries. *Trends in Cardiovascular Medicine*, *22*(8), 219–223. doi:10.1016/j.tcm.2012.08.001 PMID:22921985

Morozova, O., & Marra, M. A. (2008). Applications of next-generation sequencing technologies in functional genomics. *Genomics*, *92*(5), 255–264. doi:10.1016/j.ygeno.2008.07.001 PMID:18703132

Need, A. C., Shashi, V., Hitomi, Y., Schoch, K., Shianna, K. V., McDonald, M. T., ... & Goldstein, D. B. (2012). Clinical application of exome sequencing in undiagnosed genetic conditions. *Journal of medical genetics*, jmedgenet-2012.

Newcomer, J. D., Griffith, S. M. L., Pugh, E. W., Ling, H., Leary, D. R., Goldstein, J. L., et al. (2013). CIDRVar: A Next-Generation Sequencing Database Linking Samples, Variants, and Annotations. Center for Inherited Disease Research (CIDR). Baltimore, MD: Institute of Genetic Medicine, Johns Hopkins University; Retrieved from http://www.cidr.jhmi.edu/nih/CIDRVar.pdf

Panda, R., & Suresh, P. K. (2014). Computational identification and analysis of functional polymorphisms involved in the activation and detoxification genes implicated in endometriosis. *Gene*, *542*(2), 89–97. doi:10.1016/j.gene.2014.03.058 PMID:24698776

Parla, J. S., Iossifov, I., Grabill, I., Spector, M. S., Kramer, M., & McCombie, W. R. (2011). A comparative analysis of exome capture. *Genome Biology*, *12*(9), R97. doi:10.1186/gb-2011-12-9-r97 PMID:21958622

Pinxten, W., & Howard, H. C. (2014). Ethical issues raised by whole genome sequencing. *Best Practice & Research. Clinical Gastroenterology*, *28*(2), 269–279. doi:10.1016/j.bpg.2014.02.004 PMID:24810188

Quail, M. A., Smith, M., Coupland, P., Otto, T. D., Harris, S. R., & Connor, T. R. et al. (2012). A tale of three next generation sequencing platforms: Comparison of Ion Torrent, Pacific Biosciences and Illumina MiSeq sequencers. *BMC Genomics*, *13*(1), 341. doi:10.1186/1471-2164-13-341 PMID:22827831

Quail, M. A., Smith, M., Coupland, P., Otto, T. D., Harris, S. R., & Connor, T. R. et al. (2012). A tale of three next generation sequencing platforms: Comparison of Ion Torrent, Pacific Biosciences and Illumina MiSeq sequencers. *BMC Genomics*, *13*(1), 341. doi:10.1186/1471-2164-13-341 PMID:22827831

Rabbani, B., Tekin, M., & Mahdieh, N. (2014). The promise of whole-exome sequencing in medical genetics. *Journal of Human Genetics*, *59*(1), 5–15. doi:10.1038/jhg.2013.114 PMID:24196381

Roberts, R. J., Carneiro, M. O., & Schatz, M. C. (2013). The advantages of SMRT sequencing. *Genome Biology*, *14*(6), 405. doi:10.1186/gb-2013-14-6-405 PMID:23822731

Roche. 454 Sequencing. Retrieved from http://www.454.com/

Sie, A. S., Prins, J. B., van Zelst-Stams, W. A. G., Veltman, J. A., Feenstra, I., & Hoogerbrugge, N. (2015). Patient experiences with gene panels based on exome sequencing in clinical diagnostics: High acceptance and low distress. *Clinical Genetics*, *87*(4), 319–326. doi:10.1111/cge.12433 PMID:24863757

Soussi, T. (2014). Locus-Specific Databases in Cancer: What Future in a Post-Genomic Era? The TP53 LSDB paradigm. *Human Mutation*, *35*(6), 643–653. doi:10.1002/humu.22518 PMID:24478183

Su, X. W., Broach, J. R., Connor, J. R., Gerhard, G. S., & Simmons, Z. (2014). Genetic heterogeneity of amyotrophic lateral sclerosis: Implications for clinical practice and research. *Muscle & Nerve*, *49*(6), 786–803. doi:10.1002/mus.24198 PMID:24488689

Taber, K. A. J., Dickinson, B. D., & Wilson, M. (2014). The promise and challenges of next-generation genome sequencing for clinical care. *JAMA Internal Medicine*, *174*(2), 275–280. doi:10.1001/jamainternmed.2013.12048 PMID:24217348

Teer, J. K., & Mullikin, J. C. (2010). Exome sequencing: The sweet spot before whole genomes. *Human Molecular Genetics*, ddq333. PMID:20705737

Wildeman, M., Van Ophuizen, E., Den Dunnen, J. T., & Taschner, P. E. (2008). Improving sequence variant descriptions in mutation databases and literature using the Mutalyzer sequence variation nomenclature checker. *Human Mutation*, *29*(1), 6–13. doi:10.1002/humu.20654 PMID:18000842

Worthey, E. A., Mayer, A. N., Syverson, G. D., Helbling, D., Bonacci, B. B., & Decker, B. et al. (2011). Making a definitive diagnosis: Successful clinical application of whole exome sequencing in a child with intractable inflammatory bowel disease. *Genetics in Medicine*, *13*(3), 255–262. doi:10.1097/GIM.0b013e3182088158 PMID:21173700

Xia, J., Wang, Q., Jia, P., Wang, B., Pao, W., & Zhao, Z. (2012). NGS catalog: A database of next generation sequencing studies in humans. *Human Mutation*, *33*(6), E2341–E2355. doi:10.1002/humu.22096 PMID:22517761

Yang, Y., Muzny, D. M., Reid, J. G., Bainbridge, M. N., Willis, A., & Ward, P. A. et al. (2013). Clinical whole-exome sequencing for the diagnosis of mendelian disorders. *The New England Journal of Medicine*, *369*(16), 1502–1511. doi:10.1056/NEJMoa1306555 PMID:24088041

Zhang, M., Chen, J., Si, D., Zheng, Y., Jiao, H., & Feng, Z. et al. (2014). Whole exome sequencing identifies a novel EMD mutation in a Chinese family with dilated cardiomyopathy. *BMC Medical Genetics*, *15*(1), 77. doi:10.1186/1471-2350-15-77 PMID:24997722

Zhang, N. N., Liu, Y. T., Ma, L., Wang, L., Hao, X. Z., Yuan, Z., ... & Shi, Y. (2014). The molecular detection and clinical significance of ALK rearrangement in selected advanced non-small cell lung cancer: ALK expression provides insights into ALK targeted therapy.

Zhang, X., Li, M., & Zhang, X. J. (2011). [Exome sequencing and its application]. *Yi chuan= Hereditas/ Zhongguo yi chuan xue hui bian ji, 33*(8), 847-856.

Zhou, X., Ren, L., Li, Y., Zhang, M., Yu, Y., & Yu, J. (2010). The next-generation sequencing technology: A technology review and future perspective. *Science China Life Sciences*, *53*(1), 44–57. doi:10.1007/s11427-010-0023-6 PMID:20596955

Zhou, X., Ren, L., Meng, Q., Li, Y., Yu, Y., & Yu, J. (2010). The next-generation sequencing technology and application. *Protein & cell*, *1*(6), 520-536.

KEY TERMS AND DEFINITIONS

Apoptosis: Is the process of programmed cell death.

CIDR: Is the short for Classless Inter-Domain Routing.

DdATP: 2',3'-Dideoxyadenosine-5'-Triphosphate.

DdCTP: 2',3'-Dideoxycytidine-5'-Triphosphate.

DdGTP: 2',3'-Dideoxyguanosine-5'-Triphosphate.

DdTTP: 2',3'-Dideoxythymidine-5'-Triphosphate.

de novo: Latin expression meaning "from the beginning."

DNA: Deoxyribonucleic acid.

Electrophoresis: Is the motion of dispersed particles relative to a fluid under the influence of a spatially uniform electric field.

Epigenetics: The study of stable, long-term alterations in the transcriptional potential of a cell that are not necessarily heritable.

Genomics: Is a discipline in genetics that applies recombinant DNA, DNA sequencing methods, and bioinformatics to sequence, assemble, and analyze the function and structure of genomes (the complete set of DNA within a single cell of an organism).

GWAS: Genome-wide association study.

Illumina: Sequencing and array technologies fuel advancements in life science research, translational and consumer genomics, and molecular diagnostics.

Nanophotonics: Is the study of the behavior of light on the nanometer scale, and of the interaction of nanometer-scale objects with light.

Nucleotide: Is one of the structural components, or building blocks, of DNA and RNA.

PCR: Polymerase chain reaction.

Proteomics: Is the large-scale study of proteins, particularly their structures and functions.

Pseudogenes: Are dysfunctional relatives of genes that have lost their protein-coding ability or are otherwise no longer expressed in the cell.

Pyrosequencing: Is a method of DNA sequencing (determining the order of nucleotides in DNA) based on the "sequencing by synthesis" principle.

RTPCR: Reverse transcription polymerase chain reaction.

SiRNA: Small interfering RNAs.

SNP: Single Nucleotide Polymorphism.

SsDNA: Single-stranded DNA.

Chapter 13
Agile Methodology of Development and How to be Compliant

Yerramalli Subramaniam
CliniOps Inc., USA

Avik Pal
CliniOps Inc., USA

Arindam Dey
HCL America Inc., USA

ABSTRACT

Given that Agile software development is preferred methodology for products and services in life science industry, in this chapter we will describe how to adopt Agile software development process and still be compliant. We will focus on few Agile methodologies and provide details on what design controls we can adopt in order for the product and process to be compliant. We will also focus on some of the tools that can be used to help put such design and process control in place where we can have complete transparency and traceability.

INTRODUCTION

Software industry started a few decades ago focusing on some core sectors of industry such as banking but now it is part of almost every industry. Historically, most of the software development was done through waterfall methodology. In this, all the requirements are made available upfront and after the development is complete, there is a dedicated testing phase, which focuses more on quality control of the software. This methodology works well when there is well-defined set of requirements. This also helps in contractual terms where there well-defined set of inputs, outputs and the results are measured against those cornerstones. But these days, because of competition, companies want to rollout products faster and want to be more responsive to voice of customers. As a result, most of the companies are adopting

DOI: 10.4018/978-1-4666-8726-4.ch013

Agile methodology where continuous improvement and feedback is a critical part of the process. Over the years, Agile software development has adopted several approaches such as extreme programming, scrum, lean software development, etc. and they all share similar philosophy of adaptive, iterative and evolutionary development.

But when it comes to compliance, there are challenges to roll out the products faster with evolutionary development while still adhering to the compliance directives. Most of the compliance directives defined in 21CFR 820 (QSR) and it provides just broad guideline for compliance instead of going into details and "prescribing" the nuts and bolts of software development process. The onus is on the software development team to provide details of their development process and convince auditors how their design controls in software development process address the compliance guidelines and deliver a quality product. Apart from the development process, FDA also looks into the software quality/safety model, software maintenance and CAPA,which could be a part of safety risk management.

Agile vs Waterfall Process

Few decades ago, waterfall process was ubiquitous and was considered the standard way of developing a software product for medical devices and other life science product. Bell, Thomas E., and T. A. Thayer originally proposed this process in 1976 and in this process each stage of software development follows a sequential pattern. Requirements are gathered upfront followed by design, implementation verification, and maintenance.

In early 2001, software community proposed a new approach for software development and published the "Manifesto for Agile Software Development ". In contrast to waterfalls' linear, structured and rigid approach, Agile process focused on incremental and iterative development. This followed the philosophy of fail early and fail often by frequent software releases and getting customers feedback for next software iteration release cycle. With short development cycles and releases, customers were able to use the product and the core philosophy was that a working version of software is a true measure of the progress of the project.

In summary, Agile methodology is more focused on the actual product than producing an exhaustive document in the requirements stage. Having an exhaustive requirements may make sense for engineering companies for building bridges, ships etc. but when it comes to software engineering, over the years companies have realized that the requirements do change because of several reasons – changing market condition, value proposition etc, and the software needs to adapt to those changes. Software professionals now realize the true value of Agile methodology because it is more focused on responding to changing market condition than sticking to an original plan. Unlike the waterfall method, the software development communities have also realized that Agile process is more about collaboration with customers and not about negotiating a contract from the requirements document. Finally, Agile puts more emphasis on individual's interactions than on process and tools. This is not to say that Agile process is unstructured. On the contrary, it is process that is more adaptable to change in requirements, strict control of software quality and getting customer's feedback on the product.

FDA

Contrary to popular belief, we should consider FDA as a partner in defining software development process and not a deterrent in process for faster product development. FDA has provided guidance on various aspects of software development but the most notable is "General Principles of Software Validation; Final Guidance for Industry and FDA". In this guideline, FDA has specifically mentioned that we should consider the least burdensome approach in all areas of medical device regulation. It also states that if you believe that an alternative approach would be less burdensome, please contact FDA so they can consider your point of view. So in essence, FDA is open to ideas to make the process leaner and faster but at the same time not compromising on the quality of software.

From FDA perspective, the regulated software is broadly classified into two categories – medical device software and non-medical device software. Medical device software includes the software that is actually a part of medical device, an accessory to medical device or the software itself is a medical device. Non-medical devices are classified as the tools used for the software development process such as quality system, software configuration management, production system or systems used to maintain records. Medical devices have a wide range of complexity. Those devices could be as simple as a thermometer and could be as complicated as pacemakers or robotic surgical instrument. For this reason, the medical device is classified into Class 1, 2 and 3. Class 1 devices have the lowest risk to human lives and class 3 devices are considered to have the highest risk. FDA regulation is based on this classification and they provide guidance on each of these classes of devices accordingly. Note that the "Medical Device" definition is fairly broad. It could be a physical hardware device or entirely software or a combination of both.

The compliance directive for medical devices is defined in 21 CFR (QSR) (FDA, 2014). There are other regulations such as 21 CFR 11 (related electronic records and electronic signatures) and 21 CFR 830 (related to device establishment and pre-market approvals) but it is beyond the scope of this chapter. Apart from CFR regulation there are ISO standards for medical devices defined in ISO 13485. There is some overlap in CFR and ISO standards and for the most part if the documentations are developed for QSR compliance it can potentially be mapped to ISO 13485.

In terms of enforcements of compliance, FDA has several powers including but not limited to the list below in two broad categories.

Administrative Powers

- Unannounced and Announced Inspections
- Warning Letters
- Adverse Publicity
- FDA-Initiated Recalls and Monitoring Company-Initiated Recalls
- Delay, Suspension, or Withdrawal of Product Approvals
- Preclusion of Government contracts
- Detention and Refusal of Entry into U.S. Commerce of Imported Products

Judicial Enforcement Powers

- Civil Enforcement Powers (Seizure)
- Criminal Enforcement Powers (Prosecution)

PROCESS COMPLIANCE

When it comes to compliance, the key thing to demonstrate to FDA is that your process must be able to demonstrate that you are "operating in a state of control". As long as all the plans and procedures are defined upfront and there is documentation available to provide evidence of adhering to those plans and procedures, there is should not be of any concern that those processes follow waterfall or Agile methodologies. The main goal of software development in validated environment is that

- All Process outputs should be maintained in a consistent state
- All Process outputs should be available when needed as input to further work on the software
- Before any medical device software is released, all process outputs should be consistent with each other.

Documentation

There is often misconception regarding Agile methodology and less documentation. For non-compliant products that may be still true but when it comes to QSR compliant software, there is no compromise on documentation. It is just that the Agile process makes the creation, update and maintenance of documents much more manageable and avoids discrepancies in documentation and implementation. Before we look in the differences between waterfall and Agile, lets a quick look at the documentation that is required for compliance. As mentioned before, QSR documentation can be mapped to ISO 13485 and the Table 1 shows the mapping of QSR 21CFR 820 with ISO 13485 and the relevant documentation that are needed for compliance.

Before we get into the details of how this documentation map to Agile process, let's take a quick look at the description of each documentation.

Integrated Project Plan and Schedule

This document contains the time required to complete the tasks and activities that are needed to develop a product that meets the product performance requirements as written in the PRDand transfer to manufacturing. Team members should consider the following areas in defining the tasks and activities: project and product scope, development, verification and validation, design transfer, operations etc. The team estimates and defines the resources needed to complete tasks and activities and creates a Project Timeline, based on the assumption that all requested resources can be provided. This IPPS is reviewed and revised, as necessary, during product development.

Table 1. List of documentation for QSR and ISO compliance .

QSR Reference	ISO Reference	Software Deliverables for Documentation
Design and Development Planning	7.3.1	Integrated Project Plan & Schedule
Design Input	7.3.2	Product Requirements Document Product Specifications Document
Design Output	7.3.3	Traceability Matrix Manufacturing Readiness Report Development Design
Risk Analysis	7.3.2	Safety Risk Management Plan Safety Risk Management Report
Design Reviews	7.3.4	Development Design Review Design Verification and Validation Review
Design Verification	7.3.5	Design Verification Test Plan Design Verification Test Report
Design Validation	7.3.6	Design Validation Test Plan Design Validation Test Report Field Confirmation Plan and Report
Design Transfer	7.5.1	Design Transfer
Design Changes	7.2.2	Change Request Order
Design History File	7.5.5	Electronic Design History File

Product Requirements Document

This document describes the list of key features for the product and what the product does to satisfy customer need. These features are the foundation of minimum viable product (MVP) without which there is no market justification to launch the product. However, the requirements document just specifies the high level requirements and does not provide any specifics of how these requirements will be fulfilled. The contents of this document are more towards "what" needs to be achieved and does not mention the "how". These requirements could optionally be prioritized based on MVP.

Product Specifications Document

This is an extension of PRD where we specify the "how" part. This document describes how the users will interact with the systems, the core features that will fulfill the requirements and what screens or software modules needs to be implemented. It can also go to the detail level of screen layouts and behavior of each user interface screens and controls.

Traceability Matrix

This is where we map all the product requirements, product specification and test cases. The intent of this document is to make sure that all requirements have been interpreted properly in PSD and that the product is working as intended. By mapping the test cases, it verifies and validates the software product. At the end of project we should have complete traceability of requirements to test cases and all those test cases should be executed and passed. Bugs that are logged should have reference to test cases so that it is easy to find which requirements or features are affected.

Manufacturing Readiness Report

If this project involves a medical device, this document should describe how the software or firmware would be deployed at device manufacturer site. This includes any checklist of prerequisite tools manufacturer need to have before the design transfer. It should also specify what approval process should be in place if there is any software update. For software only project, it should describe how distribution media (usb, cd, dvdetc) will be manufactured and a checklist for readiness. If no media is involved and is following SaaS, document should provide detail description of how the software will be deployed and the approval process for updating software for future changes.

Safety Risk Management Plan and Report

These Work Instructions describe how to assess and manage product risk from development through commercialization. Risks addressed in this document are related to unintended use of the software, patient safety, public health, and the environment. Prepare initial risk analysis containing, at a minimum, the following elements - known or foreseeable hazards, safety risks, severity, frequency, risk acceptability and mitigation.

Determine if residual risk is acceptable. If risk(s) are unacceptable and controls are impractical, additional mitigations must be implemented until residual risk is acceptable. Write the SRMR, which includes, at a minimum, a description of the product being analyzed, date of the analysis, the intended use statement, and the completed risk assessment. The SRMR document is approved, at a minimum, by the designated regulatory compliance/quality representative, a development representative, and the core team leader.

Table 2. shown below summarizes the method used to evaluate acceptability of a risk associated with a hazard. After determining the severity and likelihood of a hazard, the acceptability of the associated risk is given by this table. It is based on ISO-14971:2007, Annex D.

Table 2. Risk classification criteria

Severity / Frequency	Minor	Moderate	Major
Improbable	BAR	ALARP	ALARP
Remote	ALARP	ALARP	ALARP
Occasional	ALARP	ALARP	INTOL
Probable	ALARP	INTOL	INTOL
Frequent	ALARP	INTOL	INTOL

Development Design Review

Here the goal is to develop a software architecture document that provides a detail overview of the entire software architecture. For a medical device this could also include the firmware and mechatronics designs as well. Decompose the software architecture as required to guide developers, facilitate adherence to coding standards, and describe how the requirements in the PRD are implemented. Detailed design may include the following, although other output may also be used- UI design specification, UML diagrams, API specifications, Database schemas, Network topologies, File format specifications, Communication protocols, Implementation algorithms, Interface formats for hardware and software components, Segregation of software components as required for risk control (for major level of concern software), interfaces between components, interfaces with external systems. This document may be updated with each development iteration. Additional detail or lower level architecture documents may be incorporated or added throughout

Design Verification Test Plan and Report

Design verification confirms that the design specifications mentioned in PSD have been met. During the development phase, assigned personnel prepare appropriate test plans and test procedures to support the required design verification activities. Necessary verification activities can be determined through the use of the traceability matrix. In addition, assigned personnel execute the test plans following defined procedures. Once all the verification activities are done (even if it is incremental), verification report is generated.

Design Validation Test Plan and Report

Design validation confirms that the design requirements in PRD conform to user needs and intended use(s). During the development and manufacturing phases, assigned personnel prepare appropriate validation test plans including pre-determined acceptance criteria and procedures to demonstrate that the system conforms to the defined user needs and intended use as documented in the PRD. Once all the validation activities are done (even if it is incremental), validation report is generated. During the manufacturing phase, assigned personnel execute the test plans following defined procedures. Validation testing is performed under actual or simulated user conditions. Completion of the validation activities is documented in the traceability matrix.

Design Verification and Validation Review

After all development is complete, in this review process all the internal stakeholders are involved and they go through the reports generated for verification and validation and provide recommendation on go/no-go decision for product launch.

Field Confirmation Plan and Report

After the product has gone through the design verification and validation review, the product is available for field tests. For this, key customers should have been identified who will use the product as if it is available for general availability. For this field test there should be a detail plan outlined for workflows that customers will go through and metrics for how we are going to measure the pass/fail criteria. At the end of field tests all those metrics needs to be captured and a report will be generated which will be archived in eDHF. The main goal of this field test is to show repeatability and scalability. This is particularly more critical for a medical device product.

Design Transfer

Design transfer ensures that the product is working as per the requirements. This is especially important for medical device where we need to check if the product is meeting the input and output criteria, which are measureable, verified, and under change control. The goal of this is to make sure that when the devices are manufactured, the same criteria can be used perform quality assurance and quality control.

Change Request Order (CRO)

After a product is released, any changes in requirements or major updates needs to go through a change request process. In this document, we capture the reasons for change, what changed and how it is going to affect the current features or functionality. Once the changes are document, if applicable, corresponding changes to PRD, PSD and test cases need to be updates also.

Electronic Design History File (eDHF).

This is a repository where all the project related documents are kept. It should allow quick search and retrieval of documents so that they are readily available to respective stakeholders for reference.

Waterfall

In the waterfall process, as shown in Figure 1, each sub-process follows a linear pattern during the software development cycle. Once all the requirements are identified in the "Requirement" stage, all the requirements are documented for the inputs to "Design" stage. Each requirement ideally, should have specifications, which are measurable and prioritized based on customers' use case. The "Design" stage starts when the various requirements are mapped to the software environment and implementation decisions are taken. This stage focuses on how the software will be built. This includes architectural design, high-level design and low-level design. Once all the development is complete, in Verification & Validation stage all the test cases are executed and documented. One of the major drawbacks of this waterfall approach is that if any of the upstream phases are not executed correctly; it is has serious consequences in the subsequent phases of SDLC. For instance, if the requirements are not defined correctly, the design phase is inaccurate and subsequently the implementation and test phase deliverables will have some gaps. These issues not only delay the project timelines but also have an exponential impact on the time, cost and resources as the problem cascades through the waterfall process.

Based on the process mentioned above, the documentation needed for each stage is shown in Table 3.

Figure 1. Waterfall process

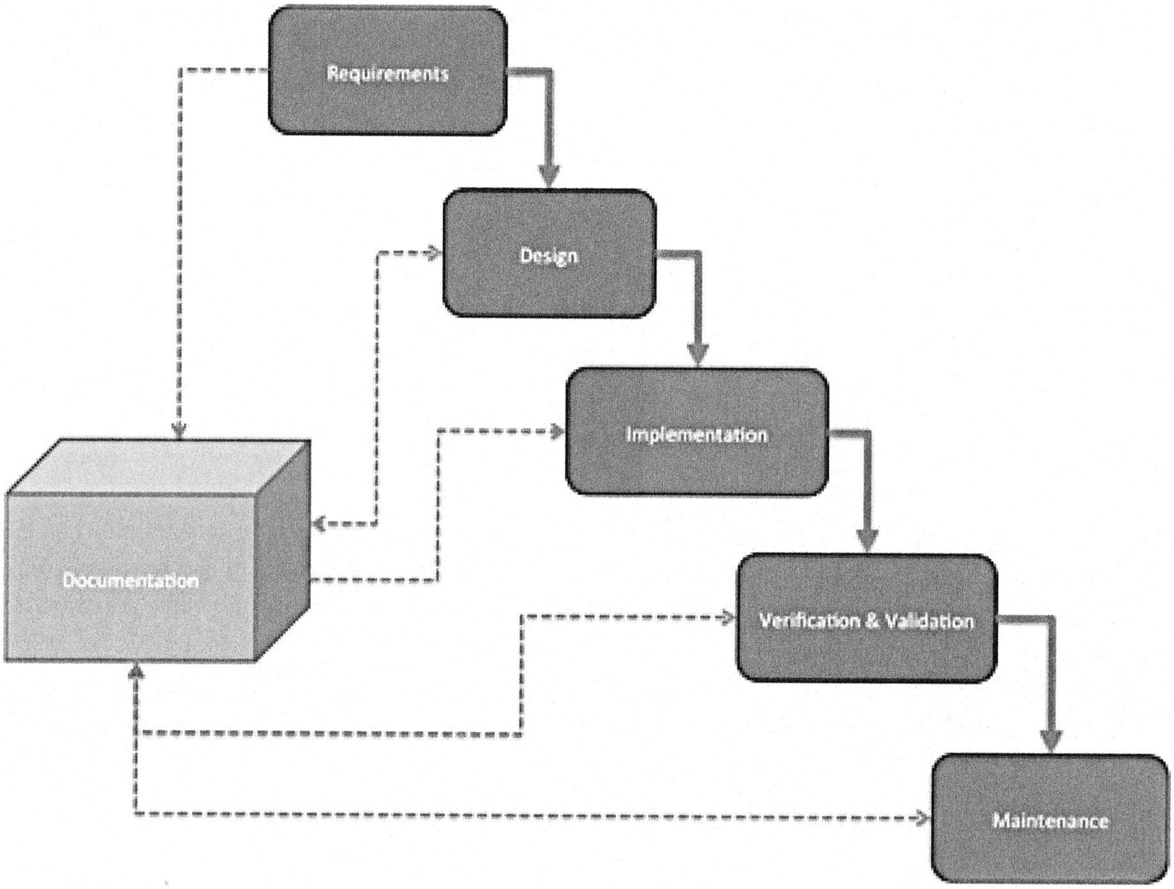

Table 3. Waterfall process documentation requirements

Stage	Documents
Requirements	Integrated Project Plan & Schedule Product Requirements Document Product Specifications Document Design Validation Test Plan
Design	Safety Risk Management Plan Development Design
Implementation	Development Design Review Design Verification Test Plan
Verification & Validation	Traceability Matrix Manufacturing Readiness Report Safety Risk Management Report Design Verification Test Report Design Validation Test Report
Maintenance	Design Verification and Validation Review Field Confirmation Plan and Report Design Transfer sections of the Product Development Plan in the IPP&S Change Request Order Electronic Design History File.

Figure 2. High level agile process

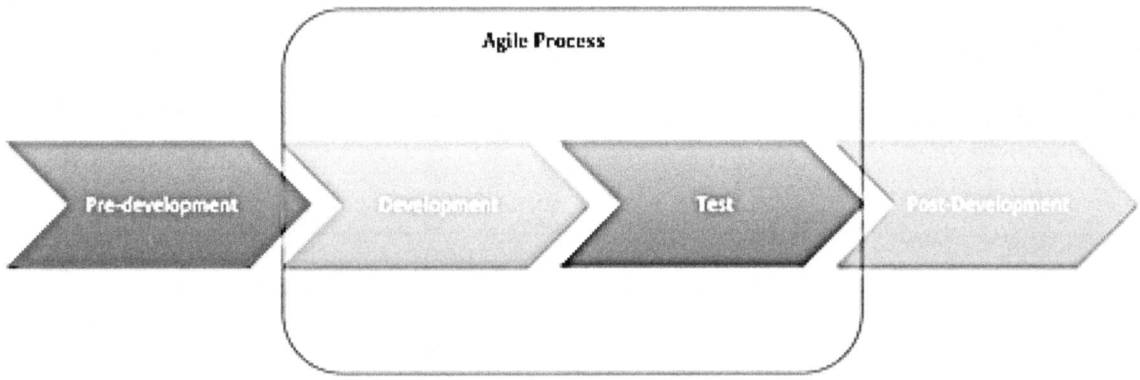

Agile

The core essence of Agile process is that it is iterative. Unlike waterfall, the entire software development process is broken down into small manageable piece. At a very high level, one such example of complete life cycle of the project could be as shown in Figure 2. The Agile process is relevant only for the development and test activities.

Pre-Development

In this stage ideas are gathered from various sources, prioritized and moved into the pipeline. Most of the ideas will come from customer feedback, focus groups and ethnography sessions. Also look at key features offered by our competitors and include them in pipeline if it fits strategic goals. In this phase team should also explore regulatory/compliance features, which are necessary or relevant. For the regulatory/compliance features, which are already implemented, we will look for revisions in guidelines. This process happens throughout the year and all ideas are captured in a repository. The ideas are then filtered based on priority and then decision is made whether it makes business sense to implement those features. These decisions could be based on conjoint analysis, strategic direction and customer adoption which could lead to increase in market share or revenue. If the decision is to not move forward with the idea, it will be archived and revisited for later releases. If needed, a prototype of the idea is built to evaluate technical options as well as risks. Once the prototype is successful, the feature moves into the formal development process.

Development& Test: In this phase we follow Agile software development methodology. Each development cycle will be broken into several iterations. Each iteration will be for 3 weeks, of which 2 weeks is spent on development followed by 1 week of testing. For each iteration, we will define epic, user stories and tasks. Developers work on their tasks to complete user stories for each iteration. Once all iterations are complete, the testing team will perform regression testing and the development team will fix issues found during testing. At the end of development cycle, we branch the code and archive all the test cases for future regression. During the transition from one development cycle to the next, the test team will work with the development team to automate as many test cases as possible.

Figure 3. Iterative agile process

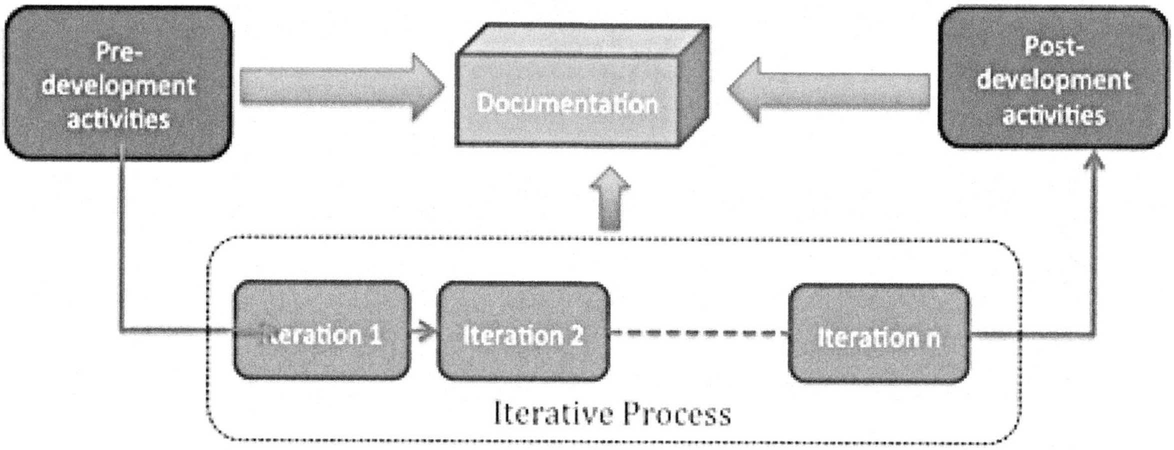

Post-development: In this stage, we perform full regression tests once again to make sure nothing this left out. At this stage all the project related documents should be up-to-date and available in DHF. After the test cycle is complete, software or firmware is available for production use. Following this, we make sure that manufacturing is ready for design transfer.

As shown in Figure 3, at the start of any project we define the high level requirements and not focus on the specifics. These requirements are something that is agreed with business leaders and product managers. For instance, if the project is to build a blood pressure monitor, we define the high level requirements of sensitivity and specificity of the device. All the specific of the design such as user interface, screen layout, usability, color etc. can be set aside for iterative development cycles. During the iterative cycles we continue to update the documents and build them incrementally. Finally in post-development activities all the documentations are finalized and archived in eDHF.

In the figure 4, we show the activities within each iteration. Each company may follow slightly different set of activities within iteration but all versions of Agile process have one thing in common- smaller and iterative development cycle. Unlike waterfall, in Agile process each iteration is small chunk of the entire development and test cycle. Each iteration cycle is usually 3-4 weeks long but such duration of iteration cycle can be defined by the software team at the start of the project. Each project is different and the software team should choose shorter or longer duration of iterative cycles based on several factors such as complexity of the project, number of developers and testers, scope of the final product deliverables, access to user acceptance etc. As shown in Figure 4, each iteration is closed feedback loop.

As shown in the Figure 4 above, the starting point of iteration is an Epic. Usually Epic is a direct translation of a requirement, which outlines the high level requirement and may contain one or more user stories.

In the iteration planning stage, this epic is broken down into manageable set of user stories. Each user story is an encapsulation of a specific feature of the product and each user story can have predecessor and dependent user stories. All the predecessor user stories have to be complete before dependent user story can begin. In iteration planning, each user story is assigned a story point. A story point is usually a measure of the complexity of the user story. The story point scale (0-5 or 1-10) can be defined by the software development team but the core essence of estimating story points is to measure how much

Figure 4. Agile process workflow

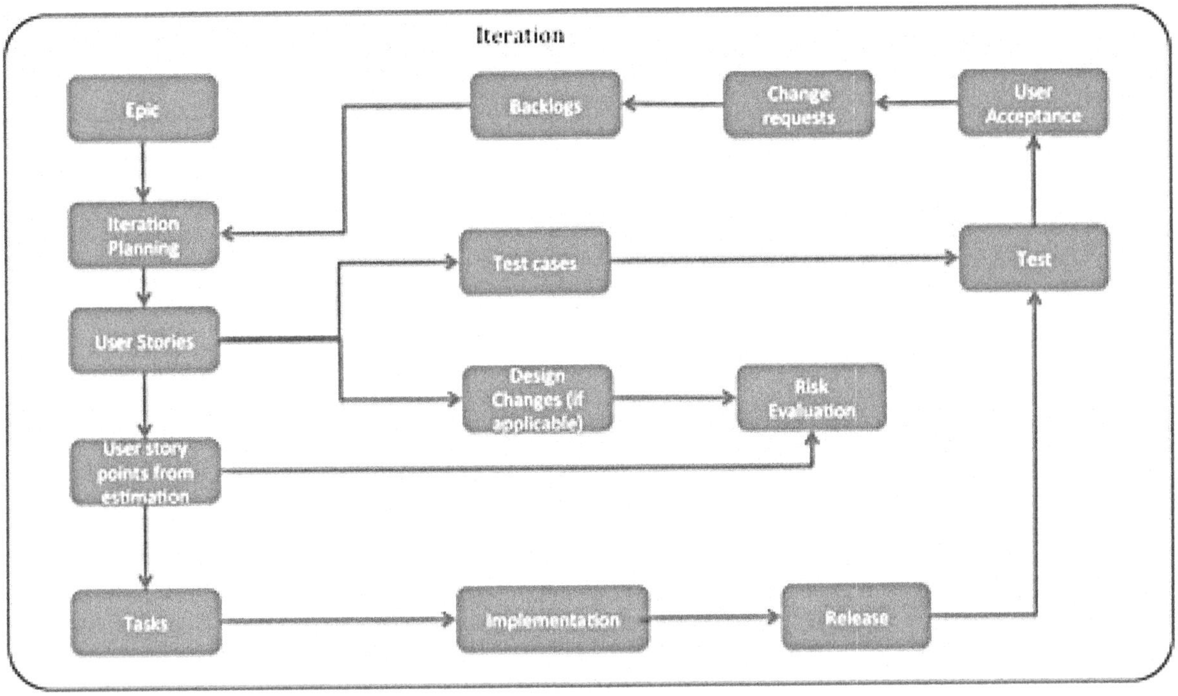

work is involved and with that story points we can have an estimate of the duration of getting that user story implemented. Usually the story points are estimated by the development team. Sometimes teams use "planning poker" for this estimation where each developer "bids" on the story points. If a developer is estimating too high or too low, they can present their rationale for such outlier estimation. Another reason for estimating story points is to see that sum of all story points should fit in the iteration timeline. Story point may also have a direct relationship with risk management; higher story points may indicate higher risks and can be used to evaluate risk for the product in incremental and iterative way. When the user stories are defined, each user story will also have a pre-defined set of inputs and outputs. With these sets of inputs and outputs, test team can start writing test cases and have acceptance criteria for these user stories when they are implemented and given to QA team for testing. Some of the user stories can have impact on the overall design architecture of the entire product. Note that implementation of an epic can potentially span over multiple iterations but each user story should be small enough to be completed in a single iteration.

Each user story contains tasks. All the tasks in the user story have to be complete before it can be given to QA for testing. Once a user story is defined, developers usually create these tasks at the start of iteration and continue to update the status or add new tasks during the iteration cycle. Sometimes these tasks may also include task for creating UI wireframes or mockup for supporting the user stories and other tasks in user stories. As a result, sometimes tasks can also have predecessor and dependent tasks. Once all the tasks are defined, developers implement the user stories and update the corresponding tasks as necessary. Towards the end of iteration cycle, the product is released to QA for testing.

QA team initially focuses on the acceptance tests and check if the core aspect of user stories has been implemented correctly. In such cases, automated test case can help perform a smoke test and generate

Table 4. Agile process documention requirements

	Stage	Documents
Pre-development activities	Idea generation	Integrated Project Plan & Schedule Product Requirements Document
	Proof of concept	Product Requirements Document
Iterative Development	Epic	Product Requirements Document
	Iteration planning	Integrated Project Plan & Schedule
	User Stories	Product Specifications Document Safety Risk Management Plan Development Design Development Design Review
	Test cases	Design Verification Test Plan Design Validation Test Plan
	Testing	Traceability Matrix
	Change requests	Product Specifications Document Change Request Order
Post-development activities	Release	Design Verification Test Report Design Validation Test Report Design Verification and Validation Review Safety Risk Management Report Field Confirmation Plan
	Manufacturing	Manufacturing Readiness Report Design Transfer Field Confirmation Report
	Commercialization	Electronic Design History File

report quickly. Once the release is accepted, QA team goes through the test cases written for the user stories and executes them. If any of the critical bugs are found, development team will have to fix the issues before closing the iteration.

After all QA activities are complete, it is given to customers who are early adopters or in some cases the customer could be the product manager. They review the features completed in the user story for the iteration and provide feedback. If there are changes requested for the current implementation, those can be prioritized and treated a backlogs for the next iteration.

Table 4. below shows what document needs to updated at various stages. Some of the documents are created in a particular stage but continue to get updated in subsequent stages. For instance, the PRD is in a draft stage in the pre-development activities. Some of the requirements will be at a high level and specifics can be added during the iteration. But once the specific have been added to the requirements, further changes in the requirements need to go through change control process.

Unlike waterfall, several documents are updated during the iteration cycle. When an Epic is defined, there may be some small updates to PRD. Subsequently, during iteration planning when the story points are estimated, it may have cascading effect on the overall project schedule so IPPS also needs to be updated. User stories have detail description of a feature or sub-feature so once a user story is defined; corresponding section of PSD also needs to be updated. Some of the complex user stories may have an impact on overall product architecture design or safety design and the corresponding document DD and SRMP should updated as necessary. There should be small design reviews if the design changes significantly and DDR document should be updated. During an iteration cycle, QA team is responsible for updating the TM, DVrTP and DVlTP. At the end of iteration cycle, if there are major changes to the requirements and functionality, it has to be formalized through CRO document and also update PSD as appropriate.

In the post-development activities, when all iterations are complete, QA team needs to generate DVrTR and DVlTR documents and create DVVR document after a formal review process. In addition development team needs create SRMR where all the issues related to risk are addressed or mitigated during the iterations. Subsequently, if the product involves a medical device, MRR and DT document should be final and after field tests are complete, FCR should be updated with the field results. Finally, prior to close of the project all documentation should be archived in eDHF.

Tools

There are several tools commercially available to help program mangers, product managers and software development team to collaborate and implement the compliance processes. A few of such tools are: IBM DOORS, HP Quality Center, JAMA, and Serena Dimension (IBM; Hewlett-Packard Development Company, 2015; JAMA, 2015; Serena Software, 2015). It should be noted that these tools just provide a framework for the process to be compliant and it is the core team who defines the process.

CONCLUSION

Several companies are embracing Agile processes to not only improve the quality of software applications but also to help improve development schedules. Unlike waterfall, Agile methodology provide an iterative development process that offers several advantages – early feedback, better project plan estimation, continuous feedback loop and improvement, always have a working version of software, avoid discrepancies in documentation and last but not the least more adaptive to changing market conditions. Specifications required for design control compliance can be integrated into Agile processes. As long as software development team has the entire process documented, follow the process that is documented and be able to provide evidence that the documented processes have been followed, there should be no issues with the compliance or audits.

REFERENCES

CFR - Code of Federal Regulations Title 21. (2014, September 1). US Food and Drug Administration Website. Retrieved from http://www.accessdata.fda.gov/scripts/cdrh/cfdocs/cfcfr/CFRSearch.cfm?CFRPart=820

Hewlett-Packard Development Company. (2015). *Quality Center Enterprise: Hewlett-Packard Development Company*. Retrieved from http://www8.hp.com/us/en/software-solutions/quality-center-quality-management/index.html?jumpid=va_R11374_us/en/large/eb/go_qc

IBM. (n. d.). *Rational DOORS: IBM*. Retrieved from http://www-03.ibm.com/software/products/en/ratidoor

JAMA. (2015). *Requirements Management: JAMA*. Retrieved from http://www.jamasoftware.com/jama-requirements-management/

Serena Software. (2015). *Serena Dimensions: Serena software*. Retrieved from http://www.serena.com

ADDITIONAL READING

U.S. Department Of Health and Human Services. FDA; Center for Devices and Radiological Health; Center for Biologics Evaluation and Research. (2002). *General Principles of Software Validation; Final Guidance for Industry and FDA Staff.* Silver Spring: FDA. Retrieved from http://www.fda.gov/downloads/MedicalDevices/DeviceRegulationandGuidance/GuidanceDocuments/ucm085371.pdf

U.S. Department of Health and Human Services. FDA; Center for Devices and Radiological Health; Office of Device Evaluation; Office of InVitro Diagnostics; Center for Biologics Evaluation and Research; Office of Blood Research and Review. (n.d.). *Guidance for the Content of Premarket Submissions for Software Contained in Medical Devices.* Silver Spring: FDA. Retrieved from http://www.fda.gov/MedicalDevices/DeviceRegulationandGuidance/GuidanceDocuments/ucm089543.htm

U.S. Department Of Health And Human Services, & the Food and Drug Administration. Center for Devices and Radiological Health; Office of Device Evaluation. (1999). *Guidance for Industry, FDA Reviewers and Compliance on Off-The-Shelf Software Use in Medical Devices.* Silver Spring: FDA. Retrieved from http://www.fda.gov/downloads/MedicalDevices/DeviceRegulationandGuidance/GuidanceDocuments/ucm073779.pdf

KEY TERMS AND DEFINITIONS

CAPA: Corrective and preventive actions.
CFR: Code federal regulation.
CRO: Change request order.
DDR: Development design Review.
DD: Development design.
DVrTP: Design Verification Test Plan.
DVlTP: Design Validation Test Plan.
DT: Design Transfer.
eDHF: Electronic design history file.
FCP: Field confirmation plan.
FCR: Field confirmation report.
FDA: Food and drug Administration (US).
IPPS: Integrated product plan and schedule.
MRR: Manufacturing readiness report.
MVP: Minimum viable product.
PRD: Product requirements document.

PSD: Product Specification document.
QA: Quality Assurance.
QSR: Quality systems regulations.
SaaS: Software as a service.
SDLC: Software development life cycle.
SRMP: Safety risk management plan.
SRMPR: Saftey risk management report.
TM: Traceability Matrix.
UI: User interface.
BAR: Broadly Accepted Range.
ALARP: As low as reasonably possible.
INTOL: Intolerable.

Compilation of References

FDA adverse event reporting system (FAERS) (n. d.). U.S. Food and Drug Administration. Retrieved from http://www.fda.gov/Drugs/GuidanceComplianceRegulatoryInformation/Surveillance/AdverseDrugEffects/default.htm

Ma, L., Mei, J., Pan, Y. (2007). Semantic Web Technologies and Data Management. Retrieved from http://www.w3.org/2007/03/RdfRDB/papers/ma.pdf

Oscarsson, M., Ingelman-Sundberg, M., Daly, A. K., & Nebert, D. W. (Eds.). (2001). Human cytochrome (p. 450). CYP. Retrieved from http://www.imm.ki.se/CYPalleles

Post approval safety data management: definitions and standards for expedited reporting. (2003). Proceedings of ICH-International Conference on Harmonisation of Technical Requirements for Registration of Pharmaceuticals for Human Use. Retrieved from http://www.ich.org/fileadmin/Public_Web_Site/ICH_Products/Guidelines/Efficacy/E2D/Step4/E2D_Guideline.pdf

Sheth, A.P. & Larson, J.A. (1990). Federated database systems for managing distributed, heterogeneous, and autonomous databases. *ACM Comput. Surv.* 22(3), 183-236. DOI=10.1145/96602.96604

We created an SMS campaign to increase HIV/AIDS awareness in Uganda. (2010). Text to Change. Retrieved from http://www.simpill.com/index.html

ABI. (2007). A History of Innovation in Genetic Analysis. Retrieved from http://tools.lifetechnologies.com/content/sfs/posters/ABI6247_SOLiD_Timeline_v4_ONLINE.pdf

ACMG Board of Directors. (2012). Points to consider in the clinical application of genomic sequencing. *Genetics in Medicine*, *14*(8), 759–761. doi:10.1038/gim.2012.74 PMID:22863877

a CRAN - package extRemes. extreme value analysis. (n. d.). Retrieved from http://cran.r-project.org/web/packages/extRemes/index.html

Affymetrix Technical Note. (2005). Guide to Probe Logarithmic Intensity Error (PLIER) Estimation, Retrieved from http://www.affymetrix.com

Agilent Technologies. (n. d.). Agilent SureSelect Target Enrichment. Retrieved from http://www.genomics.agilent.com/article.jsp?pageId=2094

Allied Health World. (2012). A tweet a day keeps the doctor away. Retrieved from http://www.mediabistro.com/alltwitter/files/2012/12/social-media-healthcare.png

AllSeq. (n. d.). Life Technologies – Ion Torrent. Retrieved from http://allseq.com/knowledgebank/sequencing-platforms/life-technologies-ion-torrent

Almenoff, J. S., DuMouchel, W., Kindman, L. A., Yang, X., & Fram, D. (2003). Disproportionality analysis using empirical bayes data mining: A tool for the evaluation of drug interactions in the post-marketing setting. *Pharmacoepidemiology and Drug Safety*, *12*(6), 517–521. doi:10.1002/pds.885 PMID:14513665

Almenoff, J., Tonning, J. M., Gould, A. L., Szarfman, A., Hauben, M., & Ouellet-Hellstrom, R. et al. (2005). Perspectives on the use of data mining in pharmacovigilance. *Drug Safety*, *28*(11), 981–1007. doi:10.2165/00002018-200528110-00002 PMID:16231953

Amin, R., & Arefin, T. (2010). The Empirical Study on the Factors Affecting Data Warehousing Success.

AMR Clinical Metrics Study. (2008). BearingPoint.

Angiuoli, S. V., Matalka, M., Gussman, A., Galens, K., Vangala, M., & Riley, D. R. et al. (2011). CloVR: A virtual machine for automated and portable sequence analysis from the desktop using cloud computing. *BMC Bioinformatics*, 12. PMID:21878105

Apidianakis, Y., Mindrinos, M. N., Xiao, W., Lau, G. W., Baldini, R. L., Davis, R. W., & Rahme, L. G. (2005). Profiling early infection responses: Pseudomonas aeruginosa eludes host defenses by suppressing antimicrobial peptide gene expression. *Proceedings of the National Academy of Sciences of the United States of America*, *102*(7), 2573–2578. doi:10.1073/pnas.0409588102 PMID:15695583

Architectural Pattern. (2015). Wikipedia. Retrieved from https://en.wikipedia.org/wiki/Architectural_pattern

Architectural Patterns and Styles (Ch. 3). (2009, Oct). MSDN Library - Application Architecture Guide (2nd ed.). Retrieved from https://msdn.microsoft.com/en-in/library/ee658117.aspx

Armstrong, S. A., Staunton, J. E., Silverman, L. B., Pieters, R., den Boer, M. L., & Minden, M. D. et al. (2002). MLL translocations specify a distinct gene expression profile that distinguishes a unique leukemia. *Nature Genetics*, *30*(1), 41–47. doi:10.1038/ng765 PMID:11731795

Auerbach, A. D., Burn, J., Cassiman, J. J., Claustres, M., Cotton, R. G., & Cutting, G. et al. (2011). Mutation (variation) databases and registries: A rationale for coordination of efforts. *Nature Reviews. Genetics*, *12*(12), 881–881. PMID:22025002

Avgeriou, P., & Zdun, U. (2005, July). Architectural patterns revisited:a pattern language. Proceedings of the 10th European Conference on Pattern Languages of Programs (EuroPlop 2005). Irsee, Germany.

Badger, L.,Grance, T.,Patt-Corner, R., & Voas, J. (2011). Cloud computing synopsis and recommendations. NIST special publication, 800, 146

Baker, R.A., Pikalov, A., Tran, Q.V., Kremenets, T., Arani, R.B., & Doraiswamy, P.M. (2009). *Atypical antipsychotic drugs and diabetes mellitus in the US food and drug administration adverse event database: A systematic Bayesian signal detection analysis.* Psychopharmacol Bull., 42(1), 11-31.

Banerjee, S. (2013). SMAC: Social, mobile, analytics, and cloud. *Cutter IT Journal*, *26*(2), 27–30. Retrieved http://www.cutter.com/content-and-analysis/journals-and-reports/cutter-it-journal/sample/itj1302/itj1302.pdf

Banks, D., Woo, E. J., Burwen, D. R., Perucci, P., Braun, M. M., & Ball, R. (2005). Comparing data mining methods on the VAERS database. *Pharmacoepidemiology and Drug Safety*, *14*(9), 601–609. doi:10.1002/pds.1107 PMID:15954077

Barbieri, C. E., Baca, S. C., Lawrence, M. S., Demichelis, F., Blattner, M., & Theurillat, J. P. et al. (2012). Exome sequencing identifies recurrent SPOP, FOXA1 and MED12 mutations in prostate cancer. *Nature Genetics*, *44*(6), 685–689. doi:10.1038/ng.2279 PMID:22610119

Bass, L., Clements, P., & Kazman, R. (2005). *Software Architecture in Practice* (2nd ed.).

Bate, A., Lindquist, M., Edwards, I. R., Olsson, S., Orre, R., Lansner, A., & De Freitas, R. M. (1998). A bayesian neural network method for adverse drug reaction signal generation. *European Journal of Clinical Pharmacology*, *54*(4), 315–321. doi:10.1007/s002280050466 PMID:9696956

Belle, D. J., & Singh, H. (2008). Genetic Factors in Drug Metabolism. *American Family Physician*, *77*(11), 1553–1560. PMID:18581835

Benjamini, Y. & Hochberg, Y. (1995). Controlling the false discovery rate: a practical and powerful approach to multiple testing. *The Journal of statistical society, 57,* 289-300.

Berry, S. M., & Berry, D. A. (2004). *Accounting for multiplicities in assessing drug safety: A three-level hierarchical mixture model*

Biesecker, L. G., Shianna, K. V., & Mullikin, J. C. (2011). Exome sequencing: The expert view. *Genome Biology*, *12*(9), 128. doi:10.1186/gb-2011-12-9-128 PMID:21920051

Bloche, G. M. (2004). Race-based therapeutics. *The New England Journal of Medicine*, *351*(20), 2035–2037. doi:10.1056/NEJMp048271 PMID:15533852

Boerma, M., van der Wees, C. G., Vrieling, H., Svensson, J. P., Wondergem, J., & van der Laarse, A. et al. (2005). Microarray analysis of gene expression profiles of cardiac myocytes and fibroblasts after mechanical stress, ionising or ultraviolet radiation. *BMC Genomics*, *6*(1), 6. doi:10.1186/1471-2164-6-6 PMID:15656902

Brown, J. S., Petronis, K., Bate, A., Zhang, F., Dashevsky, I., & Kulldorff, M. et al. (2011). Comparing two methods for detecting adverse event signals in observational data: Empirical bayes gamma poisson shrinker vs. tree-based scan statistic. *Pharmacoepidemiology and Drug Safety*, *20*, S144.

Burke, W., Trinidad, S. B., & Clayton, E. W. (2013). Seeking genomic knowledge: The case for clinical restraint. *The Hastings Law Journal*, *64*(6), 1650. PMID:24688162

Buschmann, Meunier, Rohnert, Sommerlad & Stal (1996). Pattern-Oriented Software Architecture: A System of Patterns.

Candore, G., Juhlin, K., Manlik, K., Thakrar, B., Quarcoo, N., Seabroke, S., & Slattery, J. (2015). *Comparison of statistical signal detection methods within and across spontaneous reporting databases. Drug Safety, Casella, G., & Berger, R. L. (2001). Statistical inference* (2nd ed.). Cengage Learning.

Caruccio, N. (2011). Preparation of next-generation sequencing libraries using Nextera technology: Simultaneous DNA fragmentation and adaptor tagging by in vitro transposition. *Methods in Molecular Biology (Clifton, N.J.)*, *733*, 241–255. doi:10.1007/978-1-61779-089-8_17 PMID:21431775

Casey, D. (2006). Federated Databases in Bioinformatics and Translational Medical Research.

Casey, D. (2006). How Federated Databases Benefit Bioinformatics Research.

Caster, O., Noren, G. N., Madigan, D., & Bate, A. (2013). *Logistic regression in signal detection: Another piece added to the puzzle*

Cawley, S., Bekiranov, S., Ng, H. H., Kapranov, P., Sekinger, E. A., & Kampa, D. et al. (2004). Unbiased mapping of transcription factor binding sites along human chromosomes 21 and 22 points to widespread regulation of noncoding RNAs. *Cell*, *116*(4), 499–509. doi:10.1016/S0092-8674(04)00127-8 PMID:14980218

CFR - Code of Federal Regulations Title 21. (2014, September 1). US Food and Drug Administration Website. Retrieved from http://www.accessdata.fda.gov/scripts/cdrh/cfdocs/cfcfr/CFRSearch.cfm?CFRPart=820

Chilamakuri, C. S. R., Lorenz, S., Madoui, M. A., Vodák, D., Sun, J., & Hovig, E. et al. (2014). Performance comparison of four exome capture systems for deep sequencing. *BMC Genomics*, *15*(1), 449. doi:10.1186/1471-2164-15-449 PMID:24912484

Choi, M., Scholl, U. I., Ji, W., Liu, T., Tikhonova, I. R., & Zumbo, P. et al. (2009). Genetic diagnosis by whole exome capture and massively parallel DNA sequencing. *Proceedings of the National Academy of Sciences of the United States of America*, *106*(45), 19096–19101. doi:10.1073/pnas.0910672106 PMID:19861545

CIOMS Working Group VIII. (2010). *Practical aspects of signal detection in pharmacovigilance.*

Clark, M. J., Chen, R., Lam, H. Y., Karczewski, K. J., Chen, R., & Euskirchen, G. et al. (2011). Performance comparison of exome DNA sequencing technologies. *Nature Biotechnology*, *29*(10), 908–914. doi:10.1038/nbt.1975 PMID:21947028

Clinical Pharmacology I. (2002). MCCQE Review Notes. Retrieved from http://www.ucl.ac.uk/anaesthesia/education/Pharmacology

Clinical Trial Phases. (n. d.). Cern Foundation. Retrieved from https://cern-foundation.org/?page_id=292

Cloud Standards Customer Council. (2013). Convergence of social, mobile and cloud: 7 steps to ensure success. Retrieved from http://www.cloudstandardscustomercouncil.org/Convergence_of_Cloud_Social%20_Mobile_Final.pdf

Collins, F. S. (2010). The Future of Personalized Medicine. *NIH Medline Plus*, *5*(1), 2–3.

Comer, J. E., Galindo, C. L., Chopra, A. K., & Peterson, J. W. (2005). GeneChip analyses of global transcriptional responses of murine macrophages to the lethal toxin of Bacillus anthracis. *Infection and Immunity*, *73*(3), 1879–1885. doi:10.1128/IAI.73.3.1879-1885.2005 PMID:15731093

Cornelius, V. R., & Evans, S. J. W. (2009). The use of time to event models in signal detection. *Drug Safety*, *32*(10), 926.

CRAN - package texmex. statistical modelling of extreme values. (n. d.). Retrieved from http://cran.r-project.org/web/packages/texmex/index.html

Crawley, & LaVera (2007). The Paradox of Race in the Bidil Debate. *Journal of the National Medical Association*, 99, 821-822.

Czernicki, B. (2011, February 7). IaaS, PaaS and SaaS terms clearly explained and defined. Retrieved from http://www.silverlighthack.com/post/2011/02/27/iaas-paas-and-saas-terms-explained-and-defined.aspx

Dalen, P., Dahl, M. L., Ruiz, M. L., Nordin, J., & Bertilsson, L. (1998). 10-Hydroxylation of nortriptyline in white persons with 0, 1, 2, 3, and 13 functional CYP2D6 genes. *Clinical Pharmacology and Therapeutics*, *63*(4), 444–452. doi:10.1016/S0009-9236(98)90040-6 PMID:9585799

David, J. (2007). Genome-wide mapping of in vivo protein–DNA interactions. *Science*, *316*(5830), 1497–1502. doi:10.1126/science.1141319 PMID:17540862

Den Dunnen, J. T., & Antonarakis, S. E. (2000). Mutation nomenclature extensions and suggestions to describe complex mutations: A discussion. *Human Mutation*, *15*(1), 7–12. doi:10.1002/(SICI)1098-1004(200001)15:1<7::AID-HUMU4>3.0.CO;2-N PMID:10612815

Derek, V. B. (2010). Cytochrome P450 2C9-CYP2C9. Pharmacogenetics and genomics. *Pharmacogenetics and Genomics*, *20*(4), 277–281. PMID:20150829

Desta, Z., Zhao, X., Shin, J.-G., & Flockhart, D. A. (2002). Clinical significance of the cytochrome P450 2C19 genetic polymorphism. Clinical pharmacokinetics. *Clinical Pharmacokinetics*, *41*(12), 913–958. doi:10.2165/00003088-200241120-00002 PMID:12222994

Dolled-Filhart, M.P., Lordemann, A., Dahl, W., & Haraksingh, R.R., Ou-yang, & Lin, J. C. H. (. (2012). Opportunities & Challenges With Personalized Exome Sequencing. *Personalized Medicine.*, *9*(8), 805–819. doi:10.2217/pme.12.97

DuMouchel, W. (1999). Bayesian data mining in large frequency tables, with an application to the FDA spontaneous reporting system. *The American Statistician*, *53*(3), 177–190.

Durbin, R. M., Altshuler, D. L., Durbin, R. M., Abecasis, G. R., Bentley, D. R., & Chakravarti, A. (2010). A map of human genome variation from population-scale sequencing. *Nature*, *467*, 1061–1073. PMID:20981092

Earls, E. (2012, July 19). Clinical trial delays: America's patient recruitment dilemma. Retrieved from http://www.drugdevelopment-technology.com/features/featureclinical-trial-patient-recruitment

Eid, J., Fehr, A., Gray, J., Luong, K., Lyle, J., & Otto, G. et al. (2009). Real-time DNA sequencing from single polymerase molecules. *Science*, *323*(5910), 133–138. doi:10.1126/science.1162986 PMID:19023044

Emory Genetics Laboratory. (2014). Six Medical EmExome Service Levels. Retrieved from http://geneticslab.web.emory.edu/about/news-and-events/news/2014/01/medical-emexome/

Ermak, G. (2013).. . *Modern Science & Future Medicine*, *2*, 164.

Eu-Adr Group. (n. d.). Retrieved from http://euadr-project.org/

European Medicines Agency (EMA). (2014). EudraVigilance. Retrieved from https://eudravigilance.ema.europa.eu/human/index.asp

European Medicines Agency. (2012). Guideline on good pharmacovigilance practices- Module VI. Retrieved from http://www.ema.europa.eu/docs/en_GB/document_library/Scientific_guideline/2012/02/WC500123203.pdf

European Medicines Agency. (2013). Reflection paper on risk based quality management in clinical trials. Retrieved from http://www.ema.europa.eu/docs/en_GB/document_library/Scientific_guideline/2011/08/WC500110059.pdf

Evans, B.L. (2003). Non Negative Matrix Factorization, Multidimensional Digital Signal Processing. Retrieved from http://www.ece.utexas.edu/~bevans/courses/ee381k/projects/spring03/

Evans, S. J. W., Waller, P. C., & Davis, S. (2001). Use of proportional reporting ratios (PRRs) for signal generation from spontaneous adverse drug reaction reports. *Pharmacoepidemiology and Drug Safety*, *10*(6), 483–486.

Evans, W. E., & Johnson, J. A. (2001). Pharmacogenomics: The inherited basis for interindividual differences in drug response. *Annual Review of Genomics and Human Genetics*, *2*(1), 9–39. doi:10.1146/annurev.genom.2.1.9 PMID:11701642

Evans, W. E., & McLeod, H. L. (2003). Pharmacogenomics – drug disposition, drug targets, and side effects. *The New England Journal of Medicine*, *348*(6), 538–549. doi:10.1056/NEJMra020526 PMID:12571262

Evans, W. E., & Relling, M. V. (1999). Pharmacogenomics: Translating functional genomics into rational therapeutics. *Science*, *286*(5439), 487–491. doi:10.1126/science.286.5439.487 PMID:10521338

Extedo. (n. d.). Retrieved from http://www.extedo.com/products/the-extedosuite-for-regulatory-information-management/

Fejes, A. P., Khodabakhshi, A. H., Birol, I., & Jones, S. J. (2011). Human variation database: An open-source database template for genomic discovery. *Bioinformatics (Oxford, England)*, *27*(8), 1155–1156. doi:10.1093/bioinformatics/btr100 PMID:21367872

Fokkema, I. F., Taschner, P. E., Schaafsma, G. C., Celli, J., Laros, J. F., & den Dunnen, J. T. (2011). LOVD v.2.0: The next generation in gene variant databases. *Human Mutation*, *32*(5), 557–563. http://www.lovd.nl/2.0/ doi:10.1002/humu.21438 PMID:21520333

Food & Drug Administration. (n. d.). fda.gov.

Food and Drug Administration. (2001). Guidance for industry: post marketing safety reporting for human drug and biological products including vaccines. Retrieved from http://www.fda.gov/downloads/BiologicsBloodVaccines/GuidanceComplianceRegulatoryInformation/Guidances/Vaccines/ucm092257.pdf

Food and Drug Administration. (2011). Guidance for industry responding to unsolicited requests for off-label information about prescription drugs and medical devices. Retrieved from http://www.fda.gov/downloads/Drugs/GuidanceComplianceRegulatoryInformation/Guidances/UCM285145.pdf

Food and Drug Administration. (2013). Mobile medical applications-guidance for industry and food and drug administration staff. Retrieved from http://www.fda.gov/downloads/MedicalDevices/.../UCM263366.pdf

Food and Drug Administration. (2013). Oversight of clinical investigations — a risk based approach to monitoring. Retrieved from http://www.fda.gov/downloads/Drugs/.../Guidances/UCM269919.pdf

Food and Drug Administration. (2014). Guidance for industry fulfilling regulatory requirements for postmarketing submissions of interactive promotional media for prescription human and animal drugs and biologics. Retrieved from http://www.fda.gov/downloads/Drugs/GuidanceComplianceRegulatoryInformation/Guidances/UCM381352.pdf

Futema, M., Plagnol, V., Whittall, R. A., Neil, H. A. W., & Humphries, S. E. (2012). Use of targeted exome sequencing as a diagnostic tool for Familial Hypercholesterolaemia. *Journal of Medical Genetics*, *49*(10), 644–649. doi:10.1136/jmedgenet-2012-101189 PMID:23054246

Gahl, W. A., Markello, T. C., Toro, C., Fajardo, K. F., Sincan, M., & Gill, F. et al. (2011). The national institutes of health undiagnosed diseases program: Insights into rare diseases. *Genetics in Medicine*, *14*(1), 51–59. doi:10.1038/gim.0b013e318232a005 PMID:22237431

Gartner (2011, November 15). Gartner says sales of mobile devices grew 5.6 percent in third quarter of 2011; smartphone sales increased 42 percent. Retrieved from http://www.gartner.com/newsroom/id/1848514

Gipson, G. (2012). A shrinkage-based comparative assessment of observed-to-expected disproportionality measures. *Pharmacoepidemiology and Drug Safety*, *21*(6), 589–596. doi:10.1002/pds.2349 PMID:22290739

Gipson, G., Schaaf, R., DuMouchel, W., Valentino, R., & Wisniewski, A. (2010). Impact of drug product litigation on safety signal detection in aers. *Pharmacoepidemiology and Drug Safety*, *19*, S224.

Gleeson, F. C., Kipp, B. R., Kerr, S. E., Voss, J. S., Graham, R. P., & Campion, M. B. et al. (2015). Kinase genotype analysis of gastric gastrointestinal stromal tumor cytology samples using targeted next-generation sequencing. *Clinical Gastroenterology and Hepatology*, *13*(1), 202–206. doi:10.1016/j.cgh.2014.06.024 PMID:24997326

Glossary of common terms. (n. d.). ClinicalTrials.gov.

Glossary of terms used in Pharmacovigilance. (n. d.). Uppsala. Retrieved from http://who-umc.org/Graphics/24729.pdf

Gombar, V.K., Silver, I.S., Zhao, Z. (2003). Primary Role of ADME characteristics in drug discovery and their in silico evaluation: in silico screening of chemicals for their metabolic stability. Retrieved from http://www.ncbi.nlm.nih.gov/pubmed/12769701

Gould, A. L. (2003). Practical pharmacovigilance analysis strategies. *Pharmacoepidemiology and Drug Safety*, *12*(7), 559–574.

Gould, A. L. (2008). Detecting potential safety issues in clinical trials by Bayesian screening. *Biometrical Journal. Biometrische Zeitschrift*, *50*(5), 837–851.

Green, R. C., Berg, J. S., Grody, W. W., Kalia, S. S., Korf, B. R., & Martin, C. L. et al. (2013). ACMG recommendations for reporting of incidental findings in clinical exome and genome sequencing. *Genetics in Medicine*, *15*(7), 565–574. doi:10.1038/gim.2013.73 PMID:23788249

Gunn, S. (1998, May 14). Support Vector Machines for Classification and Regression. http://homepages.cae.wisc.edu/~ece539/software/svmtoolbox/svm.pdf

Hall, A. M., & Wilkins, M. R. (2005). Warfarin: A case history in pharmacogenetics, Heart. *BMJ (Clinical Research Ed.)*, *91*, 563–564.

Harjulampi, V. (2013). *Adopting cloud computing and hosted services in pharmaceutical industry*. Jyväskylä, Finland: Jamk University of Applied Sciences.

Hearst, M. (1999). Untangling text data mining. http://www.sims.berkeley.edu/~hearst/papers/acl99/acl99-tdm.html

Hearst, M. (2003, October 17). What is text mining? Retrieved from http://www.sims.berkeley.edu/~hearst/text-mining.html

Heffernan, J. E., & Tawn, J. A. (2004). A conditional approach for multivariate extreme values. *Journal of the Royal Statistical Society. Series A, (Statistics in Society)*, *66*(3), 497–546. doi:10.1111/j.1467-9868.2004.02050.x

Hewlett-Packard Development Company. (2015). *Quality Center Enterprise: Hewlett-Packard Development Company*. Retrieved from http://www8.hp.com/us/en/software-solutions/quality-center-quality-management/index.html?jumpid=va_R11374_us/en/large/eb/go_qc

Huang, Y. F., Chen, S. C., Chiang, Y. S., Chen, T. H., & Chiu, K. P. (2012). Palindromic sequence impedes sequencing-by-ligation mechanism. *BMC Systems Biology*, *6*(Suppl 2), S10. doi:10.1186/1752-0509-6-S2-S10 PMID:23281822

Hurd, P. J., & Nelson, C. J. (2009). Advantages of next-generation sequencing versus the microarray in epigenetic research. *Briefings in Functional Genomics*, elp013. PMID:19535508

IBM. (n. d.). *Rational DOORS: IBM*. Retrieved from http://www-03.ibm.com/software/products/en/ratidoor

ICH MedDRA Maintenance and Support Services Organization (MSSO). (n. d.). MedDRA Retrieved from http://www.meddra.org/

Illumina a. Nextera Rapid Capture Exome and Expanded Exome Kits. (n. d.). Retrieved from http://www.illumina.com/products/nextera-rapid-capture-exome-kits.html

Illumina b. Systems. (n. d.). Retrieved from http://www.illumina.com/systems.html

Illumina c. Truseq Exome Enrichment Kit. (n. d.). Retrieved from http://support.illumina.com/sequencing/sequencing_kits/truseq_exome_enrichment_kit.html

Ingelman-Sundberg, M. (2004). Pharmacogenetics of cytochrome P450 and its applications in drug therapy: The past, present and future. *Trends in Pharmacological Sciences*, *25*(4), 193–200. doi:10.1016/j.tips.2004.02.007 PMID:15063083

Innovative Medicines Initiative PROTECT. (n. d.). Retrieved from http://www.imi-protect.eu/

Insight, C. C. S. (2013, June 10). Mobile phone sales will hit 1.86 billion in 2013 as strong smartphone growth continues. Retrieved February 15, 2014 from http://www.ccsinsight.com/press/company-news/1655-mobile-phone-sales-will-hit-186-billion-in-2013-as-strong-smartphone-growth-continues

International Conference on Harmonisation. (1994). *ICH harmonised tripartite guideline - clinical safety data management: Definitions and standards for expedited reporting E2A*. ().ICH. Retrieved from http://www.ich.org/fileadmin/Public_Web_Site/ICH_Products/Guidelines/Efficacy/E2A/Step4/E2A_Guideline.pdf

International Conference on Harmonisation. (1996). *ICH harmonised tripartite guideline on good clinical practice E6(R1)*. Retrieved from http://www.ich.org/fileadmin/Public_Web_Site/ICH_Products/Guidelines/Efficacy/E6_R1/Step4/E6_R1__Guideline.pdf

International Conference on Harmonisation. (2004 November, 18). *ICH harmonised tripartite guideline pharmacovigilance planning E2E*. Retrieved from http://www.ich.org/fileadmin/Public_Web_Site/ICH_Products/Guidelines/Efficacy/E2E/Step4/E2E_Guideline.pdf

International Health Terminology Standards Development Organisation (IHTSDO). (n. d.). SNOMED CT - ihtsdo.org. Retrieved from http://www.ihtsdo.org/snomed-ct/

JAMA. (2015). *Requirements Management: JAMA*. Retrieved from http://www.jamasoftware.com/jama-requirements-management/

Jiao, Y., Shi, C., Edil, B. H., de Wilde, R. F., Klimstra, D. S., & Maitra, A. et al. (2011). DAXX/ATRX, MEN1, and mTOR pathway genes are frequently altered in pancreatic neuroendocrine tumors. *Science*, *331*(6021), 1199–1203. doi:10.1126/science.1200609 PMID:21252315

Jothi, R., Cuddapah, S., Barski, A., Cui, K., & Zhao, K. (2008). Genome-wide identification of in vivo protein-DNA binding sites from ChIP-Seq data. *Nucleic Acids Research*, *36*(16), 5221–5231. doi:10.1093/nar/gkn488 PMID:18684996

Juhlin, K., Ye, X., Star, K., & Noren, G. N. (2013). Outlier removal to uncover patterns in adverse drug reaction surveillance - a simple unmasking strategy. *Pharmacoepidemiology and Drug Safety*, *22*(10), 1119–1129. PMID:23832706

Kapranov, P., Chen, L., Dederich, D., Dong, B., He, J., & Steinmann, K. E. et al. (2012). Native molecular state of adeno-associated viral vectors revealed by single-molecule sequencing. *Human Gene Therapy*, *23*(1), 46–55. doi:10.1089/hum.2011.160 PMID:21875357

Karlin, S. (2014). Adverse events in social media: FDA expects signal detection "Revolution". *The Pink Sheet*.

Kato, K. (2009). Impact of the next generation DNA sequencers. *International journal of clinical and experimental medicine*, *2*(2), 193.

Kent, W. J., Sugnet, C. W., Furey, T. S., Roskin, K. M., Pringle, T. H., Zahler, A. M., & Haussler, D. (2002). The human genome browser at UCSC (online advance copy). *Genome Research*, *12*(6), 996–1006. doi:10.1101/gr.229102. Article published online before print in May 2002 PMID:12045153

Kilzer, J., Xun, X., Bodeau, J., Breu, H., & Harris, A. (2010). A balanced barcoding system for multiplexed DNA library and SOLiD SAGE Sequencing. *Journal of Biomolecular Techniques*, *21*, 528.

Kircher, M., & Kelso, J. (2010). High-throughput DNA sequencing–concepts and limitations. *BioEssays*, *32*(6), 524–536. doi:10.1002/bies.200900181 PMID:20486139

Krawitz, P., Rödelsperger, C., Jager, M., Jostins, L., Bauer, S., & Robinson, P. N. (2010). Microindel detection in short-read sequence data. *Bioinformatics (Oxford, England)*, *26*(6), 722–729. doi:10.1093/bioinformatics/btq027 PMID:20144947

Kurz, X., Slattery, J., Addis, A., Durand, J., Segec, A., Skibicka, I., & Hidalgo-Simon, A. et al. (2010). The eudravigilance database of spontaneous adverse reactions as a tool for H1N1 vaccine safety monitoring. *Pharmacoepidemiology and Drug Safety*, *19*, S330–S331.

Lander, E. S., Linton, L. M., Birren, B., Nusbaum, C., Zody, M. C., & Baldwin, J. et al. (2001). Inter- national human genome, initial sequencing and analysis of the human genome. *Nature*, *409*(6822), 860–921. doi:10.1038/35057062 PMID:11237011

Laney, D. (2001). 3D data management: controlling data volume, velocity and variety. Retrieved from http://blogs.gartner.com/doug-laney/files/2012/01/ad949-3D-Data-Management-Controlling-Data-Volume-Velocity-and-Variety.pdf

Langmead, B., & Salzberg, S. L. (2012). Fast gapped-read alignment with Bowtie 2. *Nature Methods*, *9*(4), 357–359. doi:10.1038/nmeth.1923 PMID:22388286

Langmead, B., Trapnell, C., Pop, M., & Salzberg, S. L. (2009). Ultrafast and memory-efficient alignment of short DNA sequences to the human genome. *Genome Biology*, *10*(3), 25–30. doi:10.1186/gb-2009-10-3-r25 PMID:19261174

Langville, A., & Meyer, C. (2005, June 9). ALS Algorithms Nonnegative Matrix Factorization Text Mining. Retrieved from http://meyer.math.ncsu.edu/Meyer/Talks/SAS_6_9_05_NmfWorkshop.pdf

Levy, R. H., Thummel, K. E., Trager, W. F., Hansten, P. D., & Eichelbuam, M. (Eds.). (2000). *Metabolic Drug Interactions* (pp. 29–30). Philadelphia: Lippincott Williams & Wilkins.

Life Technologies. (n. d.). Ion AmpliSeq™ Exome RDY - OT2 Kit 1x8. Retrieved from http://www.lifetechnologies.com/order/catalog/product/4489837

Li, H., & Durbin, R. (2008). Mapping short DNA sequencing reads and calling variants using mapping quality scores. *Genome Research*, *18*(11), 1851–1858. doi:10.1101/gr.078212.108 PMID:18714091

Li, H., & Durbin, R. (2009). Fast and accurate short read alignment with Burrows-Wheeler transform. *Bioinformatics (Oxford, England)*, *25*(14), 1754–1760. doi:10.1093/bioinformatics/btp324 PMID:19451168

Lindquist, M., Edwards, I. R., Bate, A., Fucik, H., Nunes, A. M., & Stahl, M. (1999). From association to alert - A revised approach to international signal analysis. *Pharmacoepidemiology and Drug Safety*, *8*(SUPPL. 1), S15; S25.

Liu, L., Li, Y., Li, S., Hu, N., He, Y., Pong, R., ... & Law, M. (2012). Comparison of next-generation sequencing systems. BioMed *Research International*, 2012.

Lu, Z., & Su, J. (n. d.). *Clinical data management: Current status, challenges, and future directions from industry perspectives Metadata Management in Clinical Research.* Retrieved From http://www.octagonresearch.com/assets/files/MDR_whitepaper_2_2012.pdf

Lyon, G. J., Jiang, T., Van Wijk, R., Wang, W., Bodily, P. M., & Xing, J. et al. (2011). Exome sequencing and unrelated findings in the context of complex disease research: Ethical and clinical implications. *Discovery Medicine*, *12*(62), 41. PMID:21794208

Machini, K., Douglas, J., Braxton, A., Tsipis, J., & Kramer, K. (2014). Genetic counselors' views and experiences with the clinical integration of genome sequencing. *Journal of Genetic Counseling*, *23*(4), 496–505. doi:10.1007/s10897-014-9709-4 PMID:24671342

Maignen, F., Hauben, M., Hung, E., Van Holle, L., & Dogne, J. (2014). *Assessing the extent and impact of the masking effect of disproportionality analyses on two spontaneous reporting systems databases*

Maignen, F., Hauben, M., Hung, E., Holle, L. V., & Dogne, J. (2013). A conceptual approach to the masking effect of measures of disproportionality. *Pharmacoepidemiology and Drug Safety*. PMID:24243699

Majewski, J., Schwartzentruber, J., Lalonde, E., Montpetit, A., & Jabado, N. (2011). What can exome sequencing do for you? *Journal of Medical Genetics*, *48*(9), 580–589. doi:10.1136/jmedgenet-2011-100223 PMID:21730106

Mardis, E. (2008). Next-generation DNA sequencing methods. *Annual Review of Genomics and Human Genetics*, *9*(1), 387–402. doi:10.1146/annurev.genom.9.081307.164359 PMID:18576944

Marian, A. J. (2012). Challenges in medical applications of whole exome/genome sequencing discoveries. *Trends in Cardiovascular Medicine*, *22*(8), 219–223. doi:10.1016/j.tcm.2012.08.001 PMID:22921985

Martin Lewis, P., Smart, G., & Webster, A. (2006). False Positive, The commercial and clinical development of pharmacogenetics. Retrieved from http://www.york.ac.uk/media/satsu/res-pgx/FalsePositive2006.pdf

McLeod, H. L., & Evans, W. E. (2001). Pharmacogenomics: Unlocking the human genome for better drug therapy. *Annual Review of Pharmacology and Toxicology*, *41*(1), 101–121. doi:10.1146/annurev.pharmtox.41.1.101 PMID:11264452

Mertes, F., Elsharawy, A., Sauer, S., Van Helvoort, J., Van, D. Z., & Franke, A. M. (2011). Targeted enrichment of genomic DNA regions for next-generation sequencing. *Briefings in Functional Genomics*, *10*(6), 374–386. doi:10.1093/bfgp/elr033 PMID:22121152

Metzker, M. L. (2010). Sequencing technologies the next generation. *Nature Reviews. Genetics*, *11*(1), 31–46. doi:10.1038/nrg2626 PMID:19997069

Meyer, U. A. (2004). Pharmacogenetics – five decades of therapeutic lessons from genetic diversity. *Nature Reviews. Genetics*, *5*(9), 669–676. doi:10.1038/nrg1428 PMID:15372089

Monitoring, S. (2013, September 23). State of mobile 2013. Retrieved from http://www.supermonitoring.com/blog/2013/09/23/state-of-mobile-2013-infographic

Morozova, O., & Marra, M. A. (2008). Applications of next-generation sequencing technologies in functional genomics. *Genomics*, *92*(5), 255–264. doi:10.1016/j.ygeno.2008.07.001 PMID:18703132

Need, A. C., Shashi, V., Hitomi, Y., Schoch, K., Shianna, K. V., McDonald, M. T., ... & Goldstein, D. B. (2012). Clinical application of exome sequencing in undiagnosed genetic conditions. *Journal of medical genetics*, jmedgenet-2012.

Neveling, K., Collin, R., Gilissen, C., Van Huet, R., Visser, L., & Kwint, M. et al. (2012). Next-generation genetic testing for retinitis pigmentosa. *Human Mutation*, *33*(6), 963–972. doi:10.1002/humu.22045 PMID:22334370

Newcomer, J. D., Griffith, S. M. L., Pugh, E. W., Ling, H., Leary, D. R., Goldstein, J. L., et al. (2013). CIDRVar: A Next-Generation Sequencing Database Linking Samples, Variants, and Annotations. Center for Inherited Disease Research (CIDR). Baltimore, MD: Institute of Genetic Medicine, Johns Hopkins University; Retrieved from http://www.cidr.jhmi.edu/nih/CIDRVar.pdf

Noy, N.F. (2004, June 25). Semantic Integration – A Survey of Ontology based approaches. Retrieved from http://web.stanford.edu/~natalya/papers/SigmodRecordReview.pdf

Oracle Help Centre-Oracle Database Online Documentation (p. 9). (n. d.). Oracle. Retrieved from http://docs.oracle.com/cd/B19306_01/server.102/b14239/concepts.htm

Owen Ryan, P. (2009). Cytochrome P450 2D6, Pharmacogenetics and genomics. *Pharmacogenetics and Genomics*, *19*(7), 559–562. doi:10.1097/FPC.0b013e32832e0e97 PMID:19512959

P450 Drug interaction Table. (2010). Indiana University Division of Clinical Pharmacology. Retrieved from http://medicine.iupui.edu/clinpharm/ddis/table.asp

Panda, R., & Suresh, P. K. (2014). Computational identification and analysis of functional polymorphisms involved in the activation and detoxification genes implicated in endometriosis. *Gene*, *542*(2), 89–97. doi:10.1016/j.gene.2014.03.058 PMID:24698776

Parla, J. S., Iossifov, I., Grabill, I., Spector, M. S., Kramer, M., & McCombie, W. R. (2011). A comparative analysis of exome capture. *Genome Biology*, *12*(9), R97. doi:10.1186/gb-2011-12-9-r97 PMID:21958622

Pettersson, E., Lundeberg, J., & Ahmadian, A. (2009). Generations of sequencing technologies. *Genomics*, *93*(2), 105–111. doi:10.1016/j.ygeno.2008.10.003 PMID:18992322

Phillips, K., Veenstra, D., Oren, E., Lee, J., & Sadee, W. (2001). Potential role of pharmacogenomics in reducing adverse drug reactions. *Journal of the American Medical Association*, *286*(18), 2270–2279. doi:10.1001/jama.286.18.2270 PMID:11710893

Pinxten, W., & Howard, H. C. (2014). Ethical issues raised by whole genome sequencing. *Best Practice & Research. Clinical Gastroenterology*, *28*(2), 269–279. doi:10.1016/j.bpg.2014.02.004 PMID:24810188

Proteus Digital Health. (2012, July 30). Proteus digital health announces FDA clearance of ingestible sensor. Retrieved from http://www.proteus.com/proteus-digital-health-announces-fda-clearance-of-ingestible-sensor

Quail, M.A. (2012). Pacific Biosciences and Illumina MiSeq sequencers. *BMC Genomics*, 13–341.

Quail, M. A., Smith, M., Coupland, P., Otto, T. D., Harris, S. R., & Connor, T. R. et al. (2012). A tale of three next generation sequencing platforms: Comparison of Ion Torrent, Pacific Biosciences and Illumina MiSeq sequencers. *BMC Genomics*, *13*(1), 341. doi:10.1186/1471-2164-13-341 PMID:22827831

R&D Pipeline Management. (n. d.). University of Wisconsin. Retrieved from http://maravelias.che.wisc.edu/?page_id=23

Rabbani, B., Tekin, M., & Mahdieh, N. (2014). The promise of whole-exome sequencing in medical genetics. *Journal of Human Genetics*, *59*(1), 5–15. doi:10.1038/jhg.2013.114 PMID:24196381

Raney, B. J., Cline, M. S., Rosenbloom, K. R., Dreszer, T. R., Learned, K., & Barber, G. P. et al. (2011). ENCODE whole-genome data in the UCSC genome browser. *Nucleic Acids Research*, *39*(Database), 871–875. doi:10.1093/nar/gkq1017 PMID:21037257

Roberts, R. J., Carneiro, M. O., & Schatz, M. C. (2013). The advantages of SMRT sequencing. *Genome Biology*, *14*(6), 405. doi:10.1186/gb-2013-14-6-405 PMID:23822731

Roche. 454 Sequencing. Retrieved from http://www.454.com/

Rouse, M. (2011). What is Data Virtualization?

Salleh, M. Z. (2013). Systematic Pharmacogenomics Analysis of a Malay Whole Genome: Proof of Concept for Personalized Medicine. *PLoS ONE*, *10*, 1371. PMID:24009664

Sanderson, S., Emery, J., & Higgins, J. (2005). CYP2C9 gene variants, drug dose, and bleeding risk in warfarin-treated patients: A HuGEnet systematic review and meta-analysis. *Genetics in Medicine*, *7*(2), 97–104. doi:10.1097/01.GIM.0000153664.65759.CF PMID:15714076

Schadt, E., Turner, S., & Kasarskis, A. (2010). A window into third-generation sequencing. *Human Molecular Genetics*, *19*(R2), R227–R240. doi:10.1093/hmg/ddq416 PMID:20858600

Scordo, M. G., Pengo, V., Spina, E., Dahl, M. L., Gusella, M., & Padrini, R. (2002). Influence of CYP2C9 and CYP2C19 genetic polymorphisms on warfarin maintenance dose and metabolic clearance. *Clinical Pharmacology and Therapeutics*, *72*(6), 702–710. doi:10.1067/mcp.2002.129321 PMID:12496751

Scott Stuart, A. (2011). PharmGKB summary: very important pharmacogene information for cytochrome P450, family 2, subfamily C, polypeptide 19. Pharmacogenetics and genomics.

Serena Software. (2015). *Serena Dimensions: Serena software*. Retrieved from http://www.serena.com

Shendure, J., & Ji, H. (2008). Next-generation DNA sequencing. *Nature Biotechnology*, *26*(10), 1135–1145. doi:10.1038/nbt1486 PMID:18846087

Sie, A. S., Prins, J. B., van Zelst-Stams, W. A. G., Veltman, J. A., Feenstra, I., & Hoogerbrugge, N. (2015). Patient experiences with gene panels based on exome sequencing in clinical diagnostics: High acceptance and low distress. *Clinical Genetics*, *87*(4), 319–326. doi:10.1111/cge.12433 PMID:24863757

Slattery, J., Candore, G., Tregunno, P., Wong, J., Seabroke, S., & Juhlin, K. et al. (2013). Comparison of disproportionality measures in eudravigilance. *Pharmacoepidemiology and Drug Safety*, *22*, 38–39.

Smith, A., Heisler, L., St.Onge, R., Farias-Hesson, E., Wallace, I., & Bodeau, J. et al. (2010). Highly-multiplexed barcode sequencing: An efficient method for parallel analysis of pooled samples. *Nucleic Acids Research*, *38*, 142. doi:10.1093/nar/gkq368 PMID:20460461

Solomon, M. J., Larsen, P. L., & Varshavsky, A. (1988). Mapping protein-DNA interactions in vivo with formaldehyde: Evidence that histone H4 is retained on a highly transcribed gene. *Cell*, *53*(6), 937–947. doi:10.1016/S0092-8674(88)90469-2 PMID:2454748

Soussi, T. (2014). Locus-Specific Databases in Cancer: What Future in a Post-Genomic Era? The TP53 LSDB paradigm. *Human Mutation*, *35*(6), 643–653. doi:10.1002/humu.22518 PMID:24478183

Southworth, H. (2014). *Predicting potential liver toxicity from phase 2 data: A case study with ximelagatran.*

Southworth, H., & Heffernan, J. E. (2012a). Extreme value modelling of laboratory safety data from clinical studies. *Pharmaceutical Statistics*, *11*(5), 361–366. doi:10.1002/pst.1510 PMID:22684727

Southworth, H., & Heffernan, J. E. (2012b). Multivariate extreme value modelling of laboratory safety data from clinical studies. *Pharmaceutical Statistics*, *11*(5), 367–372. doi:10.1002/pst.1531 PMID:22888093

Sucher, N. J., Hennell, J. R., & Carles, M. C. (2011). DNA fingerprinting, DNA barcoding, and next generation sequencing technology in plants. *Methods in Molecular Biology (Clifton, N.J.)*, *862*, 1322. PMID:22419485

Su, X. W., Broach, J. R., Connor, J. R., Gerhard, G. S., & Simmons, Z. (2014). Genetic heterogeneity of amyotrophic lateral sclerosis: Implications for clinical practice and research. *Muscle & Nerve*, *49*(6), 786–803. doi:10.1002/mus.24198 PMID:24488689

Swerdlow, H., & Gesteland, R. (1990). Capillary gel electrophoresis for rapid, high resolution DNA sequencing. *Nucleic Acids Research*, *18*(6), 1415–1419. doi:10.1093/nar/18.6.1415 PMID:2326186

Szarfman, A., Machado, S. G., & O'Neill, R. T. (2002). Use of screening algorithms and computer systems to efficiently signal higher-than-expected combinations of drugs and events in the US FDA's spontaneous reports database. *Drug Safety*, *25*(6), 381–392. doi:10.2165/00002018-200225060-00001 PMID:12071774

Taber, K. A. J., Dickinson, B. D., & Wilson, M. (2014). The promise and challenges of next-generation genome sequencing for clinical care. *JAMA Internal Medicine*, *174*(2), 275–280. doi:10.1001/jamainternmed.2013.12048 PMID:24217348

Takahashi, H., & Echizen, H. (2003). Pharmacogenetics of CYP2C9 and interindividual variability in anticoagulant response to warfarin. *The Pharmacogenomics Journal*, *3*(4), 202–214. doi:10.1038/sj.tpj.6500182 PMID:12931134

Teer, J. K., & Mullikin, J. C. (2010). Exome sequencing: The sweet spot before whole genomes. *Human Molecular Genetics*, ddq333. PMID:20705737

The Observational Medical Outcomes Partnership (OMOP). Observational medical outcomes partnership Retrieved from http://omop.org/

Tranah, G. J., Chan, A. T., Giovannucci, E., Ma, J., Fuchs, C., & Hunter, D. J. (2005). Epoxide hydrolase and CYP2C9 polymorphisms, cigarette smoking, and risk of colorectal carcinoma in the Nurses' Health Study and the Physicians' Health Study. *Moleclar Carcinogenics*, *44*(1), 21–30. doi:10.1002/mc.20112 PMID:15924351

Tregunno, P. M., Fink, D. B., Fernandez-Fernandez, C., Lazaro-Bengoa, E., & Noren, G. N. (2014). *Performance of probabilistic method to detect duplicate individual case safety reports.*

Tregunno, P. M., Fink, D. B., Fernandez, C., & Noren, N. G. (2013). Hit-miss model detects duplicates missed by rule-based screening of individual case safety reports. *Pharmacoepidemiology and Drug Safety*, *22*, 101–102.

U.S. Food and Drug Administration. (FDA). (2014). Sentinel initiative. Retrieved from http://www.fda.gov/Safety/FDASSentinelInitiative/default.htm

University of California Los Angeles. (n.d). Clinical and Translational Science Institute, UCLA.

Uppsala Monitoring Centre (UMC). (n. d. a). VigiBase®. Retrieved from http://www.umc-products.com/DynPage.aspx?id=73590&mn1=1107&mn2=1132

Uppsala Monitoring Centre (UMC). (n. d. b). Welcome to WHO-ART. Retrieved from http://www.umc-products.com/DynPage.aspx?id=73589&mn1=1107&mn2=1664

Van Holle, L., & Bauchau, V. (2014). Signal detection on spontaneous reports of adverse events following immunisation: A comparison of the performance of a disproportionality-based algorithm and a time-to-onset-based algorithm. *Pharmacoepidemiology and Drug Safety*, *23*(2), 178–185. doi:10.1002/pds.3502 PMID:24038719

Van Puijenbroek, E. P., Bate, A., Leufkens, H. G. M., Lindquist, M., Orre, R., & Egberts, A. C. G. (2002). A comparison of measures of disproportionality for signal detection is spontaneous reporting systems for adverse drug reactions. *Pharmacoepidemiology and Drug Safety*, *11*(1), 3–10. doi:10.1002/pds.668 PMID:11998548

Venter, J. C., Adams, M. D., Myers, E. W., Li, P. W., Mural, R. J., & Sutton, G. G. (2001). The sequence of the human genome. *Science*, *291*(5507), 1304–1351. doi:10.1126/science.1058040 PMID:11181995

Vitalari, N., Strain, W., & Shaughnessy, H. (2012). Creating elastic digital architectures. Retrieved from http://www.cognizant.com/InsightsWhitepapers/Creating-Elastic-Digital-Architectures.pdf

Weinshilboum, R. (2003). Inheritance and drug response. *The New England Journal of Medicine*, *348*(6), 529–537. doi:10.1056/NEJMra020021 PMID:12571261

What is a clinical trial? (n. d.). Michigan Institute for Clinical and Health Research. Retrieved from https://www.michr.umich.edu/about/whatisaclinicaltrial

Wikoff, W. R., Frye, R. F., Zju, H., Gong, Y., Boyle, S., Churchill, E., & Cooper-Dehoff, R. M. (2013). Pharmacometabolomics reveals racial differences in response to atenolol treatment. *PLoS ONE*, *8*(3), 1–8. doi:10.1371/journal.pone.0057639 PMID:23536766

Wildeman, M., Van Ophuizen, E., Den Dunnen, J. T., & Taschner, P. E. (2008). Improving sequence variant descriptions in mutation databases and literature using the Mutalyzer sequence variation nomenclature checker. *Human Mutation*, *29*(1), 6–13. doi:10.1002/humu.20654 PMID:18000842

Wisniewski, A. F. Z., Juhlin, K., Laursen, M., Macia, M. M., Manlik, K., & Pinkston, V. K. et al. (2012). Characterisation of databases (DBS) used for signal detection (SD): Results of a survey of imi protect work package (WP) 3 participants. *Pharmacoepidemiology and Drug Safety*, *21*, 233–234. PMID:21786364

Worthey, E. A., Mayer, A. N., Syverson, G. D., Helbling, D., Bonacci, B. B., & Decker, B. et al. (2011). Making a definitive diagnosis: Successful clinical application of whole exome sequencing in a child with intractable inflammatory bowel disease. *Genetics in Medicine*, *13*(3), 255–262. doi:10.1097/GIM.0b013e3182088158 PMID:21173700

Wu, M. (2011). What is gamification, really. Retrieved from http://community.lithium.com/t5/Science-of-Social-blog/What-is-Gamification-Really/ba-p/30447

Xia, J., Wang, Q., Jia, P., Wang, B., Pao, W., & Zhao, Z. (2012). NGS catalog: A database of next generation sequencing studies in humans. *Human Mutation*, *33*(6), E2341–E2355. doi:10.1002/humu.22096 PMID:22517761

Yang, Y., Muzny, D. M., Reid, J. G., Bainbridge, M. N., Willis, A., & Ward, P. A. et al. (2013). Clinical whole-exome sequencing for the diagnosis of mendelian disorders. *The New England Journal of Medicine*, *369*(16), 1502–1511. doi:10.1056/NEJMoa1306555 PMID:24088041

Yeung, A., & Hall, G. (2007). *Spatial database systems design, implementation and project management* (Vol. 87, p. 566). Dordrecht: Springer.

Zhang, N. N., Liu, Y. T., Ma, L., Wang, L., Hao, X. Z., Yuan, Z., ... & Shi, Y. (2014). The molecular detection and clinical significance of ALK rearrangement in selected advanced non-small cell lung cancer: ALK expression provides insights into ALK targeted therapy.

Zhang, X., Li, M., & Zhang, X. J. (2011). [Exome sequencing and its application]. *Yi chuan= Hereditas/Zhongguo yi chuan xue hui bian ji, 33*(8), 847-856.

Zhang, M., Chen, J., Si, D., Zheng, Y., Jiao, H., & Feng, Z. et al. (2014). Whole exome sequencing identifies a novel EMD mutation in a Chinese family with dilated cardiomyopathy. *BMC Medical Genetics*, *15*(1), 77. doi:10.1186/1471-2350-15-77 PMID:24997722

Zhang, Y., Liu, T., Meyer, C. A., Eeckhoute, J., Johnson, D. S., & Bernstein, B. E. et al. (2008). Model-based analysis of ChIP-Seq (MACS). *Genome Biology*, *9*(9), 137–145. doi:10.1186/gb-2008-9-9-r137 PMID:18798982

Zhou, X., Ren, L., Meng, Q., Li, Y., Yu, Y., & Yu, J. (2010). The next-generation sequencing technology and application. *Protein & cell, 1*(6), 520-536.

Zhou, X., Ren, L., Li, Y., Zhang, M., Yu, Y., & Yu, J. (2010). The next-generation sequencing technology: A technology review and future perspective. *Science China Life Sciences*, *53*(1), 44–57. doi:10.1007/s11427-010-0023-6 PMID:20596955

About the Contributors

Partha Chakraborty is the Sr, Director & Head of Global Delivery R&D Practice in Cognizant Life Sciences (www.cognizant.com). He is a BTech from Indian Institute of Technology (IIT), Kharagpur. He has worked with large healthcare providers and multiple large global pharmaceuticals globally and led IT transformation in Clinical & Safety as well as the implementation of critical engagements. He has presented his point of view in FDA Science Congress, CDISC Interchange, International Society of Pharmacovigilance (ISOP) and DIA. He is instrumental in creating the 1st certificate course, in Pharmacovigilance in India, launched by Symogen, UK. He has written a chapter, called "Role of Information Technology on Drug Safety" in a recently published book "Elements of Pharmacovigilance"; "Statistical Methods Applied in Drug Safety" in a recently published book "Handbook of Research on Pharmacoinformatics."

Amit Nagal, is a bioinformatics scientist who has worked with large scale data management and analysis on HPC clusters. He is involved in developing tools/databases and giving bioinformatics approaches to drive experimental discoveries.

***

Anu Acharya is the CEO of Mapmygenome India Limited, an Indian genomics company providing a range of prognostics, diagnostics, and brain wellness solutions. She is the Founder of Ocimum Biosolutions, which she led as the CEO from 2000 to March 2013. Ocimum combines its bioinformatics products, reference databases, data generation, and data analytics on "big data" using a LIMS backbone through its proprietary platform called RaaS. Ms. Acharya was named a "Young Global Leader" by the World Economic Forum (WEF) for its class of 2011 and serves on the WEF's Global Agenda Council as a member of the Personalized and Precision Medicine. She currently serves as a governing board member at CSIR (Council for Scientific and Industrial Research, the premier R&D Organization in India) and a governing board member at the NIBMG (National Institute for Biomedical Genomics). Prior to founding Ocimum Biosolutions, Ms. Acharya has had rich experience in the Telecom, IT and entrepreneurship arenas. She worked for a start-up in the telecommunications space called Mantiss Information and a consulting firm called SEI Information where she helped create a social networking site for entrepreneurs. Her experience is backed by education at premier institutions such as the Indian Institute of Technology at Kharagpur, India (IIT) and University of Illinois where she has two Post Graduate degrees in Physics and MIS. She was selected by Red Herring Magazine to the list of "25 Tech Titans under 35" in 2006. She has been named Biospectrum "Entrepreneur of the Year" for 2008 and also has been awarded the

Astia Life Science Innovators Award in the same year. Ms. Acharya has been a past President of the Hyderabad chapter of the Entrepreneurs Organization and currently serves on the boards of ABLE (Association of Biotech Led Enterprises) and TIE Hyderabad.

Dr. Ramin B. Arani, Ph.D is a Statistical Science Director in the Advanced Analytics Centre, Biometrics and Information Sciences, at the AstraZeneca Pharmaceutical Company. He has over 20 year of research experience in clinical trials and methodologies. He has published in numerous peer reviewed journals in areas of Safety Signal detection, Survival Analysis, Multiplicity, Artificial Intelligence, Change-point models and Quantitative Benefit-Risk methods.

Ayan Choudhury is an IT professional with 17 years of experience. Ayan is designated as 'Associate Director' and plays the role of a lead solution architect in Cognizant Technology Solution's (CTS) Life Sciences practice for the last 9 years. He has strong experience in information systems pertaining to Drug development with specific focus on Clinical and Safety Information Management, Automation of Clinical Trial Operations and Regulatory Information Management and submissions. He has strong system architecture and project delivery skills. He has experience in leading and managing cross functional, multi-vendor and onsite-offshore teams.

Arindam Dey has overall 12 years of experience with rich program management experience across several clinical projects. Currently taking care of multiple projects for HCL across several US Pharma majors(Managing team of 70+) Actively involved in Business Development activities- Existing Clients plus New Clients. Represented company on different quality and process certifications like Six Sigma, ISO, BS7799,CMMi*

Chandrakant Ekkirala, is working as Associate Director with Cognizant Technologies Limited, Hyderabad in the LS division.

udayaraja GK, IGB Bioinformatics Support Specialist for the Saudi Human Genome Project

Shibichakravarthy Kannan, is a cancer research scientist with experience in computational biology, bioinformatics, next generation sequencing and clinical sequencing. After graduating from medical college he entered the fascinating world of clinical research and never looked back. His PhD thesis was on host versus pathogen response and the role of lipid raft signaling in innate immunity. The focus of his post doctoral fellowship at MD Anderson Cancer Center was on tumor immunotherapy and the role of tumor microenvironment in Follicular Lymphoma pathogenesis. Currently, he is in the in vitro diagnostics industry developing novel genetic test panels for screening diseases with genetic predisposition or inherited mutations. Going forward he is actively involved in translating research to bedside using Next Generation Sequencing technology.

Jasmine Khurana is a Research Associate at Mapmygenome.in. She has completed her Master's in Computational Natural Sciences and Bioinformatics from IIIT-Hyderabad and holds a B.tech degree in Biotechnology from U.I.E.T, Maharashi Dayanand University, Rohtak, Haryana. Her research interest includes personal genomics that further contributes to prediction of predisposed diseases and pharmacogenomics as well.

Brajendra Kumar is Sr. Bioinformatics Analyst in the Bio-IT Dept. at Ocimum BioSolutions Pvt. Ltd, Hyderabad. He received his M.Sc in Bioinformatics from University of Madras in year 2006 and started his career as Bioinformatics programmer at the Genome Life Sciences Pvt.Ltd, Chennai (www.genome.com) . Over the period of six years in GLS, he worked on various Bioinformatics research and development projects,including applications development and NGS data analysis services. He got his first authorship for his contribution in development of 'RoBuST', published in journal "BMC Plant Biology" (http://www.biomedcentral.com/1471-2229/10/161). Brajendra has contributed in the design and development of "Genome Explorer"(a GIC platform for NGS data Analysis), playing a techno-functional role. He moved to Ocimum BioSolutions Pvt. Ltd. in the year 2013 and is currently leading the Bio-IT consulting team for NGS data analysis services. His professional interest lies in applications development for Bioinformatics and is keen interested to contribute to clinical research projects.

Kanishka Mukherjee, joined GVK BIO as AVP - Head of IT and Analytics in 2014 and is based at Hyderabad, India. Kanishka has more than 19 years of diverse experience of working for Healthcare and Life Sciences companies across US, Europe and APAC. He has led the global delivery for multiple large Pharmaceutical accounts and has experience in heading of a very large transformation programs for multiple Organizations. Kanishka hold Bachelor degrees in Physics and Law; is a Post-graduate in Computer Science and has a Certificate on Global Business Leadership from U21 Global, Singapore

Avik Kumar Pal, has extensive experience in enterprise applications, managing global delivery at various fortune 500 clients across North America, Europe, Latin America & Asia. Before starting Clin-iOps, he managed client relationships and solutioning at NTT DATA. Prior to that, he worked with two successful start-ups, FocusFrame (acquired by Hexaware Technologies) and Euclid (acquired by Persistent Systems). He is a Founding Board member at iKure, a social entrepreneurship healthcare startup in India. He is also the President of 'IIT Foundation', San Francisco Bay Area Chapter, the alumni association of Indian Institute of Technology, Kharagpur. Avik holds a B. Tech (Honors) from IIT Kharagpur, and an MBA from University of San Francisco, where he was awarded the 'McLaren Fellowship'. Avik is passionate about Social Entrepreneurship and Impact. He also enjoys travelling and has travelled to 25+ countries worldwide, including an amazing expedition to Antarctica.

Sushma Patil, is currently working as a genetic counsellor for Mapmygenome, India on personalised genomics products. She holds a Postgraduate Diploma certificate degree in Genetic Counselling from Kamineni Education Trust, Hyderabad, India.

Sowmyanarayan Srinivasan, has over 15 years of experience in the area of R&D informatics. He is also focused on building capabilities in the emerging transformation areas including Translational Medicine & NGS. He has spent the early part of his career in supporting bioinformatics product development and marketing.

Yerramalli Subramaniam (Subbu) has over 15 years of experience in the life science industry. Prior to joining CliniOps, Subbu worked at Applied Biosystems (acquired by Life Technologies/Thermo Fisher) where he led the design and development of several RUO and diagnostic instruments. In this position, he interacted extensively with various pharma companies, core facilities, genome centers and PI labs to understand customer needs and provide custom solutions. Subbu has in-depth knowledge of

CE and FDA regulations and has launched multiple instrument and software products with these regulatory compliances that received several industry awards. Prior to working at Applied Biosystems, Subbu worked at Sanofi as part of their supply chain automation effort. He also has a publication and several patents to his credit in the area of qPCR and next generation sequencing. Subbu holds a B. Tech (Honors) from IIT Kharagpur, and an MBA from Haas School of Business, University of California, Berkeley.

Geethanjali Tanikella is the Sr. Manager of Corporate Communication at Mapmygenome, where she handles all the content to be published - scientific, technical, marketing, and more. She is a writing and editing professional with 12+ years of experience. Her areas of expertise include Corporate Communication, Marketing Communication, Scientific Writing, Technical Communication, XML, and Content Editing.

Manu Venugopal is a Lead Business Consultant in Life Sciences R&D at Accenture, India. With over 12 years of industry experience, his primary focus has been on providing business analysis and consulting expertise in building technology solutions in the area of Clinical Development and Pharmacovigilance. He has worked with the R&D organizations of many of the Top 15 pharmaceutical companies of the world, playing the role of a functional consultant. His recent interest lies in exploring how the new age technologies such as social media, mobility and analytics can bring about disruptive transformations in Life Sciences R&D space. In the past two years, he has been instrumental in building several technology and knowledge assets using social media and analytical tools.

Antoni F.Z. Wisniewski is Systems Area Lead for Patient Safety Analytics at AstraZeneca and is based in the UK. He is a biologist by training, worked in clinical research at the University of Nottingham for eleven years and has worked in the field of pharmacovigilance at AstraZeneca since 1998. His main interest is the practical application of technology to aid analysis and evaluation of safety data related to the use of medicinal products.

Index

N

O

P

Q

R

S

T

U

W

CPSIA information can be obtained at www.ICGtesting.com
Printed in the USA
BVOW09*1258100416

443419BV00003B/4/P

9 781466 687264